**Plant Gene Silencing: Mechanisms and Applications**

FSC
www.fsc.org
MIX
Paper from responsible sources
FSC® C013604

## CABI Biotechnology Series

Biotechnology, in particular the use of transgenic organisms, has a wide range of applications including agriculture, forestry, food and health. There is evidence that it could make a major impact in producing plants and animals that are able to resist stresses and diseases, thereby increasing food security. There is also potential to produce pharmaceuticals in plants through biotechnology, and provide foods that are nutritionally enhanced. Genetically modified organisms can also be used in cleaning up pollution and contamination. However, the application of biotechnology has raised concerns about biosafety, and it is vital to ensure that genetically modified organisms do not pose new risks to the environment or health. To understand the full potential of biotechnology and the issues that relate to it, scientists need access to information that not only provides an overview of and background to the field, but also keeps them up to date with the latest research findings.

This series, which extends the scope of CABI's successful Biotechnology in Agriculture Series, addresses all topics relating to biotechnology including transgenic organisms, molecular analysis techniques, molecular pharming, *in vitro* culture, public opinion, economics, development and biosafety. Aimed at researchers, upper-level students and policy makers, titles in the series provide international coverage of topics related to biotechnology, including both a synthesis of facts and discussions of future research perspectives and possible solutions.

### Titles available

*Animal Nutrition with Transgenic Plants*
Edited by G. Flachowsky

*Plant-derived Pharmaceuticals: Principles and Applications for Developing Countries*
Edited by K.L. Hefferon

*Transgenic Insects: Techniques and Applications*
Edited by M.Q. Benedict

*Bt Resistance: Characterization and Strategies for GM Crops Producing Bacillus thuringiensis Toxins*
Edited by M. Soberón, Y. Gao and A. Bravo

*Plant Gene Silencing: Mechanisms and Applications*
Edited by T. Dalmay

# Plant Gene Silencing: Mechanisms and Applications

*Edited by*

**Tamas Dalmay**

*School of Biological Sciences, University of East Anglia, Norwich, UK*

CABI

**CABI is a trading name of CAB International**

CABI
Nosworthy Way
Wallingford
Oxfordshire OX10 8DE
UK

Tel: +44 (0)1491 832111
Fax: +44 (0)1491 833508
E-mail: info@cabi.org
Website: www.cabi.org

CABI
745 Atlantic Avenue
8th Floor
Boston, MA 02111
USA

Tel: +1 (617)682-9015
E-mail: cabi-nao@cabi.org



**British Library Cataloguing-in-Publication Data**

A catalogue record for this book is available from the British Library, London, UK.

**Library of Congress Cataloging-in-publication data**

Names: Dalmay, Tamas, editor.
Title: Plant gene silencing : mechanisms and applications / edited by Tamas Dalmay.
Other titles: Plant gene silencing (Dalmay) | CABI biotechnology series ; 5.
Description: Wallingford, Oxfordshire, UK ; Boston, MA : CABI, [2017] | Series: CABI biotechnology series ; 5 | Includes bibliographical references.
Identifiers: LCCN 2016057958 (print) | LCCN 2017009210 (ebook) | ISBN 9781780647678 (hbk : alk. paper) | ISBN 9781780647685 (ePDF) | ISBN 9781780647692 (ePub)
Subjects: LCSH: Plant gene silencing. | MESH: Gene Silencing | Gene Expression Regulation, Plant
Classification: LCC QK981.4 .P5794 2017 (print) | LCC QK981.4 (ebook) | NLM QK 981.4 | DDC 572.8/652--dc23
LC record available at https://lccn.loc.gov/2016057958

ISBN: 978 1 78064 767 8 (hardback)
978 1 78064 768 5 (e-book)
978 1 78064 769 2 (e-pub)

Commissioning editor: David Hemming
Editorial assistant: Emma McCann
Production editors: Alan Worth/Lauren Povey

Typeset by SPi, Pondicherry, India.
Printed and bound in the UK by CPI Group (UK) Ltd, Croydon, CR0 4YY.

# Contents

# Contributors

**Alberto Carbonell**, Instituto de Biología Molecular y Celular de Plantas (CSIC-UPV), Campus UPV, Valencia, Spain. E-mail: acarbonell@ibmcp.upv.es

**Heriberto Cerutti**, School of Biological Sciences and Center for Plant Science Innovation, University of Nebraska-Lincoln, Lincoln, USA. E-mail: hcerutti1@unl.edu

**Tamas Dalmay**, School of Biological Sciences, University of East Anglia, Norwich NR4 7TJ, UK. E-mail: t.dalmay@uea.ac.uk

**Soumita Das**, National Centre for Biological Sciences, Bellary Road, Bangalore 560065, India. E-mail: soumitad@ncbs.res.in

**Asis Datta**, National Institute of Plant Genome Research, New Delhi 110067, India. E-mail: asisdatta@hotmail.com

**Emilie Elvira-Matelot**, Institut Jean-Pierre Bourgin, INRA, AgroParisTech, CNRS, Université Paris-Saclay, RD10, 78000 Versailles, France. E-mail: emilie.elvira-matelot@inserm.fr

**John Fosu-Nyarko**, Plant Biotechnology Research Group, Western Australian State Agricultural Biotechnology Centre, School of Veterinary and Life Sciences, Murdoch University, Perth, Western Australia. E-mail: j.fosu-nyarko@murdoch.edu.au

**Juan Antonio García**, National Centre for Biotechnology CNB, CSIC, Cantoblanco, Madrid, Spain. E-mail: jagarcia@cnb.csic.es

**Sumit Ghosh**, CSIR-Central Institute of Medicinal and Aromatic Plants, Lucknow 226015, India. E-mail: sumitghosh@cimap.res.in

**Vincenza Ilardi**, Consiglio per la Ricerca in Agricoltura e l'Analisi dell'Economia Agraria – Centro di Ricerca per la Patologia Vegetale (CREA-PAV), Rome, Italy. E-mail: vincenza.ilardi@crea.gov.it

**Sadia Iqbal**, Plant Biotechnology Research Group, Western Australian State Agricultural Biotechnology Centre, School of Veterinary and Life Sciences, Murdoch University, Perth, Western Australia. E-mail: s.iqbal@murdoch.edu.au

**Mohammad Irfan**, National Institute of Plant Genome Research, New Delhi 110067, India. E-mail: irfanqadri01@gmail.com

**Michael G.K. Jones**, Plant Biotechnology Research Group, Western Australian State Agricultural Biotechnology Centre, School of Veterinary and Life Sciences, Murdoch University, Perth, Western Australia. E-mail: m.jones@murdoch.edu.au

**Eun-Jeong Kim**, School of Biological Sciences and Center for Plant Science Innovation, University of Nebraska-Lincoln, Lincoln, USA. E-mail: sassyej@gmail.com

**Aline Koch**, Institute for Phytopathology, Centre for BioSystems, Land Use and Nutrition, Justus Liebig University, Giessen, Germany. E-mail: aline.koch@agrar.uni-giessen.de

**Karl-Heinz Kogel**, Institute for Phytopathology, Centre for BioSystems, Land Use and Nutrition, Justus Liebig University, Giessen, Germany. E-mail: karl-heinz.kogel@agrar.uni-giessen.de

**Juan José López-Moya**, Centre for Research in Agricultural Genomics CRAG, CSIC-IRTA-UAB-UB, Campus Bellaterra, Barcelona, Spain. E-mail: juanjose.lopez@cragenomica.es

**Alessandra Lucioli**, Agenzia Nazionale per le Nuove Tecnologie, l'Energia e lo Sviluppo Economico Sostenibile (ENEA), C.R. Casaccia, Rome, Italy. E-mail: alessandra.lucioli@enea.it

**Xinrong Ma**, School of Biological Sciences and Center for Plant Science Innovation, University of Nebraska-Lincoln, Lincoln, USA. E-mail: luckymaxinrong@yahoo.com

**Ángel Emilio Martínez de Alba**, Institut Jean-Pierre Bourgin, INRA, AgroParisTech, CNRS, Université Paris-Saclay, RD10, 78000 Versailles, France. E-mail: aemartidealba@gmail.com

**Ares Mingot**, Centre for Research in Agricultural Genomics CRAG, CSIC-IRTA-UAB-UB, Campus Bellaterra, Barcelona, Spain. E-mail: ares.mingot@cragenomica.es

**Leandro Peña**, Laboratorio de Biotecnología de Cítricos, Instituto de Biología Molecular y Celular de Plantas (IBMCP), Consejo Superior de Investigaciones Científicas/Universidad Politécnica de Valencia, Valencia, Spain; Laboratorio de Biotecnologia Vegetal, Pesquisa & Desenvolvimento, Fundo de Defensa da Citricultura (Fundecitrus), Araraquara, Sao Paulo, Brazil. E-mail: lpenya@fundecitrus.com.br

**Elsa Pons**, Laboratorio de Biotecnología de Cítricos, Instituto de Biología Molecular y Celular de Plantas (IBMCP), Consejo Superior de Investigaciones Científicas/Universidad Politécnica de Valencia, Valencia, Spain; Laboratorio de Biotecnologia Vegetal, Pesquisa & Desenvolvimento, Fundo de Defesa da Citricultura (Fundecitrus), Araraquara, Sao Paulo, Brazil. E-mail: elponba@ibmcp.upv.es

**Rosa M. Ruiz-Vázquez**, University of Murcia, Murcia, Spain. E-mail: rmruiz@um.es

**P.V. Shivaprasad**, National Centre for Biological Sciences, Bellary Road, Bangalore 560065, India. E-mail: shivaprasad@ncbs.res.in

**Rahul Raj Singh**, National Centre for Biological Sciences, Bellary Road, Bangalore 560065, India. E-mail: rahulrs@ncbs.res.in

**Mario Tavazza**, Agenzia Nazionale per le Nuove Tecnologie, l'Energia e lo Sviluppo Economico Sostenibile (ENEA), C.R. Casaccia, Rome, Italy. E-mail: mario.tavazza@enea.it

**Santiago Torres-Martínez**, University of Murcia, Murcia, Spain. E-mail: storres@um.es

**Adrian Valli**, National Centre for Biotechnology CNB, CSIC, Cantoblanco, Madrid, Spain. E-mail: avalli@cnb.csic.es

**Hervé Vaucheret**, Institut Jean-Pierre Bourgin, INRA, AgroParisTech, CNRS, Université Paris-Saclay, RD10, 78000 Versailles, France. E-mail: herve.vaucheret@versailles.inra.fr

# Preface

It is more than 25 years ago that gene silencing was described for the first time, although at that stage the mechanism was not understood (Napoli *et al*., 1990; van der Krol *et al*., 1990). The progress during the last 25 years or so has been simply remarkable: we now do not just understand how gene silencing happens down to a very detailed molecular level, but many strategies have been developed to harness this phenomenon. This book describes both the theory of gene silencing and also the application.

The first five chapters discuss different aspects of the gene silencing mechanism. Since the silencing pathways are particularly diverse in plants, a whole chapter is dedicated to describe these (Chapter 1). It is a generally accepted view that gene silencing has evolved in plants as a defence mechanism against viruses, therefore Chapter 2 discusses the 'arms race' between plants and viruses, how viruses trigger silencing and also evolved proteins that can suppress it. Another aspect of gene silencing is the epigenetic changes caused by silencing. This, and how we can direct epigenetic changes, is described in Chapter 3. Finally, the theoretical part is closed by two chapters on how gene silencing works in algae (Chapter 4) and fungi (Chapter 5), two groups of organisms related to plants.

The second part of the book is dedicated to application of gene silencing. Small non-coding RNAs are key molecules in the mechanism and Chapter 6 discusses various strategies to produce small artificial RNAs. The following chapters describe the application of gene silencing to influence specific, agronomically important traits in plants, including traits for industrial use (Chapter 7) and nutritional value (Chapter 8). The last three chapters review the use of gene silencing to provide resistance against different types of pathogens including fungi (Chapter 9), nematodes (Chapter 10) and viruses (Chapter 11).

## References

Napoli, C., Lemieux, C. and Jorgensen, R. (1990) Introduction of a chimeric chalcone synthase gene into petunia results in reversible co-suppression of homologous genes in trans. *Plant Cell* 2, 279–289.

van der Krol, A.R., Mur, L.A., Beld, M., Mol, J.N.M. and Stuitje, A.R. (1990) Flavonoid genes in petunia: addition of a limited number of gene copies may lead to a suppression of gene expression. *Plant Cell* 2, 291–299.

# 1 Diversity of RNA Silencing Pathways in Plants

**Emilie Elvira-Matelot, Ángel Emilio Martínez de Alba and Hervé Vaucheret***

*Institut Jean-Pierre Bourgin, INRA, AgroParisTech, CNRS, Université Paris-Saclay, Versailles, France*

## 1.1 Introduction

RNA silencing is a manifestation of eukaryote defences against exogenous invading nucleic acids. Indeed, infection by pathogens, including fungi, bacteria, viruses or viroids, generally results in the production of pathogen-specific short interfering RNAs (siRNAs), the hallmark of RNA silencing (Hamilton and Baulcombe, 1999; Navarro *et al.*, 2008). When loaded onto ARGONAUTE (AGO) proteins, these siRNAs guide the cleavage of the long RNAs naturally encoded by the invader (Vaucheret, 2008). However, despite the highly sequence-specific effect of siRNAs, pathogen-derived RNAs generally are not eliminated because most pathogens encode proteins that counteract the biogenesis or the action of siRNAs (Pumplin and Voinnet, 2013; Csorba *et al.*, 2015).

RNA silencing is also used to control endogenous invading nucleic acids such as transposable elements (TE). In fact, TE silencing is mandatory to prevent uncontrolled expansion of these elements within the genome and avoiding subsequent deleterious effects, including gene disruption, gene activation or internal recombination. Unlike viruses, TEs generally do not encode proteins that have the capacity to block RNA silencing. Therefore, TEs generally are efficiently controlled by RNA silencing. Nevertheless, the protection of TE RNAs by TE proteins has been reported (Mari-Ordonez *et al.*, 2013). Moreover, TE silencing can be erased under certain stress conditions (for example heat stress), leading to transient expression of TE RNAs and possible TE movement (Pecinka *et al.*, 2010; Ito *et al.*, 2011).

In contrast to pathogens and TEs, endogenous protein-coding genes generally are not a source for siRNA production and therefore are not subjected to RNA silencing. Indeed, only a handful of endogenous genes, in particular varieties, have been shown to produce siRNAs at levels that allow blocking transcription (transcriptional gene silencing or TGS) or degrading mRNAs (post-transcriptional gene silencing or PTGS), depending whether the siRNAs derive from the promoter or the transcribed region. Remarkably, these varieties exhibit genomic rearrangements, involving either duplication events or TEs inserted within or adjacent to the gene, whereas

* Corresponding author e-mail: Herve.Vaucheret@versailles.inra.fr

regular varieties that lack such rearrangements do not produce siRNAs and do not show silencing (Coen and Carpenter, 1988; Bender and Fink, 1995; Cubas *et al.*, 1999; Clough *et al.*, 2004; Tuteja *et al.*, 2004; Della Vedova *et al.*, 2005; Manning *et al.*, 2006; Martin *et al.*, 2009; Tuteja *et al.*, 2009; Durand *et al.*, 2012). It is assumed that genomic rearrangements resulting in the silencing of endogenous protein-coding genes are tolerated because they affect dispensable genes, and that cells undergoing genomic rearrangements that provoke the silencing of essential genes do not survive. This hypothesis implies that, during evolution, endogenous protein-coding genes are shaped to avoid producing siRNAs and undergoing silencing.

## 1.2 Transgene-based Genetic Screens to Unravel Silencing Pathways

The situation of endogenous protein-coding genes contrasts sharply to that of transgenes, which often undergo RNA silencing, although they are designed to structurally resemble and function like endogenous protein-coding genes. Note that RNA silencing was actually discovered as an unintended consequence of plant transformation (Matzke *et al.*, 1989; Napoli *et al.*, 1990; Smith *et al.*, 1990; van der Krol *et al.*, 1990). Indeed, it is now known that introduction of transgenes in the form of naked DNA, or by infection with disarmed bacteria such as *Agrobacterium*, always activates the production of siRNAs (Llave *et al.*, 2002). Following stable integration in the genome, transgenes are either expressed or silenced. Nevertheless, silencing sometimes occurs after a period of normal expression that can last several generations. The reasons why certain transformants express a transgene whereas others undergo silencing by TGS or PTGS remain not well understood, and this raises important issues about the reliability of transgene expression. Importantly, when the transgene undergoing silencing carries sequences derived from an endogenous gene, transgene-derived siRNAs also affect the endogenous copy or copies, a phenomenon referred to as co-suppression (Napoli *et al.*, 1990).

The fact that transgenes frequently undergo silencing whereas endogenous protein-coding genes do not, indicates that transgenes are often perceived as invaders that need to be silenced like pathogens or TEs. During the transient phase of extra-chromosomal expression, transgenes are generally present in high copy number, which may result in abnormally high levels of RNAs, thus mimicking what happens with invader RNAs during an infection, and activation of RNA silencing. Following integration in genomic areas allowing high levels of transcription, transgenes can still continue to produce high levels of RNAs, thus maintaining RNA silencing active against them. Supporting this hypothesis, transgenes that carry strongly expressed promoters are generally more prone to undergo silencing than transgenes that carry weakly expressed promoters. Stable integration of several transgene copies within the genome can also activate anti-transposons RNA silencing. Supporting this second hypothesis, transgenic plants exhibiting high transgene copy numbers are generally more prone to undergo silencing than plants carrying single copies.

Almost 20 years ago, the first forward genetic screens based on the reactivation of silenced transgenes identified the core components of the PTGS and TGS pathways. Enhancer screens were then set up, revealing cellular functions that antagonize silencing. More recently refined genetic screens, including sensitized screens and suppressor screens, have allowed identification of a variety of regulatory components. So far, 12 and 18 forward genetic screens dedicated to PTGS and TGS, respectively, have been published. The outcome of these screens is described in Table 1.1 and Table 1.2. Because transgenes only serve as excellent reporters of endogenous functions, we do not describe further how each transgene locus is silenced. In the next sections, we describe what transgene-based genetic screens have told us about natural silencing pathways.

## 1.3 PTGS Pathways

### 1.3.1 Antiviral PTGS

Antiviral PTGS starts by the processing of virus-derived dsRNA into 21- and 22-nt primary siRNAs by DICER-LIKE 4 (DCL4) and DCL2, respectively (Bouche *et al.*, 2006; Deleris *et al.*, 2006; Fusaro *et al.*, 2006). Virus-derived dsRNA molecules represent either: (i) the natural form of dsRNA viruses; (ii) intermediate forms of the replication of ssRNA viruses; (iii) partially folded viral ssRNAs; or (iv) molecules resulting from the action of RNA-DEPENDENT-RNA-POLYMERASE (RDR) enzymes on aberrant or subgenomic viral ssRNA. Primary siRNAs are methylated at their 3′ end by the methyltransferase HUA ENHANCER 1 (HEN1) (Boutet *et al.*, 2003; Li *et al.*, 2005) before loading onto AGO proteins, mainly AGO1 and AGO2 but also AGO5 or AGO7 (Morel *et al.*, 2002; Qu *et al.*, 2008; Harvey *et al.*, 2011; Wang *et al.*, 2011b; Brosseau and Moffett, 2015) to guide the cleavage of viral ssRNA through sequence homology. AGO-mediated cleavage generates RNA fragments that escape degradation due to the protective activity of SUPPRESSOR OF GENE SILENCING 3 (SGS3) (Mourrain *et al.*, 2000; Yoshikawa *et al.*, 2013). With the assistance of the putative RNA export protein SILENCING-DEFECTIVE (SDE5) (Hernandez-Pinzon *et al.*, 2007), SGS3-protected cleavage products are transformed into dsRNA by RDR6 (Mourrain *et al.*, 2000). These dsRNA are processed into siRNA duplexes by DCL4 to produce secondary siRNAs that reinforce AGO-mediated RNA cleavage, thus creating an amplification loop (Fig. 1.1). Such a process should eliminate viral RNA; however, most viruses have developed strategies to handle PTGS by expressing proteins called VSR (viral suppressors of RNA silencing), which block one or other of the steps of the PTGS pathway (Pumplin and Voinnet, 2013; Csorba *et al.*, 2015).

This antiviral PTGS model also explains how PTGS is activated against sense transgenes that are not supposed to produce dsRNAs. Accordingly, transgenes that produce aberrant RNAs in sufficient amounts to escape degradation by nuclear and cytoplasmic RNA quality control (RQC) pathways (see below) are transformed into dsRNA by RDR6. The nature of transgene aberrant RNAs has long remained a mystery until the recent identification of uncapped transgene RNAs resulting from the 3′ end processing of readthrough transcripts (Parent *et al.*, 2015b). RDR6-derived transgene dsRNAs are processed into 21-nt and 22-nt primary by DCL4 and DCL2 (Parent *et al.*, 2015a), and loaded onto AGO1, which cleaves complementary target RNAs (Morel *et al.*, 2002; Baumberger and Baulcombe, 2005). Transgene RNA cleavage fragments are transformed into

Table 1.1. Mutants identified in PTGS genetic screens.

| Reporter gene/ transgene | Basis of the screen | Mutants nomenclature | Mutants | Gene product |
|---|---|---|---|---|
| *p35S:rolB* | Wild-type phenotype | *enhancer of gene silencing (egs)* | *egs1* | Unknown |
| | | | *egs2* | Unknown |
| *p35S:GUS* (*L1*) | GUS expression | *supressor of gene silencing (sgs)* | *sgs1* | Transcription factor NAC52 |
| | | | *sgs2* | RNA dependent RNA polymerase RDR6 |
| | | | *sgs3* | RNA binding protein SGS3 |
| | | | *sgs4* | RNA slicer AGO1 |
| | | | *sgs5* | RNA methylase HEN1 |
| | | | *sgs6* | DNA methyltransferase MET1 |
| | | | *sgs7* | RNA export protein SDE5 |
| | | | *sgs8* | H3K4me2/3 demethylase JMJ14 |
| | | | *sgs9* | RNA traficking protein HPR1 |
| *p35S:NIA2* (*2a3*) | Wild-type growth | *supressor of gene silencing (sgs)* | *sgs2* | RNA dependent RNA polymerase RDR6 |
| | | | *sgs3* | RNA binding protein SGS3 |
| | | | *sgs13* | RNA helicase protein SDE3 |
| | | | *sgs14* | Small nuclear protein SmD1 |
| | | | *sgs17* | Cyp40 like gene SQUINT |
| *p35S:GUS (L1)* in *ago1-27* or *ago1-33* | Absence of GUS activity | none | *fry1* | Inositol polyphosphoric acid 1-phosphatase FRY1 |
| | | | *ski3* | Exosome cofactor SUPERKILLER3 |
| *p35S:MP17-GFP* | Wild-type phenotype | *increased transgene silencing (its)* | *xrn4* | 5´-3´ exoribonuclease XRN4 |
| | | | *dcp2* | mRNA decapping enzyme DCP2 |
| *p35S:GFP* x *p35S: PVX-GFP* (*GxA*) | GFP expression | *silencing defective (sde)* | *sde1* | RNA-dependent RNA polymerase RDR6 |
| | | | *sde2* | RNA binding protein SGS3 |
| | | | *sde3* | RNA helicase protein SDE3 |
| | | | *sde4* | PolIV subunit NRPD1 |
| | | | *sde5* | RNA export protein SDE5 |
| | | | *sde6* | RNA-dependent RNA polymerase RDR2 |
| *p35S:PVX-GFP* in *rdr6* | enhanced GFP expression | | *dcl4* | DICER-LIKE protein DCL4 |
| | | | *hen1* | RNA methylase HEN1 |
| | | | *upf1* | NMD factor UPFRAMESHIFT1 |

| | | | | |
|---|---|---|---|---|
| *p35S:PVX-PDS* | Increase in *PDS* silencing | *enhanced silencing phenotype (esp)* | *esp1* | mRNA 3′ formation CstF64-like |
| | | | *esp3* | RNA helicase splicing factor PRP2 |
| | | | *esp4* | mRNA 3′ formation symplekin |
| | | | *esp5* | mRNA 3′ formation CPSF100 |
| *p35S:GFP-miR171* (*GFP171.1*) | GFP expression | *microrna biogenesis deficient (mbd)* | *mbd1* | DICER-LIKE protein DCL1 |
| | | | *mbd2* | RNA methylase HEN1 |
| | | *microrna action deficient (mad)* | *mad3* | HMGR enzyme |
| | | | *mad4* | Sterol C-8 isomerase |
| | | | *mad5* | Microtubule severing protein KTN1 |
| | | | *mad7* | Exosome cofactor SUPERKILLER2 |
| *p35S:LUC* and *p35S:miR-LUC* | LUC expression | none | *cpl1* | C-TERMINAL DOMAIN PHOSPHATASE-LIKE 1 |
| *p35S:TAS3aPDS* (*TAS3-syn*) | Decreased photobleaching | none | *ago7* | RNA slicer AGO7 |
| | | | *dcl4* | DICER-LIKE protein DCL4 |
| | | | *rdr6* | RNA-dependent RNA polymerase RDR6 |
| | | | *sgs3* | RNA binding protein SGS3 |
| | | | *mir390* | *MIR390* |
| *pSUC2:SUL* | Decreased photobleaching | *silencing movement-deficient (smd)* | *smd1* | RNA dependent RNA polymerase RDR2 |
| | | | *smd2* | PolIV largest subunit NRPD1 |
| | | | *dcl4* | DICER-LIKE protein DCL4 |
| *pSUC2:PDS* (*JAP3*) | Decreased photobleaching | none | *clsy1* | SNF2 domain–containing protein CLSY1 |
| | | | *nrpd1* | Nuclear RNA polymerase D1 |
| | | | *rdr2* | RNA-dependent RNA polymerase RDR2 |
| | | | *dcl3* | DICER-LIKE protein DCL3 |
| | | | *nrpd2* | Nuclear RNA polymerase D2 |
| | | | *fca* | mRNA processing protein FCA |
| | | | *fpa* | mRNA processing protein FPA |
| | | | *jmj14* | H3K4me2/3 demethylase JMJ14 |
| | | | *tex1* | RNA trafficking protein TEX1 |

**Table 1.2.** Mutants identified in TGS genetic screens.

| Reporter gene/transgene | Basis of the screen | Mutant nomenclature | Mutants | Gene product |
|---|---|---|---|---|
| *p35S:CHS-pNos:NptII (C)* | Decreased anthocyanin | *homology dependent gene silencing (hog)* | *hog1* | S-Adenosyl-homocysteine hydrolase SAHH1 |
| | Kanamycin resistance | *modifier of silencing (sil)* | *sil1* | Rpd3-like histone deacetylase HDA6 |
| *p35S:HPT (A)* | Hygromycin resistance | *somniferous effect (som)* | *som1* | Chromatin remodelling factor DDM1 |
| | | | *mom1* | Chromatin remodelling factor MOM1 |
| *pUBQ3:LUC in mom1* | LUC expression | *mom1 enhancer (moe)* | *moe1* | PolV subunit NRPE1 |
| *SUP in clk (clk-st)* | Wild-type flowers | *none* | *cmt3* | Cytosine methyltransferase CMT3 |
| | | | *kyp* | H3K9 methyltransferase SUVH4 |
| | | | *ago4* | RNA slicer AGO4 |
| *Endogenous PAI1-PAI4 in Ws* | Reduced UV fluorescence | *none* | *cmt3* | Cytosine methyl transferase CMT3 |
| | | | *suvh4* | H3K9 methyltransferase SUVH4 |
| *p35S::GUS (L5)* | GUS expression | *none* | *rpa2* | Replication protein A 2nd subunit RPA2 |
| *pRD29A:LUC* | Impaired LUC expression | *repressor of silencing (ros)* | *ros1* | Glycosylase ROS1 |
| | | | *ros3* | Regulator of DNA demethylation ROS3 |
| | | | *ros4* | Acetyltransferase protein IDM1 |
| *pRD29A:LUC in ros1* | LUC expression | *rna-directed dna methylation (rdm)* | *rdm1* | RNA-directed DNA methylation RDM1 |
| | | | *rdm2* | PolIV/PolV subunits NRPD4/NRPE4 |
| | | | *rdm3* | KOW domain-containing transcription factor KTF1 |
| | | | *rdm4* | Transcription factor for both PolII and PolIV |
| | | | *rdm5* | PolIV subunit NRPD1 |
| | | | *rdm6* | PolV subunit NRPE1 |
| | | | *rdm7* | PolIV subunit NRPD2a |
| | | | *rdm8* | RNA slicer AGO4 |
| | | | *rdm9* | RNA methylase HEN1 |
| | | | *rdm10* | Chromatin remodelling factor DRD1 |

| | | | | |
|---|---|---|---|---|
| | | | *rdm11* | Rpd3-like histone deacetylase HDA6 |
| | | | *rdm12* | Similarities with SGS3/ partner for RDR2 RDM12/IDN2 |
| | | | *rdm16* | Splicing factor PRP3 |
| | | | *rpa2* | Replication protein A 2nd subunit RPA2 |
| | | | *ago6* | RNA slicer AGO6 |
| | | | *ubp26* | Ubiquitin protease UBP26 |
| | | | *dtf1* | Putative DNA binding transcription factor DTF1 |
| | | | *zop1* | Zinc finger (ZnF) and OCRE domain-containing protein ZOP1 |
| | | | *dtf1* | Sawadee Homeodomain Homolog protein SHH1 |
| | | | *prp31* | Splicing factor PRP31 |
| *Ectopic FWA copies* | Late flowering phenotype | *involved in de novo DNA methylation (idn)* | *idn1* | Similar to the hinge-domain region of structural maintenance of chromosomes |
| | | | *idn2* | SGS3-like partner for RDR2 RDM12/ IDN2 |
| | | | *sr45* | Splicing factor Serine-arginine rich SR45 |
| | | | *drm2* | Domains rearranged methyltransferase DRM2 |
| | | | *dcl3* | DICER-LIKE protein DCL3 |
| | | | *rdr2* | RNA dependent RNA polymerase RDR2 |
| | | | *ago4* | RNA slicer AGO4 |
| | | | *ago6* | RNA slicer AGO6 |
| | | | *nrpe1* | PolV subunit NRPE1 |
| | | | *drd1* | Putative chromatin remodelling factor DRD1 |
| | | | *ktf1* | KOW domain-containing transcription factor KTF1 |

*Continued*

**Table 1.2.** Continued.

| Reporter gene/transgene | Basis of the screen | Mutant nomenclature | Mutants | Gene product |
|---|---|---|---|---|
| | | | *clsy1* | SNF2 domain-containing protein CLSY1 |
| | | | *nrpd/e2* | PolIV/PolV subunit NRPD2/NRPE2 |
| *p35S:NOS x pNOS:NPTII (KxH)* | Kanamycin resistance | *rna-mediated transcriptional silencing (rts)* | *rts1* | Rpd3-like histone deacetylase HDA6 |
| | | | *rts2* | DNA methyltransferase MET1 |
| *p35S:α x pα:GFP* | GFP expression | *defective in rna-directed dna methylation (drd)* | *drd1* | Chromatin remodelling factor DRD1 |
| | | | *drd2* | PolIV/PolV subunit NRPD2/NRPE2 |
| | | | *drd3* | PolV subunit NRPE1 |
| *p35S:ENH x pENH:GFP (SxT)* | GFP expression | *defective in meristem silencing (dms)* | *dms1* | Putative chromatin remodelling factor DRD1 |
| | | | *dms2* | PolIV subunit NRPD2a |
| | | | *dms3* | Similar to the hinge-domain region of structural maintenance of chromosomes |
| | | | *dms4* | Putative transcription factor IWR1 |
| | | | *dms5* | PolIV subunit NRPD1b |
| | | | *dms6* | DICER-LIKE protein DCL3 |
| | | | *dms7* | RNA-directed DNA methylation RDM1 |
| | | | *dms8* | Domains rearranged methyltransferase DRM2 |
| | | | *dms9* | RNA slicer AGO6 |
| | | | *dms10* | SGS3-like partner for RDR2 RDM12/IDN2 |
| | | | *dms11* | ATPase DMS11/GHKL |
| *pNOS:35S x p35S:GFP* | GFP expression | *rna-directed dna methylation defective (rmd)* | *rmd1* | PolIV subunit NRPD1b |

| | | | | |
|---|---|---|---|---|
| | | | *rmd3* | PolIV subunit NRPD1a |
| *pSDC:GFP* | GFP expression | *none* | *morc1* | ATPase-containing Microrchidia1 (MORC1) |
| | | | *morc6* | ATPase-containing Microrchidia1 (MORC6) |
| *p35S:LUC-AP2 in rdr6 (LUCH)* | Increase in LUC expression | *none* | *ago4* | RNA slicer AGO4 |
| | | | *drd1* | Chromatin remodelling factor DRD1 |
| | | | *drm2* | Domains rearranged methyltransferase DRM2 |
| | | | *mom1* | Chromatin remodelling factor MOM1 |
| | | | *hen1* | RNA methylase HEN1 |
| | | | *amp1* | ER-associated ALTERED MERISTEM PROGRAM1 |
| | | | *suvh1* | H3K4 methyltransferase SUVH1 |
| *p35S:LUC-AP2 in rdr6 (LUCL)* | Increase in LUC expression | *none* | *top1a* | DNA topoisomerase 1a |
| *p35S:SUC2* | Restored root growth | *asi (anti-silencing)* | *asi1* | RNA binding protein IMB2 |
| | | | *ros1* | Glycosylase ROS1 |
| | | | *ros4* | Acetyltransferase protein IDM1 |
| *pAPUM9:GFP (Silex)* | Restored GFP expression | *epic (epigenetic control)* | *hda6* | Rpd3-like histone deacetylase HDA6 |

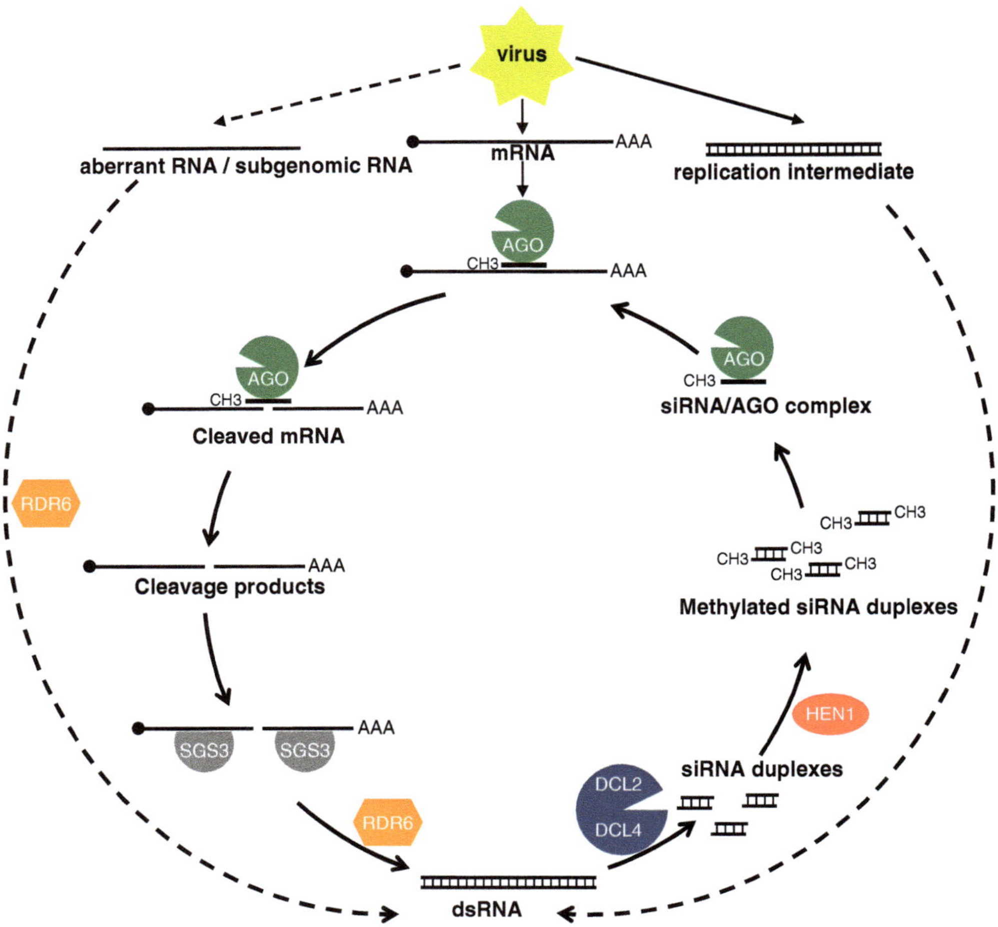


**Fig. 1.1.** Model for antiviral PTGS. Dashed arrows indicate putative initiation routes. Plain arrows indicate the amplification step. See section 1.3.1 of the text for details on the mechanisms and for additional actors involved.

dsRNA through the action of SGS3, SDE5 and RDR6 (Mourrain *et al.*, 2000; Jauvion *et al.*, 2010) and processed into siRNA duplexes by DCL4 to produce secondary 21-nt siRNAs that reinforce the cleavage of transgene mRNA through AGO1. Additional factors contribute to the efficiency of transgene PTGS, for example the RNA helicase SDE3 that binds to AGO1 (Dalmay *et al.*, 2001; Garcia *et al.*, 2012), or the RNA trafficking protein HYPER RECOMBINATION 1 (HPR1), which is likely to play a role in bringing RNA molecules to the right place during PTGS (Hernandez-Pinzon *et al.*, 2007; Jauvion *et al.*, 2010; Yelina *et al.*, 2010). In addition, the nuclear ribonucleoprotein SmD1 is likely to facilitate PTGS by protecting transgene aberrant RNAs from degradation by the RQC machinery in the nucleus, thus increasing the amount of transgene aberrant RNAs that succeed in entering siRNA-bodies in the cytoplasm to eventually activate PTGS (Elvira-Matelot *et al.*, 2016b).

### 1.3.2 RQC as a first layer of defence limiting PTGS

RQC encompasses RNA decay pathways that ensure the elimination of error-bearing RNAs. RQC should therefore eliminate the aberrant RNAs that activate PTGS. However, PTGS is regularly activated against pathogens and transgenes, probably because the amount of aberrant RNAs produced by viruses and transgenes exceeds the capacity of RQC pathways.

RQC generally involves the removal of the 5′ cap and/or the 3′ poly(A) tail. The removal of either modification is initiated when RNAs are not properly processed or translated. For example, when translation is arrested either owing to the presence of a premature termination codon or owing to excessive 3′ untranslated region (UTR) length, a process referred to as nonsense-mediated decay (NMD) is activated (Belostotsky and Sieburth, 2009). NMD generally involves the recruitment and activation of conserved UPFRAMESHIFT 1 (UPF1), UPF2 and UPF3 proteins to defective transcripts that are translationally stalled. This recruitment, either by invoking decapping and deadenylation pathways or via endonucleolytic cleavage, generates aberrant RNAs that are subsequently degraded through exonucleolytic cleavage. In *Arabidopsis*, the removal of the cap structure is catalysed by a set of conserved proteins that constitute the decapping complex, including DECAPPING1 (DCP1), DCP2, DCP5, VARICOSE (VCS) and DEAD BOX HELICASE HOMOLOG1 (DHH1) (Xu *et al.*, 2006; Goeres *et al.*, 2007; Iwasaki *et al.*, 2007). On the other hand, the shortening of the 3′ poly(A) tail (deadenylation) is catalysed by the conserved 3′-to-5′ POLY(A)-SPECIFIC RIBONUCLEASE (PARN) as well as by the conserved CARBON CATABOLITE REPRESSOR 4 (CCR4) complex (Belostotsky and Sieburth, 2009). The 5′-to-3′ XRN exoribonucleases degrade RNA with unprotected 5′ ends (Kastenmayer and Green, 2000), whereas the multimeric exosome complex contains 3′-to-5′ exoribonucleases that degrade RNA with unprotected 3′ ends (Chekanova *et al.*, 2007). *Arabidopsis* expresses three XRN proteins, the nucleolar XRN2, the nucleoplasmic XRN3, and cytoplasmic XRN4 (Kastenmayer and Green, 2000). Biochemical and molecular characterization of the *Arabidopsis* exosome core complex identified nine subunits: RIBOSOMAL RNA PROCESSING4 (RRP4), RRP40, RRP41, RRP42, RRP43, RRP45, RRP46, CENTROMERE ENHANCER OF POSITION EFFECT1 SYNTHETIC LETHAL PROTEIN4 (CLS4) and mRNA TRANSPORT REGULATOR3 (MTR3) (Chekanova *et al.*, 2007), plus specific cofactors that confer subcellular specialization; for example, MTR4 in the nucleolus, HEN2 in the nucleoplasm and the SUPERKILLER (SKI) complex in the cytoplasm (Lange *et al.*, 2014; Yu *et al.*, 2015; Zhang *et al.*, 2015).

Whereas RQC and PTGS were originally considered as exclusive pathways, eliminating endogenous aberrant RNAs and exogenous RNAs, respectively, it turned out that RQC generally serves as a first layer of defence against aberrant RNAs of both origins, and that PTGS is activated when RQC is unable to eliminate these aberrant RNAs. Indeed, compromising NMD factors UPF1 or UPF3, decapping enzymes DCP1, DCP2 or VCS, 5′-to-3′ exoribonucleases XRN2, XRN3 or XRN4, exosome core subunits RRP4 or RRP41, or exosome cofactors RRP6L1, MTR4, HEN2 or SKI3 enhance transgene PTGS (Gazzani *et al.*, 2004; Gy *et al.*, 2007; Thran *et al.*, 2012; Moreno *et al.*, 2013; Lange *et al.*, 2014; Zhang *et al.*, 2015; Hematy *et al.*, 2016), indicating that RQC limits the efficiency of PTGS. Moreover, mutations in XRN4 or

UPF1 also affected the efficiency of antiviral PTGS (Gy *et al.*, 2007; Garcia *et al.*, 2014). It is likely that aberrant RNAs are first exposed to degradation by RQC, and only if RQC is compromised or saturated do aberrant RNAs enter into siRNA-bodies where they are transformed into double-stranded RNA (dsRNA) by cellular RDR, thus allowing the production of siRNAs and the sequence-specific degradation of both functional and dysfunctional homologous mRNAs. Supporting this hypothesis, transgene loci that spontaneously trigger PTGS were found to produce uncapped RNAs at much higher levels than transgene loci that do not spontaneously trigger PTGS (Parent *et al.*, 2015b). Moreover, mutating XRN4 results in increased levels of uncapped RNAs from non-spontaneously triggering loci and subsequent triggering of PTGS (Parent *et al.*, 2015b). Also supporting the hypothesis that PTGS is triggered when RQC capacity is exceeded, P-bodies (where decapping enzymes reside) and siRNA-bodies (where cellular RDR6 resides) were found to constitute two distinct but adjacent foci (Jouannet *et al.*, 2012; Moreno *et al.*, 2013; Martínez de Alba *et al.*, 2015), suggesting that after saturating the degradation capacity of P-bodies, aberrant RNAs can move to siRNA-bodies to activate PTGS.

Remarkably, compromising decapping in *dcp2* and *vcs* mutants, or compromising both 5′-to-3′ and 3′-to-5′ RNA degradation in the *xrn4 ski2* double mutant provokes the entry of hundreds of endogenous mRNAs into the PTGS pathway and the production of siRNAs referred to as RNA quality control-specific siRNAs (rqc-siRNAs) or coding transcripts siRNAs (ct-siRNAs), respectively (Martínez de Alba *et al.*, 2015; Zhang *et al.*, 2015). In the conditions tested, ~1800 endogenous mRNAs produce rqc-siRNAs (Martínez de Alba *et al.*, 2015), while ~450 endogenous mRNAs produce ct-siRNAs (Zhang *et al.*, 2015), among which ~200 are common. Most of the ct-siRNAs identified in the *xrn4 ski2* double mutant depend on RDR6 for their production (441 out of 456), whereas only part of the rqc-siRNAs identified in *dcp2* and *vcs* mutants depend on RDR6 (350 out of 1785). Since rqc-siRNAs come from both strands, it is likely that another cellular RDR is at play for the production of certain rqc-siRNAs. RDR1 is a good candidate. Indeed, RDR1 has been recently implicated in the production of another category of endogenous siRNAs, called virus-activated siRNAs (vasiRNAs), which are produced from ~1200 endogenous protein-coding genes when plants are infected by viruses (Cao *et al.*, 2014). Most of the vasiRNAs identified in virus-infected plants depend on RDR1 for their production (1068 out of 1172). Remarkably, ~350 genes producing vasiRNAs in virus-infected plants produce rqc-siRNAs in *dcp2* and *vcs* mutants, supporting the hypothesis that RDR1 participates in the production of rqc-siRNAs in decapping mutants. These results also suggest that viruses could provoke the production of siRNAs from endogenous protein-coding genes by inhibiting RQC mechanisms, or by stimulating the production of aberrant RNAs up to a level that saturates the RQC pathway and triggers their entry into the PTGS pathway.

### 1.3.3 Specialized PTGS pathways directed against certain endogenous mRNA

As shown above, endogenous mRNAs are usually not targeted by PTGS because RQC pathways have evolved to efficiently eliminate aberrant RNAs produced by endogenous genes without producing siRNAs that could destroy functional mRNAs.

Nevertheless, plants and other eukaryotes have evolved specialized PTGS pathways to selectively regulate the abundance of certain endogenous mRNAs through the action of particular small RNAs, namely microRNAs (miRNAs), trans-acting siRNAs (ta-siRNAs) and natural antisense siRNAs (nat-siRNAs) (Fig. 1.2).

*MIR* genes are transcribed by PolII into long single-stranded primary transcripts (pri-miRNA), which exhibit typical PolII cap structures at their 5′ end and poly(A) tails at their 3′ end, and often contain introns (Jones-Rhoades *et al.*, 2006). They adopt a fold-back stem-loop structure that is processed into a mature miRNA duplex by DCL1 in *Arabidopsis* (Park *et al.*, 2002; Reinhart *et al.*, 2002; Kurihara and Watanabe, 2004). Accurate maturation and processing of pri-miRNA also requires the Cap-binding protein 20 (CBP20) and CBP80/ABH1 (Gregory *et al.*, 2008; Kim *et al.*, 2008; Laubinger *et al.*, 2008), the zinc finger protein SERRATE (SE) (Lobbes *et al.*, 2006; Yang *et al.*, 2006), the dsRNA binding protein/HYPONASTIC LEAVES 1 (DRB1/HYL1) (Han *et al.*, 2004; Vazquez *et al.*, 2004a), the Forkhead-associated (FHA) domain-containing protein DAWDLE (DDL) (Yu *et al.*, 2008), the TOUGH protein (TGH) (Ren *et al.*, 2012), the Proline-rich protein SICKLE (SIC) (Zhan *et al.*, 2012) and the RNA-binding protein MODIFIER OF SNC1, 2 (MOS2) (Wu *et al.*, 2013). miRNAs are methylated at their 3′ terminal nucleotide by the RNA methyltransferase HEN1 (Boutet *et al.*, 2003; Li *et al.*, 2005; Yu *et al.*, 2005) and most are exported to the cytoplasm by the exportin-5 homologue HASTY (HST) (Park *et al.*, 2005). One strand of the miRNA duplex acts as a guide strand and is selectively loaded onto an AGO protein, whereas the other strand, the passenger strand (miRNA*) is discarded from the complex and rapidly degraded. Most miRNAs associate to AGO1. However, specific association of miR408 or miR393* with AGO2, of miR390 with AGO7 and of miR165/166 with AGO10 have been reported (Mi *et al.*, 2008; Montgomery *et al.*, 2008a; Takeda *et al.*, 2008; Zhu *et al.*, 2011). Plant miRNAs promote the cleavage of their target RNA, to which they bind perfectly or near-perfectly, by employing mostly AGO1 as the RNA slicer. Therefore, cleavage is assumed as the common approach for miRNA-mediated gene regulation in plants (Rhoades *et al.*, 2002; Baumberger and Baulcombe, 2005; Schwab *et al.*, 2005). However, in addition to regulating RNA degradation, miRNAs sometimes direct DNA methylation (Bao *et al.*, 2004) or inhibit translation (Aukerman and Sakai, 2003; Chen, 2004; Gandikota *et al.*, 2007; Brodersen *et al.*, 2008; Lanet *et al.*, 2009; Mallory *et al.*, 2009). Although AGO1 per se is sufficient to promote RNA cleavage (Baumberger and Baulcombe, 2005), *in vivo* AGO1 activity appears modulated, directly or indirectly, by several cellular effectors, including the plant orthologue of Cyclophilin 40 SQUINT (SQN), the Heat Shock Protein 90 (HSP90) (Smith *et al.*, 2009), the F-Box protein FBW2 (Earley *et al.*, 2010), the importin b protein ENHANCED miRNA ACTIVITY (EMA1)//SUPER SENSITIVE TO ABA AND DROUGHT 2 (SAD2) (Wang *et al.*, 2011a), the GW-proteins SILENCING DEFECTIVE 3 (SDE3) (Garcia *et al.*, 2012) and SUO (Yang *et al.*, 2012). Moreover the amount of AGO1 mRNA is regulated by AGO1 (Vaucheret *et al.*, 2004; Mallory and Vaucheret, 2009) and AGO10 (Mallory *et al.*, 2009).

*TRANS ACTING siRNA* (*TAS*) genes are transcribed by PolII into long single-stranded RNAs that contain specific miRNA binding sites (Vazquez *et al.*, 2004b; Allen *et al.*, 2005; Vaucheret, 2005; Yoshikawa *et al.*, 2005; Rajagopalan *et al.*, 2006).

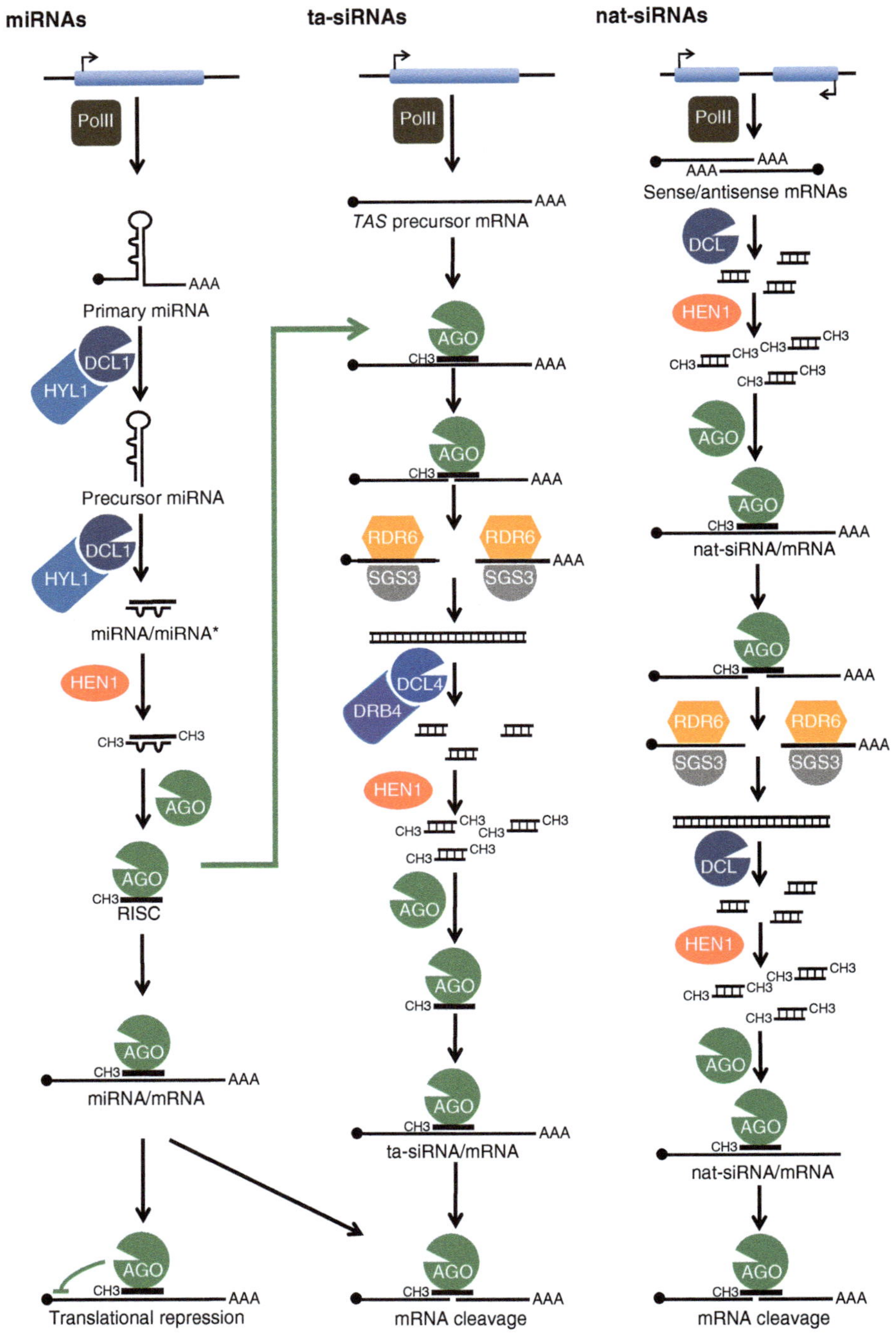


**Fig. 1.2.** Endogenous miRNA, ta-siRNA and nat-siRNA pathways. See section 1.3.3 of the text for details on the mechanisms and for additional actors involved.

It is likely that *TAS* RNAs are transferred by the THO/TREX complex to miRNA/AGO catalytic centres (Jauvion *et al.*, 2010; Yelina *et al.*, 2010). After cleavage, the RNA-binding SGS3 protein stabilizes the cleavage products, which probably prevents their degradation, allowing recruiting RDR6 which, assisted by the putative RNA export factor SDE5, catalyses the synthesis of a second complementary RNA strand (Yoshikawa *et al.*, 2005; Hernandez-Pinzon *et al.*, 2007; Elmayan *et al.*, 2009; Jauvion *et al.*, 2010). Next, DCL4 assisted by its interacting partner DsRNA BINDING PROTEIN 4 (DRB4) processes the dsRNA to generate a population of 21-nt ta-siRNAs in phase with the miRNA guided cleavage site (Gasciolli *et al.*, 2005; Xie *et al.*, 2005; Nakazawa *et al.*, 2007). Thus, the initial cleavage site guided by the miRNA determines the ta-siRNAs sequence and subsequently its targets (Vazquez *et al.*, 2004b; Allen *et al.*, 2005; Vaucheret, 2005; Yoshikawa *et al.*, 2005; Axtell *et al.*, 2006; Rajagopalan *et al.*, 2006; Montgomery *et al.*, 2008b). Similar to most miRNAs, ta-siRNAs duplexes are methylated by HEN1 (Li *et al.*, 2005) and one strand of the duplex associates with AGO1 to guide cleavage of target mRNAs (Allen and Howell, 2010).

If they are co-expressed, genes that are transcribed from complementary DNA strands at the same locus produce overlapping sense/antisense transcripts. Despite the fact that dsRNAs can result from the annealing of sense/antisense transcripts, the production of siRNAs referred to as nat-siRNAs not only requires a DCL, but also the activity of PolIV, RDR6 and SGS3 (Borsani *et al.*, 2005; Katiyar-Agarwal *et al.*, 2006). Primary nat-siRNAs are loaded onto a yet unidentified AGO protein to direct the cleavage of the constitutively expressed complementary transcript. In a second step, the cleaved transcript is converted into dsRNA in a PolIV-, RDR6- and SGS3-dependent manner (Borsani *et al.*, 2005). This RNA amplification step may extend beyond the overlapping region to form siRNAs outside the overlapping region. Further processing of the newly synthesized dsRNA in a DCL1-dependent fashion would generate 21-nt nat-siRNAs, which target the constitutive expressed transcripts (Borsani *et al.*, 2005). The RNA methyltransferase mutant *hen1* reduces the level of nat-siRNAs accumulation (Katiyar-Agarwal *et al.*, 2006), indicating that nat-siRNAs are methylated by HEN1 like other siRNAs. The involvement of so many factors in the biogenesis of nat-siRNAs implies that multiple layers of control exist and that the formation of the NAT pair may be necessary but not sufficient for the generation of nat-siRNAs. Recent genome-wide analyses showed the widespread existence of overlapping sense/antisense transcripts, which raises the possibility that nat-siRNAs could be major effectors of gene regulation. Although it is still unclear how many of these converging transcripts lead to RNA silencing, a fast and controlled production of nat-siRNAs could govern a plant-adaptive protection mechanism in response to either abiotic or biotic stress (Borsani *et al.*, 2005; Katiyar-Agarwal *et al.*, 2006).

### 1.3.4 PTGS pathways directed against transposons

Besides protein-coding genes, plant genomes contain many repeated sequences, including transposons, which need to be silenced to avoid inducing mutations and chromosome instability if multiplying within the genome. The way transposons and repeats are maintained in a transcriptionally silent state has been well deciphered (see TGS section below). However, how transposon silencing is initiated against active transposons is only starting to be understood. In *met1* mutants, loss of DNA methylation allowed reactivation of an intact *ATCOPIA93* family representative, *EVD18*. Crossing out *met1* allowed following the fate of this element, revealing that transposon mRNAs are first targeted by the PTGS machinery (RDR6, DCL4) to produce 21-nt siRNAs. However, these siRNAs fail to guide cleavage of transposon mRNAs because EVD encodes a nucleocapside that protects EVD mRNAs. Multiplication of the transposon leads to saturation of DCL4 and subsequent production of 24-nt by DCL3. These 24-nt siRNAs guide DNA methylation through AGO4, first within the EVD transcribed sequences, then spreading into the LTR (promoter) region, leading to TGS initiation (Mari-Ordonez *et al.*, 2013).

An alternative pathway was revealed when looking at the fate of an Athila6A element reactivated in *ddm1* mutants. Indeed, 21- and 22-nt siRNAs resulting from the degradation of Athila6A mRNAs by the PTGS machinery (RDR6, DCL2, DCL4, AGO1) can be directly incorporated into AGO6 to guide DNA methylation (McCue *et al.*, 2012; Nuthikattu *et al.*, 2013; McCue *et al.*, 2015) (Fig. 1.3).

Moreover, specific genomic loci, including TEs, were shown to undergo DNA methylation through atypical 21-22-nt siRNAs (Pontier *et al.*, 2012). This alternative TGS pathway is independent of RdDM components (RDR2, AGO4), but depends on classical PTGS pathway components, such as RDR6 and AGO2 (Pontier *et al.*, 2012). Moreover, this 21-22-nt-mediated DNA methylation pathway requires NEEDED FOR RDR2 INDEPENDENT DNA METHYLATION (NERD), a member of the GW repeat protein family, which generally binds to AGO proteins. NERD is thought to bind unmethylated histone H3 lysine 9 at specific genomic target loci and direct DNA methylation via its interaction with AGO2 bound to 21-22-nt siRNAs (Pontier *et al.*, 2012) (Fig. 1.3).

At last, transposon mRNA were shown to be targeted by PTGS-derived 21-nt siRNAs in cells where DDM1 is naturally not expressed, i.e. in the vegetative nucleus of pollen grains and in dedifferentiated plant cell cultures. Indeed, 21-nt siRNAs – referred to as epigenetically activated small interfering RNAs – (easiRNAs) are produced by thousands of transposon transcripts that are specifically targeted by more than 50 miRNAs. Similar to ta-siRNAs, easiRNAs result from the transformation of cleavage products into dsRNA by RDR6 and processing by DCL4. Therefore, miRNA-directed easiRNA production appears as a mechanism that specifically targets transposons when they are epigenetically reactivated during reprogramming of the germ line (Creasey *et al.*, 2014).

## 1.4 TGS Pathways

### 1.4.1 PolIV-RdDM pathway

Maintenance of transcriptional silencing at transposons and repeats involves PolIV-RdDM, i.e. RNA-directed DNA methylation mediated by PolIV-dependent

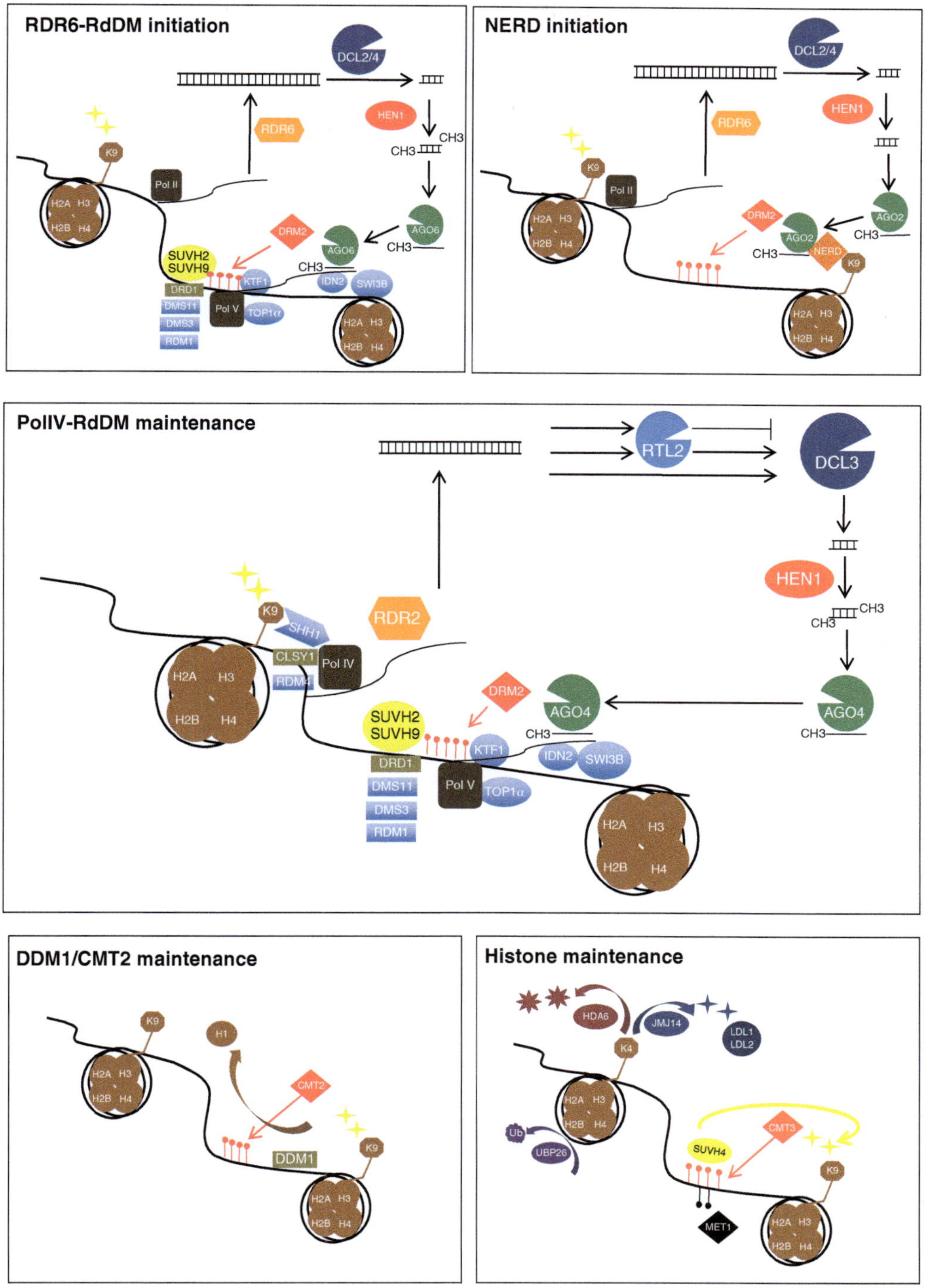


Fig. 1.3. Initiation and maintenance silencing pathways controlling transposons. See sections 1.3.4, 1.4.1 and 1.4.2 of the text for details on the mechanisms and for additional actors involved.

24-nt siRNAs (Henderson and Jacobsen, 2007; Zaratiegui *et al.*, 2007). The biogenesis of most 24-nt siRNAs depends on the plant-specific DNA-dependent RNA polymerase PolIV, a derivative of PolII, which shares several subunits with PolII but also exhibits specialized subunits, including NRPD1. PolIV is assisted by CLASSY1 (CLSY1), a SNF2-like chromatin remodelling factor (Herr *et al.*, 2005; Kanno *et al.*, 2005; Onodera *et al.*, 2005; Pontier *et al.*, 2005) and SAWADEE HOMEODOMAIN HOMOLOG1 (SHH1), a homeodomain protein which recognizes H3K9me2 (He *et al.*, 2009a; Law *et al.*, 2011; Johnson *et al.*, 2014). Single-stranded transcripts, referred to as P4-RNAs, are short, usually 27- to 45-nt in length (Blevins *et al.*, 2015; Zhai *et al.*, 2015). These precursors are depleted in both *polIV* and *rdr2* mutants, indicating that they are dependent on RDR2, reinforcing the idea that PolIV and RDR2 activities are coupled to produce P4R2 RNAs (Law *et al.*, 2011; Haag *et al.*, 2012). P4R2 RNAs are subsequently transformed into dsRNA through the action of RDR2 (Xie *et al.*, 2004), in partnership with INVOLVED IN DE NOVO 2 (IDN2)/RNA DIRECTED DNA METHYLATION 12 (RDM12), an RNA binding protein similar to SGS3 (Zheng *et al.*, 2010). Because of their short size, P4R2-RNAs each give rise to only one 24-nt siRNA, originating from either the 5′ or the 3′ end. These results fit well with the fact that DCL3 preferentially cleaves short precursors (Nagano *et al.*, 2014). Before being processed by DCL3, a fraction of P4R2 RNAs is cleaved by RNASE THREE LIKE 2 (RTL2). Depending on the P4R2 RNA considered, RTL2-mediated cleavage either reduces or enhances the production of 24-nt siRNAs, indicating that RTL2 acts as a modulator of 24-nt siRNA production at defined loci (Elvira-Matelot *et al.*, 2016a) (Fig. 1.3).

Like other small RNAs, DCL3-derived 24-nt siRNAs duplexes are methylated by HEN1 (Xie *et al.*, 2004). One strand of the duplex is loaded into a RISC-like complex containing AGO4, AGO6 or AGO9 (Havecker *et al.*, 2010). The complex formed by AGO proteins and 24-nt siRNA interacts with PolV-derived scaffold transcripts. PolV-dependent transcripts originate from the same regions as P4 RNAs, measure around 200bp long and have a cap structure at their 5′ end but lack the 3′ poly(A) tail (Wierzbicki *et al.*, 2008). PolV transcription involves the DDR complex (Law *et al.*, 2010), which contains DEFECTIVE IN RNA DIRECTED DNA METHYLATION 1 (DRD1) (Kanno *et al.*, 2004), a SNF2 chromatin remodelling factor, which acts in cooperation with RDM1, DEFECTIVE IN MERISTEM SILENCING 3 (DMS3) and the DMS11/GHKL ATPase (Lorkovic *et al.*, 2012). This PolV transcript serves as a scaffold molecule to recruit the *de novo* DNA methyltransferase DRM2 by interacting with the AGO-siRNA complex factor (Wierzbicki *et al.*, 2009) through a link made by RDM3/KTF1, a conserved PolII transcription elongation factor similar to SUPPRESSOR OF TY INSERTION 5 (SPT5) (He *et al.*, 2009b). RDM12/IDN2 contains an XS domain known to bind dsRNA with 5′ overhangs and is therefore hypothesized to stabilize interactions between AGO bound 24-nt siRNAs and PolV-derived scaffold transcripts (Ausin *et al.*, 2009). Very recently, IDN2 was also shown to mediate the interaction between PolV transcripts and SWI3B, a newly identified subunit of the SWI/SNF DNA remodelling complex, which regulates nucleosomes positioning (Zhu *et al.*, 2013). IDN2 forms a complex with IDN2 PARALOG 1 (IDP1) and IDP2 (IDN2-IDP1/IDP2 complex) (Zhang *et al.*, 2012) also identified as FACTOR OF DNA METHYLATION 1 (FDM1) and FDM2 in another study (Xie *et al.*, 2012) (Fig. 1.3).

Besides the core components of the RdDM pathway described above, there are many factors that contribute to regulating RdDM activity or RdDM components. For example, many components of the splicing machinery were also shown to modulate RdDM. It is unclear whether mutations in the SR 45 splicing factor directly affect RdDM or the splicing of members of the RdDM pathway (Ausin *et al.*, 2012). However, ZINC FINGER (ZnF) AND OCRE DOMAIN-CONTAINING PROTEIN 1 (ZOP1), MOS4 ASSOCIATED COMPLEX 3 (MAC3), MOS4, MOS12, MOS14 and several splicing factors such as PRECURSOR RNA PROCESSING (PRP3), PRP6 and PRP31 directly affect RdDM (Huang *et al.*, 2013; Zhang *et al.*, 2013; Du *et al.*, 2015), indicating that the splicing machinery can participate to a certain extent in non-coding RNA processing at the RdDM target loci.

The activity of PolIV and PolV polymerases is regulated by INTERACT WITH RNA POLII 1 (IWR1)/RDM4/DMS4, a transcription factor conserved in yeast (He *et al.*, 2009a; Kanno *et al.*, 2010; Law *et al.*, 2011). Consequently, *iwr1/rdm4/dms4* mutants show a reduced amount of PolIV-dependent primary 24-nt siRNAs, as well as a reduced amount of PolV-dependent intergenic transcripts, resulting in a loss of DNA methylation at the target loci. MORPHEUS' MOLECULE 1 (MOM1), which shares homology with the ATPase domain of the SWI2/SNF2 chromatin remodelling proteins, also regulates TGS via a complex interplay with polymerases PolIV or PolV (Yokthongwattana *et al.*, 2010). PolV activity also requires the action of TOPOISOMERASE ALPHA (TOP1α), via the release of DNA topological tension generated by transcription (Dinh *et al.*, 2014). Lastly, PolIV and PolV recruitment depends on factors that modify histones. Indeed, PolIV recruitment requires SHH1, which preferentially binds methylated H3K9 but not methylated H3K4 (Law *et al.*, 2011; Law *et al.*, 2013), whereas PolV requires SUVH2 and SUVH9 histone methyltransferases proteins for its recruitment to methylated DNA, through their SRA (SET and RING associated) domains. However, this recruitment does not depend on the histone modification activities of SUVH2 and SUVH9 because these two enzymes lack methyltransferase activity. Rather, immunoprecipitation experiments showed that SUVH2 could interact with the DDR complex (Johnson *et al.*, 2014) (Fig. 1.3).

DNA methylation is also influenced by chromatin factors. Histone modifications actually play a role in the DNA methylation maintenance through self-reinforcing loops between DNA methylation and histone methylation, mostly through the action of the histone methyltransferases SUVH2, SUVH4 (also called KYP), SUVH5 and SUVH6 (Jackson *et al.*, 2002; Ebbs *et al.*, 2005; Ebbs and Bender, 2006). For example, the histone methyltransferase SUVH4 is recruited to CHG methylation through its SRA domain and, in turn, CMT3 binds SUVH4-derived methylated H3K9 through its chromodomain, thus reinforcing CHG methylation (Johnson *et al.*, 2007). On the other hand, other histone-modifying enzymes remove active marks, thus promoting H3K9 methylation. Such histone-modifying enzymes include the histone deacetylase HDA6 (Aufsatz *et al.*, 2002), the histone demethylase JMJ14 (Deleris *et al.*, 2010) and the LYSINE-SPECIFIC HISTONE DEMETHYLASE 1 (LDL1) and LDL2 (Greenberg *et al.*, 2013). Finally, deubiquitylation of Histone 2B by UBP26 is required for heterochromatic histone H3 methylation and DNA methylation (Sridhar *et al.*, 2007).

Chromatin organization also plays a role in RdDM, independently of DNA methylation and histone modifications. For example, TEs are upregulated in *morc1* and *morc6* single or double mutants, but 24-nt siRNAs accumulation, DNA methylation and H3K9 methylation are unaffected. These mutants are characterized by a decondensation of chromocenters (Moissiard *et al.*, 2012; Moissiard *et al.*, 2014). Similarly, MOM1 affects RdDM independently on DNA methylation and histone modifications. However, it is likely to act differently from MORC1/6 because *mom1 morc6* double mutants show synergistic effects with *mom1* on common loci (Moissiard *et al.*, 2014).

Finally, the maintenance of methylation and TGS at RdDM targets is counterbalanced by several demethylating DNA glycosylases, including DEMETER (DME), DME-LIKE 2 (DML2) and DML3 (Penterman *et al.*, 2007; Ortega-Galisteo *et al.*, 2008). Moreover, active demethylation occurs through ROS1, which is guided by RNAs bound to the RNA-binding protein ROS3 (Zheng *et al.*, 2008). DNA demethylation can also be mediated through ROS4/IDM1, an acetyltransferase protein that binds to unmethylated histone H3K4, and generates acetylated marks on histone H3 (Qian *et al.*, 2012).

The majority of endogenous loci naturally targeted by RdDM are transposons and repeats. Nevertheless, a small fraction of protein-coding genes (PCG) produce 24-nt siRNAs, and this raises the question whether they contribute to regulating gene expression. Interestingly, 13% of the loci producing P4R2 RNAs that are regulated by RTL2 are located on PCG. DNA methylation and 24-nt siRNAs inversely correlate with mRNA accumulation at such loci, indicating that RdDM participates in modulating gene expression, at least at RTL2-regulated loci (Elvira-Matelot *et al.*, 2016a). DNA viruses also are targeted by RdDM. For example, like other plant viruses, geminiviruses are targeted by RNA silencing; but, unlike RNA viruses, the viral genome is targeted by small-RNA-directed methylation. This is probably because geminiviruses produce double-stranded DNA intermediates that associate with cellular histone proteins to form minichromosomes. Mutations in AGO4, CMT3, DRM2, SUVH4 or PolIV increase sensitivity to geminivirus infection, indicating that plants use chromatin methylation as a defence against DNA viruses. Nevertheless, geminiviruses counteract this defence by expressing proteins that interfere with the proper functioning of the plant DNA methylation cycle (Raja *et al.*, 2008).

### 1.4.2 DDM1/CMT2 pathway

Extensive CHH methylation is still observed in mutants impaired in PolIV-RdDM, implying that other pathway(s) contribute to maintaining CHH methylation. Such a pathway has recently been identified. It involves the SWI2/SNF2-Like protein DDM1 and the previously uncharacterized chromomethyltransferase CMT2. DDM1 is a chromatin-remodeller that binds nucleosomes and promotes nucleosome repositioning *in vitro* (Brzeski and Jerzmanowski, 2003). DDM1 has been identified as a critical factor for maintenance of DNA methylation as its loss of function leads to a 70% loss of methylation in TEs (Jeddeloh *et al.*, 1999). DDM1 and RdDM were long considered as independent pathways, as they synergistically silence rDNA loci (Blevins *et al.*, 2009). This hypothesis was recently confirmed by showing that DDM1 and RdDM, through DRD1, also have synergistic effects in silencing almost all TEs in *Arabidopsis* (Zemach *et al*., 2013). These authors showed that residual CHH methylation in *ddm1* correlates with residual methylation observed in *cmt2*, and anti-correlated with that in RdDM mutants, thus identifying CMT2 as part of the DDM1 pathway. DDM1/CMT2 and RdDM target different TEs and domains, and thus act synergistically to silence almost all TEs in *Arabidopsis*. The DDM1/CMT2 pathway preferentially targets long heterochromatic TEs, whereas the RdDM pathway is more efficient in targeting preferentially short TEs located at the vicinity of genes, in euchromatic regions. This is in agreement with the fact that RdDM requires transcription by PolIV and PolV. Therefore, DDM1/CMT2 can counteract the influence of linker histone H1, thus favouring methyltransferases access to heterochromatic regions (Zemach *et al*., 2013) (Fig. 1.3).

## 1.5 Conclusions

Regulatory mechanisms have been put in place to control invading nucleic acids from endogenous (mainly transposons) or exogenous (mainly viruses) origins. Silencing is mediated by siRNAs, and can occur at either transcriptional or post-transcriptional level. From these defence mechanisms, specialized pathways have emerged to control certain endogenous genes through the action of new classes of small RNAs: miRNA, ta-siRNA or nat-siRNA. Interestingly, the development of transformation methods has revealed that transgenes are more prone to silencing than are endogenous genes, suggesting that the position and arrangement of genes within the genome is not random, and that genes cannot be moved around without perturbing expression.

## References

Allen, E. and Howell, M.D. (2010) miRNAs in the biogenesis of trans-acting siRNAs in higher plants. *Seminars in Cell and Developmental Biology* 21, 798–804.

Allen, E., Xie, Z., Gustafson, A.M. and Carrington, J.C. (2005) microRNA-directed phasing during trans-acting siRNA biogenesis in plants. *Cell* 121, 207–221.

Aufsatz, W., Mette, M.F., van der Winden, J., Matzke, M. and Matzke, A.J. (2002) HDA6, a putative histone deacetylase needed to enhance DNA methylation induced by double-stranded RNA. *The EMBO Journal* 21, 6832–6841.

Aukerman, M.J. and Sakai, H. (2003) Regulation of flowering time and floral organ identity by a MicroRNA and its APETALA2-like target genes. *Plant Cell* 15, 2730–2741.

Ausin, I., Mockler, T.C., Chory, J. and Jacobsen, S.E. (2009) IDN1 and IDN2 are required for *de novo* DNA methylation in *Arabidopsis thaliana*. *Nature Structural and Molecular Biology* 16, 1325–1327.

Ausin, I., Greenberg, M.V., Li, C.F. and Jacobsen, S.E. (2012) The splicing factor SR45 affects the RNA-directed DNA methylation pathway in *Arabidopsis*. *Epigenetics* 7, 29–33.

Axtell, M.J., Jan, C., Rajagopalan, R. and Bartel, D.P. (2006) A two-hit trigger for siRNA biogenesis in plants. *Cell* 127, 565–577.

Bao, N., Lye, K.W. and Barton, M.K. (2004) MicroRNA binding sites in *Arabidopsis* class III HD-ZIP mRNAs are required for methylation of the template chromosome. *Developmental Cell* 7, 653–662.

Baumberger, N. and Baulcombe, D.C. (2005) *Arabidopsis* ARGONAUTE1 is an RNA Slicer that selectively recruits microRNAs and short interfering RNAs. *Proceedings of the National Academy of Sciences of the United States of America* 102, 11928–11933.

Belostotsky, D.A. and Sieburth, L.E. (2009) Kill the messenger: mRNA decay and plant development. *Current Opinion in Plant Biology* 12, 96–102.

Bender, J. and Fink, G.R. (1995) Epigenetic control of an endogenous gene family is revealed by a novel blue fluorescent mutant of *Arabidopsis*. *Cell* 83, 725–734.

Blevins, T., Pontes, O., Pikaard, C.S. and Meins, F. Jr (2009) Heterochromatic siRNAs and DDM1 independently silence aberrant 5S rDNA transcripts in *Arabidopsis*. *PLoS One* 4, e5932.

Blevins, T., Podicheti, R., Mishra, V., Marasco, M., Tang, H. and Pikaard, C.S. (2015) Identification of Pol IV and RDR2-dependent precursors of 24 nt siRNAs guiding *de novo* DNA methylation in *Arabidopsis*. *Elife* 4, e09591.

Borsani, O., Zhu, J., Verslues, P.E., Sunkar, R. and Zhu, J.K. (2005) Endogenous siRNAs derived from a pair of natural cis-antisense transcripts regulate salt tolerance in *Arabidopsis*. *Cell* 123, 1279–1291.

Bouche, N., Lauressergues, D., Gasciolli, V. and Vaucheret, H. (2006) An antagonistic function for *Arabidopsis* DCL2 in development and a new function for DCL4 in generating viral siRNAs. *The EMBO Journal* 25, 3347–3356.

Boutet, S., Vazquez, F., Liu, J., Beclin, C., Fagard, M., Gratias, A., Morel, J.B., Crete, P., Chen, X. and Vaucheret, H. (2003) *Arabidopsis* HEN1: a genetic link between endogenous miRNA controlling development and siRNA controlling transgene silencing and virus resistance. *Current Biology* 13, 843–848.

Brodersen, P., Sakvarelidze-Achard, L., Bruun-Rasmussen, M., Dunoyer, P., Yamamoto, Y.Y., Sieburth, L. and Voinnet, O. (2008) Widespread translational inhibition by plant miRNAs and siRNAs. *Science* 320, 1185–1190.

Brosseau, C. and Moffett, P. (2015) Functional and genetic analysis identify a role for *Arabidopsis* ARGONAUTE5 in antiviral RNA silencing. *Plant Cell* 27, 1742–1754.

Brzeski, J. and Jerzmanowski, A. (2003) Deficient in DNA methylation 1 (DDM1) defines a novel family of chromatin-remodeling factors. *The Journal of Biological Chemistry* 278, 823–828.

Cao, M., Du, P., Wang, X., Yu, Y.Q., Qiu, Y.H., Li, W., Gal-On, A., Zhou, C., Li, Y. and Ding, S.W. (2014) Virus infection triggers widespread silencing of host genes by a distinct class of endogenous siRNAs in *Arabidopsis*. *Proceedings of the National Academy of Sciences of the United States of America* 111, 14613–14618.

Chekanova, J.A., Gregory, B.D., Reverdatto, S.V., Chen, H., Kumar, R., Hooker, T., Yazaki, J., Li, P., Skiba, N., Peng, Q., Alonso, J., Brukhin, V., Grossniklaus, U., Ecker, J.R. and Belostotsky, D.A. (2007) Genome-wide high-resolution mapping of exosome substrates reveals hidden features in the *Arabidopsis* transcriptome. *Cell* 131, 1340–1353.

Chen, X. (2004) A microRNA as a translational repressor of APETALA2 in *Arabidopsis* flower development. *Science* 303, 2022–2025.

Clough, S.J., Tuteja, J.H., Li, M., Marek, L.F., Shoemaker, R.C. and Vodkin, L.O. (2004) Features of a 103-kb gene-rich region in soybean include an inverted perfect repeat cluster of CHS genes comprising the I locus. *Genome* 47, 819–831.

Coen, E.S. and Carpenter, R. (1988) A semi-dominant allele, niv-525, acts in trans to inhibit expression of its wild-type homologue in *Antirrhinum majus*. *The EMBO Journal* 7, 877–883.

Creasey, K.M., Zhai, J., Borges, F., Van Ex, F., Regulski, M., Meyers, B.C. and Martienssen, R.A. (2014) miRNAs trigger widespread epigenetically activated siRNAs from transposons in *Arabidopsis*. *Nature* 508, 411–415.

Csorba, T., Kontra, L. and Burgyan, J. (2015) Viral silencing suppressors: tools forged to fine-tune host-pathogen coexistence. *Virology* 479–480, 85–103.

Cubas, P., Vincent, C. and Coen, E. (1999) An epigenetic mutation responsible for natural variation in floral symmetry. *Nature* 401, 157–161.

Dalmay, T., Horsefield, R., Braunstein, T.H. and Baulcombe, D.C. (2001) SDE3 encodes an RNA helicase required for post-transcriptional gene silencing in *Arabidopsis*. *The EMBO Journal* 20, 2069–2078.

Deleris, A., Gallego-Bartolome, J., Bao, J., Kasschau, K.D., Carrington, J.C. and Voinnet, O. (2006) Hierarchical action and inhibition of plant dicer-like proteins in antiviral defense. *Science* 313, 68–71.

Deleris, A., Greenberg, M.V., Ausin, I., Law, R.W., Moissiard, G., Schubert, D. and Jacobsen, S.E. (2010) Involvement of a Jumonji-C domain-containing histone demethylase in DRM2-mediated maintenance of DNA methylation. *EMBO Reports* 11, 950–955.

Della Vedova, C.B., Lorbiecke, R., Kirsch, H., Schulte, M.B., Scheets, K., Borchert, L.M., Scheffler, B.E., Wienand, U., Cone, K.C. and Birchler, J.A. (2005) The dominant inhibitory chalcone synthase allele C2-Idf (inhibitor diffuse) from *Zea mays* (L.) acts via an endogenous RNA silencing mechanism. *Genetics* 170, 1989–2002.

Dinh, T.T., Gao, L., Liu, X., Li, D., Li, S., Zhao, Y., O'Leary, M., Le, B., Schmitz, R.J., Manavella, P.A., Weigel, D., Pontes, O., Ecker, J.R. and Chen, X. (2014) DNA topoisomerase 1alpha promotes transcriptional silencing of transposable elements through DNA methylation and histone lysine 9 dimethylation in *Arabidopsis*. *PLoS Genetics* 10, e1004446.

Du, J.L., Zhang, S.W., Huang, H.W., Cai, T., Li, L., Chen, S. and He, X.J. (2015) The splicing factor PRP31 is involved in transcriptional gene silencing and stress response in *Arabidopsis*. *Molecular Plant* 8, 1053–1068.

Durand, S., Bouche, N., Perez Strand, E., Loudet, O. and Camilleri, C. (2012) Rapid establishment of genetic incompatibility through natural epigenetic variation. *Current Biology* 22, 326–331.

Earley, K., Smith, M., Weber, R., Gregory, B. and Poethig, R. (2010) An endogenous F-box protein regulates ARGONAUTE1 in *Arabidopsis thaliana*. *Silence* 1, 15.

Ebbs, M.L. and Bender, J. (2006) Locus-specific control of DNA methylation by the *Arabidopsis* SUVH5 histone methyltransferase. *Plant Cell* 18, 1166–1176.

Ebbs, M.L., Bartee, L. and Bender, J. (2005) H3 lysine 9 methylation is maintained on a transcribed inverted repeat by combined action of SUVH6 and SUVH4 methyltransferases. *Molecular and Cellular Biology* 25, 10507–10515.

Elmayan, T., Adenot, X., Gissot, L., Lauressergues, D., Gy, I. and Vaucheret, H. (2009) A neomorphic sgs3 allele stabilizing miRNA cleavage products reveals that SGS3 acts as a homodimer. *FEBS J* 276, 835–844.

Elvira-Matelot, E., Hachet, M., Shamandi, N., Comella, P., Saez-Vasquez, J., Zytnicki, M. and Vaucheret, H. (2016a) *Arabidopsis* RNASE THREE LIKE2 modulates the expression of protein-coding genes via 24-Nucleotide small interfering RNA-directed DNA methylation. *Plant Cell* 28, 406–425.

Elvira-Matelot, E., Bardou, F., Ariel, F., Jauvion, V., Bouteiller, N., Le Masson, I., Cao, J., Crespi, M.D. and Vaucheret, H. (2016b) The nuclear ribonucleoprotein SmD1 interplays with splicing, RNA quality control, and posttranscriptional gene silencing in *Arabidopsis*. *Plant Cell* 28, 426–438.

Fusaro, A.F., Matthew, L., Smith, N.A., Curtin, S.J., Dedic-Hagan, J., Ellacott, G.A., Watson, J.M., Wang, M.B., Brosnan, C., Carroll, B.J. and Waterhouse, P.M. (2006) RNA interference-inducing hairpin RNAs in plants act through the viral defence pathway. *EMBO Reports* 7, 1168–1175.

Gandikota, M., Birkenbihl, R.P., Hohmann, S., Cardon, G.H., Saedler, H. and Huijser, P. (2007) The miRNA156/157 recognition element in the 3' UTR of the *Arabidopsis* SBP box gene SPL3 prevents early flowering by translational inhibition in seedlings. *The Plant Journal* 49, 683–693.

Garcia, D., Garcia, S., Pontier, D., Marchais, A., Renou, J.P., Lagrange, T. and Voinnet, O. (2012) Ago hook and RNA helicase motifs underpin dual roles for SDE3 in antiviral defense and silencing of nonconserved intergenic regions. *Molecular Cell* 48, 109–120.

Garcia, D., Garcia, S. and Voinnet, O. (2014) Nonsense-mediated decay serves as a general viral restriction mechanism in plants. *Cell Host and Microbe* 16, 391–402.

Gasciolli, V., Mallory, A.C., Bartel, D.P. and Vaucheret, H. (2005) Partially redundant functions of *Arabidopsis* DICER-like enzymes and a role for DCL4 in producing trans-acting siRNAs. *Current Biology* 15, 1494–1500.

Gazzani, S., Lawrenson, T., Woodward, C., Headon, D. and Sablowski, R. (2004) A link between mRNA turnover and RNA interference in *Arabidopsis*. *Science* 306, 1046–1048.

Goeres, D.C., Van Norman, J.M., Zhang, W., Fauver, N.A., Spencer, M.L. and Sieburth, L.E. (2007) Components of the *Arabidopsis* mRNA decapping complex are required for early seedling development. *Plant Cell* 19, 1549–1564.

Greenberg, M.V., Deleris, A., Hale, C.J., Liu, A., Feng, S. and Jacobsen, S.E. (2013) Interplay between active chromatin marks and RNA-directed DNA methylation in *Arabidopsis thaliana*. *PLoS Genetics* 9, e1003946.

Gregory, B.D., O'Malley, R.C., Lister, R., Urich, M.A., Tonti-Filippini, J., Chen, H., Millar, A.H. and Ecker, J.R. (2008) A link between RNA metabolism and silencing affecting *Arabidopsis* development. *Developmental Cell* 14, 854–866.

Gy, I., Gasciolli, V., Lauressergues, D., Morel, J.B., Gombert, J., Proux, F., Proux, C., Vaucheret, H. and Mallory, A.C. (2007) *Arabidopsis* FIERY1, XRN2, and XRN3 are endogenous RNA silencing suppressors. *Plant Cell* 19, 3451–3461.

Haag, J.R., Ream, T.S., Marasco, M., Nicora, C.D., Norbeck, A.D., Pasa-Tolic, L. and Pikaard, C.S. (2012) In vitro transcription activities of Pol IV, Pol V, and RDR2 reveal coupling of Pol IV and RDR2 for dsRNA synthesis in plant RNA silencing. *Molecular Cell* 48, 811–818.

Hamilton, A.J. and Baulcombe, D.C. (1999) A species of small antisense RNA in posttranscriptional gene silencing in plants. *Science* 286, 950–952.

Han, M.H., Goud, S., Song, L. and Fedoroff, N. (2004) The *Arabidopsis* double-stranded RNA-binding protein HYL1 plays a role in microRNA-mediated gene regulation. *Proceedings of the National Academy of Sciences of the United States of America* 101, 1093–1098.

Harvey, J.J., Lewsey, M.G., Patel, K., Westwood, J., Heimstadt, S., Carr, J.P. and Baulcombe, D.C. (2011) An antiviral defense role of AGO2 in plants. *PLoS One* 6, e14639.

Havecker, E.R., Wallbridge, L.M., Hardcastle, T.J., Bush, M.S., Kelly, K.A., Dunn, R.M., Schwach, F., Doonan, J.H. and Baulcombe, D.C. (2010) The *Arabidopsis* RNA-directed DNA methylation argonautes functionally diverge based on their expression and interaction with target loci. *Plant Cell* 22, 321–334.

He, X.J., Hsu, Y.F., Pontes, O., Zhu, J., Lu, J., Bressan, R.A., Pikaard, C., Wang, C.S. and Zhu, J.K. (2009a) NRPD4, a protein related to the RPB4 subunit of RNA polymerase II, is a component of RNA polymerases IV and V and is required for RNA-directed DNA methylation. *Genes and Development* 23, 318–330.

He, X.J., Hsu, Y.F., Zhu, S., Liu, H.L., Pontes, O., Zhu, J., Cui, X., Wang, C.S. and Zhu, J.K. (2009b) A conserved transcriptional regulator is required for RNA-directed DNA methylation and plant development. *Genes and Development* 23, 2717–2722.

Hematy, K., Bellec, Y., Podicheti, R., Bouteiller, N., Anne, P., Morineau, C., Haslam, R.P., Beaudoin, F., Napier, J.A., Mockaitis, K., Gagliardi, D., Vaucheret, H., Lange, H. and Faure, J.D. (2016) The Zinc-finger protein SOP1 is required for a subset of the nuclear exosome functions in *Arabidopsis*. *PLoS Genetics* 12, e1005817.

Henderson, I.R. and Jacobsen, S.E. (2007) Epigenetic inheritance in plants. *Nature* 447, 418–424.

Hernandez-Pinzon, I., Yelina, N.E., Schwach, F., Studholme, D.J., Baulcombe, D. and Dalmay, T. (2007) SDE5, the putative homologue of a human mRNA export factor, is required for transgene silencing and accumulation of trans-acting endogenous siRNA. *The Plant Journal* 50, 140–148.

Herr, A.J., Jensen, M.B., Dalmay, T. and Baulcombe, D.C. (2005) RNA polymerase IV directs silencing of endogenous DNA. *Science* 308, 118–120.

Huang, C.F., Miki, D., Tang, K., Zhou, H.R., Zheng, Z., Chen, W., Ma, Z.Y., Yang, L., Zhang, H., Liu, R., He, X.J. and Zhu, J.K. (2013) A Pre-mRNA-splicing factor is required for RNA-directed DNA methylation in *Arabidopsis*. *PLoS Genetics* 9, e1003779.

Ito, H., Gaubert, H., Bucher, E., Mirouze, M., Vaillant, I. and Paszkowski, J. (2011) An siRNA pathway prevents transgenerational retrotransposition in plants subjected to stress. *Nature* 472, 115–119.

Iwasaki, S., Takeda, A., Motose, H. and Watanabe, Y. (2007) Characterization of *Arabidopsis* decapping proteins AtDCP1 and AtDCP2, which are essential for post-embryonic development. *FEBS Letters* 581, 2455–2459.

Jackson, J.P., Lindroth, A.M., Cao, X. and Jacobsen, S.E. (2002) Control of CpNpG DNA methylation by the KRYPTONITE histone H3 methyltransferase. *Nature* 416, 556–560.

Jauvion, V., Elmayan, T. and Vaucheret, H. (2010) The conserved RNA trafficking proteins HPR1 and TEX1 are involved in the production of endogenous and exogenous small interfering RNA in *Arabidopsis*. *Plant Cell* 22, 2697–2709.

Jeddeloh, J.A., Stokes, T.L. and Richards, E.J. (1999) Maintenance of genomic methylation requires a SWI2/SNF2-like protein. *Nature Genetics* 22, 94–97.

Johnson, L.M., Bostick, M., Zhang, X., Kraft, E., Henderson, I., Callis, J. and Jacobsen, S.E. (2007) The SRA methyl-cytosine-binding domain links DNA and histone methylation. *Current Biology* 17, 379–384.

Johnson, L.M., Du, J., Hale, C.J., Bischof, S., Feng, S., Chodavarapu, R.K., Zhong, X., Marson, G., Pellegrini, M., Segal, D.J., Patel, D.J. and Jacobsen, S.E. (2014) SRA- and SET-domain-containing proteins link RNA polymerase V occupancy to DNA methylation. *Nature* 507, 124–128.

Jones-Rhoades, M.W., Bartel, D.P. and Bartel, B. (2006) MicroRNAS and their regulatory roles in plants. *Annual Review of Plant Biology* 57, 19–53.

Jouannet, V., Moreno, A.B., Elmayan, T., Vaucheret, H., Crespi, M.D. and Maizel, A. (2012) Cytoplasmic *Arabidopsis* AGO7 accumulates in membrane-associated siRNA bodies and is required for ta-siRNA biogenesis. *The EMBO Journal* 31, 1704–1713.

Kanno, T., Mette, M.F., Kreil, D.P., Aufsatz, W., Matzke, M. and Matzke, A.J. (2004) Involvement of putative SNF2 chromatin remodeling protein DRD1 in RNA-directed DNA methylation. *Current Biology* 14, 801–805.

Kanno, T., Huettel, B., Mette, M.F., Aufsatz, W., Jaligot, E., Daxinger, L., Kreil, D.P., Matzke, M. and Matzke, A.J. (2005) Atypical RNA polymerase subunits required for RNA-directed DNA methylation. *Nature Genetics* 37, 761–765.

Kanno, T., Bucher, E., Daxinger, L., Huettel, B., Kreil, D.P., Breinig, F., Lind, M., Schmitt, M.J., Simon, S.A., Gurazada, S.G., Meyers, B.C., Lorkovic, Z.J., Matzke, A.J. and Matzke, M. (2010) RNA-directed DNA methylation and plant development require an IWR1-type transcription factor. *EMBO Reports* 11, 65–71.

Kastenmayer, J.P. and Green, P.J. (2000) Novel features of the XRN-family in *Arabidopsis*: evidence that AtXRN4, one of several orthologs of nuclear Xrn2p/Rat1p, functions in the cytoplasm. *Proceedings of the National Academy of Sciences of the United States of America* 97, 13985–13990.

Katiyar-Agarwal, S., Morgan, R., Dahlbeck, D., Borsani, O., Villegas, A. Jr, Zhu, J.K., Staskawicz, B.J. and Jin, H. (2006) A pathogen-inducible endogenous siRNA in plant immunity. *Proceedings of the National Academy of Sciences of the United States of America* 103, 18002–18007.

Kim, S., Yang, J.Y., Xu, J., Jang, I.C., Prigge, M.J. and Chua, N.H. (2008) Two cap-binding proteins CBP20 and CBP80 are involved in processing primary MicroRNAs. *Plant and Cell Physiology* 49, 1634–1644.

Kurihara, Y. and Watanabe, Y. (2004) *Arabidopsis* micro-RNA biogenesis through Dicer-like 1 protein functions. *Proceedings of the National Academy of Sciences of the United States of America* 101, 12753–12758.

Lanet, E., Delannoy, E., Sormani, R., Floris, M., Brodersen, P., Crete, P., Voinnet, O. and Robaglia, C. (2009) Biochemical evidence for translational repression by *Arabidopsis* microRNAs. *Plant Cell* 21, 1762–1768.

Lange, H., Zuber, H., Sement, F.M., Chicher, J., Kuhn, L., Hammann, P., Brunaud, V., Berard, C., Bouteiller, N., Balzergue, S., Aubourg, S., Martin-Magniette, M.L., Vaucheret, H. and Gagliardi, D. (2014) The RNA helicases AtMTR4 and HEN2 target specific subsets of nuclear transcripts for degradation by the nuclear exosome in *Arabidopsis thaliana*. *PLoS Genetics* 10, e1004564.

Laubinger, S., Sachsenberg, T., Zeller, G., Busch, W., Lohmann, J.U., Ratsch, G. and Weigel, D. (2008) Dual roles of the nuclear cap-binding complex and SERRATE in pre-mRNA splicing and microRNA processing in *Arabidopsis thaliana*. *Proceedings of the National Academy of Sciences of the United States of America* 105, 8795–8800.

Law, J.A., Ausin, I., Johnson, L.M., Vashisht, A.A., Zhu, J.K., Wohlschlegel, J.A. and Jacobsen, S.E. (2010) A protein complex required for polymerase V transcripts and RNA- directed DNA methylation in *Arabidopsis*. *Current Biology* 20, 951–956.

Law, J.A., Vashisht, A.A., Wohlschlegel, J.A. and Jacobsen, S.E. (2011) SHH1, a homeodomain protein required for DNA methylation, as well as RDR2, RDM4, and chromatin remodeling factors, associate with RNA polymerase IV. *PLoS Genetics* 7, e1002195.

Law, J.A., Du, J., Hale, C.J., Feng, S., Krajewski, K., Palanca, A.M., Strahl, B.D., Patel, D.J. and Jacobsen, S.E. (2013) Polymerase IV occupancy at RNA-directed DNA methylation sites requires SHH1. *Nature* 498, 385–389.

Li, J., Yang, Z., Yu, B., Liu, J. and Chen, X. (2005) Methylation protects miRNAs and siRNAs from a 3'-end uridylation activity in *Arabidopsis*. *Current Biology* 15, 1501–1507.

Llave, C., Kasschau, K.D., Rector, M.A. and Carrington, J.C. (2002) Endogenous and silencing-associated small RNAs in plants. *Plant Cell* 14, 1605–1619.

Lobbes, D., Rallapalli, G., Schmidt, D.D., Martin, C. and Clarke, J. (2006) SERRATE: a new player on the plant microRNA scene. *EMBO Reports* 7, 1052–1058.

Lorkovic, Z.J., Naumann, U., Matzke, A.J. and Matzke, M. (2012) Involvement of a GHKL ATPase in RNA-directed DNA methylation in *Arabidopsis thaliana*. *Current Biology* 22, 933–938.

Mallory, A.C. and Vaucheret, H. (2009) ARGONAUTE 1 homeostasis invokes the coordinate action of the microRNA and siRNA pathways. *EMBO Reports* 10, 521–526.

Mallory, A.C., Hinze, A., Tucker, M.R., Bouche, N., Gasciolli, V., Elmayan, T., Lauressergues, D., Jauvion, V., Vaucheret, H. and Laux, T. (2009) Redundant and specific roles of the ARGONAUTE proteins AGO1 and ZLL in development and small RNA-directed gene silencing. *PLoS Genetics* 5, e1000646.

Manning, K., Tor, M., Poole, M., Hong, Y., Thompson, A.J., King, G.J., Giovannoni, J.J. and Seymour, G.B. (2006) A naturally occurring epigenetic mutation in a gene encoding an SBP-box transcription factor inhibits tomato fruit ripening. *Nature Genetics* 38, 948–952.

Mari-Ordonez, A., Marchais, A., Etcheverry, M., Martin, A., Colot, V. and Voinnet, O. (2013) Reconstructing *de novo* silencing of an active plant retrotransposon. *Nature Genetics* 45, 1029–1039.

Martin, A., Troadec, C., Boualem, A., Rajab, M., Fernandez, R., Morin, H., Pitrat, M., Dogimont, C. and Bendahmane, A. (2009) A transposon-induced epigenetic change leads to sex determination in melon. *Nature* 461, 1135–1138.

Martínez de Alba, A.E., Moreno, A.B., Gabriel, M., Mallory, A.C., Christ, A., Bounon, R., Balzergue, S., Aubourg, S., Gautheret, D., Crespi, M.D., Vaucheret, H. and Maizel, A. (2015) In plants, decapping prevents RDR6-dependent production of small interfering RNAs from endogenous mRNAs. *Nucleic Acids Research* 43, 2902–2913.

Matzke, M.A., Primig, M., Trnovsky, J. and Matzke, A.J. (1989) Reversible methylation and inactivation of marker genes in sequentially transformed tobacco plants. *The EMBO Journal* 8, 643–649.

McCue, A.D., Nuthikattu, S., Reeder, S.H. and Slotkin, R.K. (2012) Gene expression and stress response mediated by the epigenetic regulation of a transposable element small RNA. *PLoS Genetics* 8, e1002474.

McCue, A.D., Panda, K., Nuthikattu, S., Choudury, S.G., Thomas, E.N. and Slotkin, R.K. (2015) ARGONAUTE 6 bridges transposable element mRNA-derived siRNAs to the establishment of DNA methylation. *The EMBO Journal* 34, 20–35.

Mi, S., Cai, T., Hu, Y., Chen, Y., Hodges, E., Ni, F., Wu, L., Li, S., Zhou, H., Long, C., Chen, S., Hannon, G.J. and Qi, Y. (2008) Sorting of small RNAs into *Arabidopsis* argonaute complexes is directed by the 5' terminal nucleotide. *Cell* 133, 116–127.

Moissiard, G., Cokus, S.J., Cary, J., Feng, S., Billi, A.C., Stroud, H., Husmann, D., Zhan, Y., Lajoie, B.R., McCord, R.P., Hale, C.J., Feng, W., Michaels, S.D., Frand, A.R., Pellegrini, M., Dekker, J., Kim, J.K. and Jacobsen, S.E. (2012) MORC family ATPases required for heterochromatin condensation and gene silencing. *Science* 336, 1448–1451.

Moissiard, G., Bischof, S., Husmann, D., Pastor, W.A., Hale, C.J., Yen, L., Stroud, H., Papikian, A., Vashisht, A.A., Wohlschlegel, J.A. and Jacobsen, S.E. (2014) Transcriptional gene silencing by *Arabidopsis* microrchidia homologues involves the formation of heteromers. *Proceedings of the National Academy of Sciences of the United States of America* 111, 7474–7479.

Montgomery, T.A., Howell, M.D., Cuperus, J.T., Li, D., Hansen, J.E., Alexander, A.L., Chapman, E.J., Fahlgren, N., Allen, E. and Carrington, J.C. (2008a) Specificity of ARGONAUTE7-miR390 interaction and dual functionality in TAS3 trans-acting siRNA formation. *Cell* 133, 128–141.

Montgomery, T.A., Yoo, S.J., Fahlgren, N., Gilbert, S.D., Howell, M.D., Sullivan, C.M., Alexander, A., Nguyen, G., Allen, E., Ahn, J.H. and Carrington, J.C. (2008b) AGO1-miR173 complex initiates phased siRNA formation in plants. *Proceedings of the National Academy of Sciences of the United States of America* 105, 20055–20062.

Morel, J.B., Godon, C., Mourrain, P., Beclin, C., Boutet, S., Feuerbach, F., Proux, F. and Vaucheret, H. (2002) Fertile hypomorphic ARGONAUTE (ago1) mutants impaired in post-transcriptional gene silencing and virus resistance. *Plant Cell* 14, 629–639.

Moreno, A.B., Martinez de Alba, A.E., Bardou, F., Crespi, M.D., Vaucheret, H., Maizel, A. and Mallory, A.C. (2013) Cytoplasmic and nuclear quality control and turnover of single-stranded RNA modulate post-transcriptional gene silencing in plants. *Nucleic Acids Research* 41, 4699–4708.

Mourrain, P., Beclin, C., Elmayan, T., Feuerbach, F., Godon, C., Morel, J.B., Jouette, D., Lacombe, A.M., Nikic, S., Picault, N., Remoue, K., Sanial, M., Vo, T.A. and Vaucheret, H. (2000) *Arabidopsis* SGS2 and SGS3 genes are required for posttranscriptional gene silencing and natural virus resistance. *Cell* 101, 533–542.

Nagano, H., Fukudome, A., Hiraguri, A., Moriyama, H. and Fukuhara, T. (2014) Distinct substrate specificities of *Arabidopsis* DCL3 and DCL4. *Nucleic Acids Research* 42, 1845–1856.

Nakazawa, Y., Hiraguri, A., Moriyama, H. and Fukuhara, T. (2007) The dsRNA-binding protein DRB4 interacts with the Dicer-like protein DCL4 in vivo and functions in the trans-acting siRNA pathway. *Plant Molecular Biology* 63, 777–785.

Napoli, C., Lemieux, C. and Jorgensen, R. (1990) Introduction of a chimeric chalcone synthase gene into petunia results in reversible co-suppression of homologous genes in trans. *Plant Cell* 2, 279–289.

Navarro, L., Jay, F., Nomura, K., He, S.Y. and Voinnet, O. (2008) Suppression of the microRNA pathway by bacterial effector proteins. *Science* 321, 964–967.

Nuthikattu, S., McCue, A.D., Panda, K., Fultz, D., DeFraia, C., Thomas, E.N. and Slotkin, R.K. (2013) The initiation of epigenetic silencing of active transposable elements is triggered by RDR6 and 21-22 nucleotide small interfering RNAs. *Plant Physiology* 162, 116–131.

Onodera, Y., Haag, J.R., Ream, T., Costa Nunes, P., Pontes, O. and Pikaard, C.S. (2005) Plant nuclear RNA polymerase IV mediates siRNA and DNA methylation-dependent heterochromatin formation. *Cell* 120, 613–622.

Ortega-Galisteo, A.P., Morales-Ruiz, T., Ariza, R.R. and Roldan-Arjona, T. (2008) *Arabidopsis* DEMETER-LIKE proteins DML2 and DML3 are required for appropriate distribution of DNA methylation marks. *Plant Molecular Biology* 67, 671–681.

Parent, J.S., Bouteiller, N., Elmayan, T. and Vaucheret, H. (2015a) Respective contributions of *Arabidopsis* DCL2 and DCL4 to RNA silencing. *The Plant Journal* 81, 223–232.

Parent, J.S., Jauvion, V., Bouche, N., Beclin, C., Hachet, M., Zytnicki, M. and Vaucheret, H. (2015b) Post-transcriptional gene silencing triggered by sense transgenes involves uncapped antisense RNA and differs from silencing intentionally triggered by antisense transgenes. *Nucleic Acids Research* 43, 8464–8475.

Park, M.Y., Wu, G., Gonzalez-Sulser, A., Vaucheret, H. and Poethig, R.S. (2005) Nuclear processing and export of microRNAs in *Arabidopsis*. *Proceedings of the National Academy of Sciences of the United States of America* 102, 3691–3696.

Park, W., Li, J., Song, R., Messing, J. and Chen, X. (2002) CARPEL FACTORY, a Dicer homolog, and HEN1, a novel protein, act in microRNA metabolism in *Arabidopsis thaliana*. *Current Biology* 12, 1484–1495.

Pecinka, A., Dinh, H.Q., Baubec, T., Rosa, M., Lettner, N. and Mittelsten Scheid, O. (2010) Epigenetic regulation of repetitive elements is attenuated by prolonged heat stress in *Arabidopsis*. *Plant Cell* 22, 3118–3129.

Penterman, J., Zilberman, D., Huh, J.H., Ballinger, T., Henikoff, S. and Fischer, R.L. (2007) DNA demethylation in the *Arabidopsis* genome. *Proceedings of the National Academy of Sciences of the United States of America* 104, 6752–6757.

Pontier, D., Yahubyan, G., Vega, D., Bulski, A., Saez-Vasquez, J., Hakimi, M.A., Lerbs-Mache, S., Colot, V. and Lagrange, T. (2005) Reinforcement of silencing at transposons and highly repeated sequences requires the concerted action of two distinct RNA polymerases IV in *Arabidopsis*. *Genes and Development* 19, 2030–2040.

Pontier, D., Picart, C., Roudier, F., Garcia, D., Lahmy, S., Azevedo, J., Alart, E., Laudie, M., Karlowski, W.M., Cooke, R., Colot, V., Voinnet, O. and Lagrange, T. (2012) NERD, a plant-specific GW protein, defines an additional RNAi-dependent chromatin-based pathway in *Arabidopsis*. *Molecular Cell* 48, 121–132.

Pumplin, N. and Voinnet, O. (2013) RNA silencing suppression by plant pathogens: defence, counter-defence and counter-counter-defence. *Nature Reviews Microbiology* 11, 745–760.

Qian, W., Miki, D., Zhang, H., Liu, Y., Zhang, X., Tang, K., Kan, Y., La, H., Li, X., Li, S., Zhu, X., Shi, X., Zhang, K., Pontes, O., Chen, X., Liu, R., Gong, Z. and Zhu, J.K. (2012) A histone acetyltransferase regulates active DNA demethylation in *Arabidopsis*. *Science* 336, 1445–1448.

Qu, F., Ye, X. and Morris, T.J. (2008) *Arabidopsis* DRB4, AGO1, AGO7, and RDR6 participate in a DCL4-initiated antiviral RNA silencing pathway negatively regulated by DCL1. *Proceedings of the National Academy of Sciences of the United States of America* 105, 14732–14737.

Raja, P., Sanville, B.C., Buchmann, R.C. and Bisaro, D.M. (2008) Viral genome methylation as an epigenetic defense against geminiviruses. *Journal of Virology* 82, 8997–9007.

Rajagopalan, R., Vaucheret, H., Trejo, J. and Bartel, D.P. (2006) A diverse and evolutionarily fluid set of microRNAs in *Arabidopsis thaliana*. *Genes and Development* 20, 3407–3425.

Reinhart, B.J., Weinstein, E.G., Rhoades, M.W., Bartel, B. and Bartel, D.P. (2002) MicroRNAs in plants. *Genes and Development* 16, 1616–1626.

Ren, G., Xie, M., Dou, Y., Zhang, S., Zhang, C. and Yu, B. (2012) Regulation of miRNA abundance by RNA binding protein TOUGH in *Arabidopsis*. *Proceedings of the National Academy of Sciences of the United States of America* 109, 12817–12821.

Rhoades, M.W., Reinhart, B.J., Lim, L.P., Burge, C.B., Bartel, B. and Bartel, D.P. (2002) Prediction of plant microRNA targets. *Cell* 110, 513–520.

Schwab, R., Palatnik, J.F., Riester, M., Schommer, C., Schmid, M. and Weigel, D. (2005) Specific effects of microRNAs on the plant transcriptome. *Developmental Cell* 8, 517–527.

Smith, C.J., Watson, C.F., Bird, C.R., Ray, J., Schuch, W. and Grierson, D. (1990) Expression of a truncated tomato polygalacturonase gene inhibits expression of the endogenous gene in transgenic plants. *Molecular and General Genetics* 224, 477–481.

Smith, M.R., Willmann, M.R., Wu, G., Berardini, T.Z., Moller, B., Weijers, D. and Poethig, R.S. (2009) Cyclophilin 40 is required for microRNA activity in *Arabidopsis*. *Proceedings of the National Academy of Sciences of the United States of America* 106, 5424–5429.

Sridhar, V.V., Kapoor, A., Zhang, K., Zhu, J., Zhou, T., Hasegawa, P.M., Bressan, R.A. and Zhu, J.K. (2007) Control of DNA methylation and heterochromatic silencing by histone H2B deubiquitination. *Nature* 447, 735–738.

Takeda, A., Iwasaki, S., Watanabe, T., Utsumi, M. and Watanabe, Y. (2008) The mechanism selecting the guide strand from small RNA duplexes is different among argonaute proteins. *Plant and Cell Physiology* 49, 493–500.

Thran, M., Link, K. and Sonnewald, U. (2012) The *Arabidopsis* DCP2 gene is required for proper mRNA turnover and prevents transgene silencing in *Arabidopsis*. *The Plant Journal* 72, 368–377.

Tuteja, J.H., Clough, S.J., Chan, W.C. and Vodkin, L.O. (2004) Tissue-specific gene silencing mediated by a naturally occurring chalcone synthase gene cluster in *Glycine max*. *Plant Cell* 16, 819–835.

Tuteja, J.H., Zabala, G., Varala, K., Hudson, M. and Vodkin, L.O. (2009) Endogenous, tissue-specific short interfering RNAs silence the chalcone synthase gene family in *Glycine max* seed coats. *Plant Cell* 21, 3063–3077.

van der Krol, A.R., Mur, L.A., Beld, M., Mol, J.N. and Stuitje, A.R. (1990) Flavonoid genes in petunia: addition of a limited number of gene copies may lead to a suppression of gene expression. *Plant Cell* 2, 291–299.

Vaucheret, H. (2005) MicroRNA-dependent trans-acting siRNA production. *Science's STKE* 2005, pe43.

Vaucheret, H. (2008) Plant ARGONAUTES. *Trends in Plant Science* 13, 350–358.

Vaucheret, H., Vazquez, F., Crete, P. and Bartel, D.P. (2004) The action of ARGONAUTE1 in the miRNA pathway and its regulation by the miRNA pathway are crucial for plant development. *Genes and Development* 18, 1187–1197.

Vazquez, F., Gasciolli, V., Crete, P. and Vaucheret, H. (2004a) The nuclear dsRNA binding protein HYL1 is required for microRNA accumulation and plant development, but not posttranscriptional transgene silencing. *Current Biology* 14, 346–351.

Vazquez, F., Vaucheret, H., Rajagopalan, R., Lepers, C., Gasciolli, V., Mallory, A.C., Hilbert, J.L., Bartel, D.P. and Crete, P. (2004b) Endogenous trans-acting siRNAs regulate the accumulation of *Arabidopsis* mRNAs. *Molecular Cell* 16, 69–79.

Wang, W., Ye, R., Xin, Y., Fang, X., Li, C., Shi, H., Zhou, X. and Qi, Y. (2011a) An importin beta protein negatively regulates MicroRNA activity in *Arabidopsis*. *Plant Cell* 23, 3565–3576.

Wang, X.B., Jovel, J., Udomporn, P., Wang, Y., Wu, Q., Li, W.X., Gasciolli, V., Vaucheret, H. and Ding, S.W. (2011b) The 21-nucleotide, but not 22-nucleotide, viral secondary small interfering RNAs direct potent antiviral defense by two cooperative argonautes in *Arabidopsis thaliana*. *Plant Cell* 23, 1625–1638.

Wierzbicki, A.T., Haag, J.R. and Pikaard, C.S. (2008) Noncoding transcription by RNA polymerase Pol IVb/Pol V mediates transcriptional silencing of overlapping and adjacent genes. *Cell* 135, 635–648.

Wierzbicki, A.T., Ream, T.S., Haag, J.R. and Pikaard, C.S. (2009) RNA polymerase V transcription guides ARGONAUTE4 to chromatin. *Nature Genetics* 41, 630–634.

Wu, X., Shi, Y., Li, J., Xu, L., Fang, Y., Li, X. and Qi, Y. (2013) A role for the RNA-binding protein MOS2 in microRNA maturation in *Arabidopsis*. *Cell Research* 23(5), 645–647.

Xie, M., Ren, G., Zhang, C. and Yu, B. (2012) The DNA- and RNA-binding protein FACTOR of DNA METHYLATION 1 requires XH domain-mediated complex formation for its function in RNA-directed DNA methylation. *The Plant Journal* 72, 491–500.

Xie, Z., Johansen, L.K., Gustafson, A.M., Kasschau, K.D., Lellis, A.D., Zilberman, D., Jacobsen, S.E. and Carrington, J.C. (2004) Genetic and functional diversification of small RNA pathways in plants. *PLoS Biology* 2, E104.

Xie, Z., Allen, E., Wilken, A. and Carrington, J.C. (2005) DICER-LIKE 4 functions in trans-acting small interfering RNA biogenesis and vegetative phase change in *Arabidopsis thaliana*. *Proceedings of the National Academy of Sciences of the United States of America* 102, 12984–12989.

Xu, J., Yang, J.Y., Niu, Q.W. and Chua, N.H. (2006) *Arabidopsis* DCP2, DCP1, and VARICOSE form a decapping complex required for postembryonic development. *Plant Cell* 18, 3386–3398.

Yang, L., Liu, Z., Lu, F., Dong, A. and Huang, H. (2006) SERRATE is a novel nuclear regulator in primary microRNA processing in *Arabidopsis*. *The Plant Journal* 47, 841–850.

Yang, L., Wu, G. and Poethig, R.S. (2012) Mutations in the GW-repeat protein SUO reveal a developmental function for microRNA-mediated translational repression in *Arabidopsis*. *Proceedings of the National Academy of Sciences of the United States of America* 109, 315–320.

Yelina, N.E., Smith, L.M., Jones, A.M., Patel, K., Kelly, K.A. and Baulcombe, D.C. (2010) Putative *Arabidopsis* THO/TREX mRNA export complex is involved in transgene and endogenous siRNA biosynthesis. *Proceedings of the National Academy of Sciences of the United States of America* 107, 13948–13953.

Yokthongwattana, C., Bucher, E., Caikovski, M., Vaillant, I., Nicolet, J., Mittelsten Scheid, O. and Paszkowski, J. (2010) MOM1 and Pol-IV/V interactions regulate the intensity and specificity of transcriptional gene silencing. *The EMBO Journal* 29, 340–351.

Yoshikawa, M., Peragine, A., Park, M.Y. and Poethig, R.S. (2005) A pathway for the biogenesis of trans-acting siRNAs in *Arabidopsis*. *Genes and Development* 19, 2164–2175.

Yoshikawa, M., Iki, T., Tsutsui, Y., Miyashita, K., Poethig, R.S., Habu, Y. and Ishikawa, M. (2013) 3' fragment of miR173-programmed RISC-cleaved RNA is protected from degradation in a complex with RISC and SGS3. *Proceedings of the National Academy of Sciences of the United States of America* 110, 4117–4122.

Yu, A., Saudemont, B., Bouteiller, N., Elvira-Matelot, E., Lepere, G., Parent, J.S., Morel, J.B., Cao, J., Elmayan, T. and Vaucheret, H. (2015) Second-site mutagenesis of a hypomorphic argonaute1 allele identifies SUPERKILLER3 as an endogenous suppressor of transgene posttranscriptional gene silencing. *Plant Physiology* 169, 1266–1274.

Yu, B., Yang, Z., Li, J., Minakhina, S., Yang, M., Padgett, R.W., Steward, R. and Chen, X. (2005) Methylation as a crucial step in plant microRNA biogenesis. *Science* 307, 932–935.

Yu, B., Bi, L., Zheng, B., Ji, L., Chevalier, D., Agarwal, M., Ramachandran, V., Li, W., Lagrange, T., Walker, J.C. and Chen, X. (2008) The FHA domain proteins DAWDLE in *Arabidopsis* and SNIP1 in humans act in small RNA biogenesis. *Proceedings of the National Academy of Sciences of the United States of America* 105, 10073–10078.

Zaratiegui, M., Irvine, D.V. and Martienssen, R.A. (2007) Noncoding RNAs and gene silencing. *Cell* 128, 763–776.

Zemach, A., Kim, M.Y., Hsieh, P.H., Coleman-Derr, D., Eshed-Williams, L., Thao, K., Harmer, S.L. and Zilberman, D. (2013) The *Arabidopsis* nucleosome remodeler DDM1 allows DNA methyltransferases to access H1-containing heterochromatin. *Cell* 153, 193–205.

Zhai, J., Bischof, S., Wang, H., Feng, S., Lee, T.F., Teng, C., Chen, X., Park, S.Y., Liu, L., Gallego-Bartolome, J., Liu, W., Henderson, I.R., Meyers, B.C., Ausin, I. and Jacobsen, S.E. (2015) A one precursor one siRNA model for Pol IV-dependent siRNA biogenesis. *Cell* 163, 445–455.

Zhan, X., Wang, B., Li, H., Liu, R., Kalia, R.K., Zhu, J.K. and Chinnusamy, V. (2012) *Arabidopsis* proline-rich protein important for development and abiotic stress tolerance is involved in microRNA biogenesis. *Proceedings of the National Academy of Sciences of the United States of America* 109, 18198–18203.

Zhang, C.J., Ning, Y.Q., Zhang, S.W., Chen, Q., Shao, C.R., Guo, Y.W., Zhou, J.X., Li, L., Chen, S. and He, X.J. (2012) IDN2 and its paralogs form a complex required for RNA-directed DNA methylation. *PLoS Genetics* 8, e1002693.

Zhang, C.J., Zhou, J.X., Liu, J., Ma, Z.Y., Zhang, S.W., Dou, K., Huang, H.W., Cai, T., Liu, R., Zhu, J.K. and He, X.J. (2013) The splicing machinery promotes RNA-directed DNA methylation and transcriptional silencing in *Arabidopsis*. *The EMBO Journal* 32, 1128–1140.

Zhang, X., Zhu, Y., Liu, X., Hong, X., Xu, Y., Zhu, P., Shen, Y., Wu, H., Ji, Y., Wen, X., Zhang, C., Zhao, Q., Wang, Y., Lu, J. and Guo, H. (2015) Plant biology. Suppression of endogenous gene silencing by bidirectional cytoplasmic RNA decay in *Arabidopsis*. *Science* 348, 120–123.

Zheng, X., Pontes, O., Zhu, J., Miki, D., Zhang, F., Li, W.X., Iida, K., Kapoor, A., Pikaard, C.S. and Zhu, J.K. (2008) ROS3 is an RNA-binding protein required for DNA demethylation in *Arabidopsis*. *Nature* 455, 1259–1262.

Zheng, Z., Xing, Y., He, X.J., Li, W., Hu, Y., Yadav, S.K., Oh, J. and Zhu, J.K. (2010) An SGS3-like protein functions in RNA-directed DNA methylation and transcriptional gene silencing in *Arabidopsis*. *The Plant Journal* 62, 92–99.

Zhu, H., Hu, F., Wang, R., Zhou, X., Sze, S.H., Liou, L.W., Barefoot, A., Dickman, M. and Zhang, X. (2011) *Arabidopsis* Argonaute10 specifically sequesters miR166/165 to regulate shoot apical meristem development. *Cell* 145, 242–256.

Zhu, Y., Rowley, M.J., Bohmdorfer, G. and Wierzbicki, A.T. (2013) A SWI/SNF chromatin-remodeling complex acts in noncoding RNA-mediated transcriptional silencing. *Molecular Cell* 49, 298–309.

# 2 Induction and Suppression of Silencing by Plant Viruses

**Ares Mingot[1], Adrian Valli[2], Juan José López-Moya[1,*] and Juan Antonio García[2,*]**

*[1]Centre for Research in Agricultural Genomics, Barcelona, Spain; [2]National Centre for Biotechnology, Madrid, Spain*

## 2.1 Introduction

RNA silencing refers to an evolutionary conserved system that controls gene expression through sequence-specific mechanisms guided by small RNAs. This system is present in most eukaryotic organisms, playing important roles in different biological processes such as development, genome integrity and host defence against viruses and transposons (Siomi and Siomi, 2009).

The antiviral role of RNA silencing was discovered in plants (Hamilton and Baulcombe, 1999), but it is also present in insects (Li *et al.*, 2002), nematodes (Lu *et al.*, 2005; Wilkins *et al.*, 2005), fungi (Segers *et al.*, 2007) and, under certain circumstances, it has also been proposed to be functional against viruses in mammals (Li *et al.*, 2013; Maillard *et al.*, 2013). This defensive strategy relies on the accumulation of viral-derived small interfering RNAs (vsiRNAs), which have the ability to promote degradation of the complementary viral genome once they are loaded in the effector complexes (Valli *et al.*, 2009). Viruses, in turn, have developed a counter-defensive strategy based on the expression of RNA silencing suppressors (RSSs) able to block or interfere with the host silencing pathway (Csorba *et al.*, 2015).

An overview of the plant antiviral silencing pathway is provided in the next section.

## 2.2 Antiviral RNA Silencing Pathway in Plants

All RNA silencing pathways, including the antiviral one, share a core machinery with several essential components that are activated by the presence of double-stranded RNAs (dsRNAs) precursors. These dsRNAs are first processed by RNase III-like enzymes belonging to the Dicer family (DCLs), which produce short (s)RNA duplexes (21–24 nt). Additionally, a second wave of these duplexes might derive from new dsRNAs, synthesized by RNA-dependent RNA polymerases (RDRs). The

* Corresponding author e-mail: juanjose.lopez@cragenomica.es
* Corresponding author e-mail: jagarcia@cnb.csic.es

  *Plant Gene Silencing: Mechanisms and Applications* (ed. T. Dalmay)

generated sRNA duplexes are recruited by Argonaute proteins (AGOs), and the so-called 'passenger strand' of the duplex is selectively displaced and degraded, while the 'guide strand' is retained to conduct the complex towards complementary RNA/DNA sequences in order to promote silencing (Bologna and Voinnet, 2014). Figure 2.1 shows a schematic view of the RNA silencing pathway that is involved in plant antiviral defence.

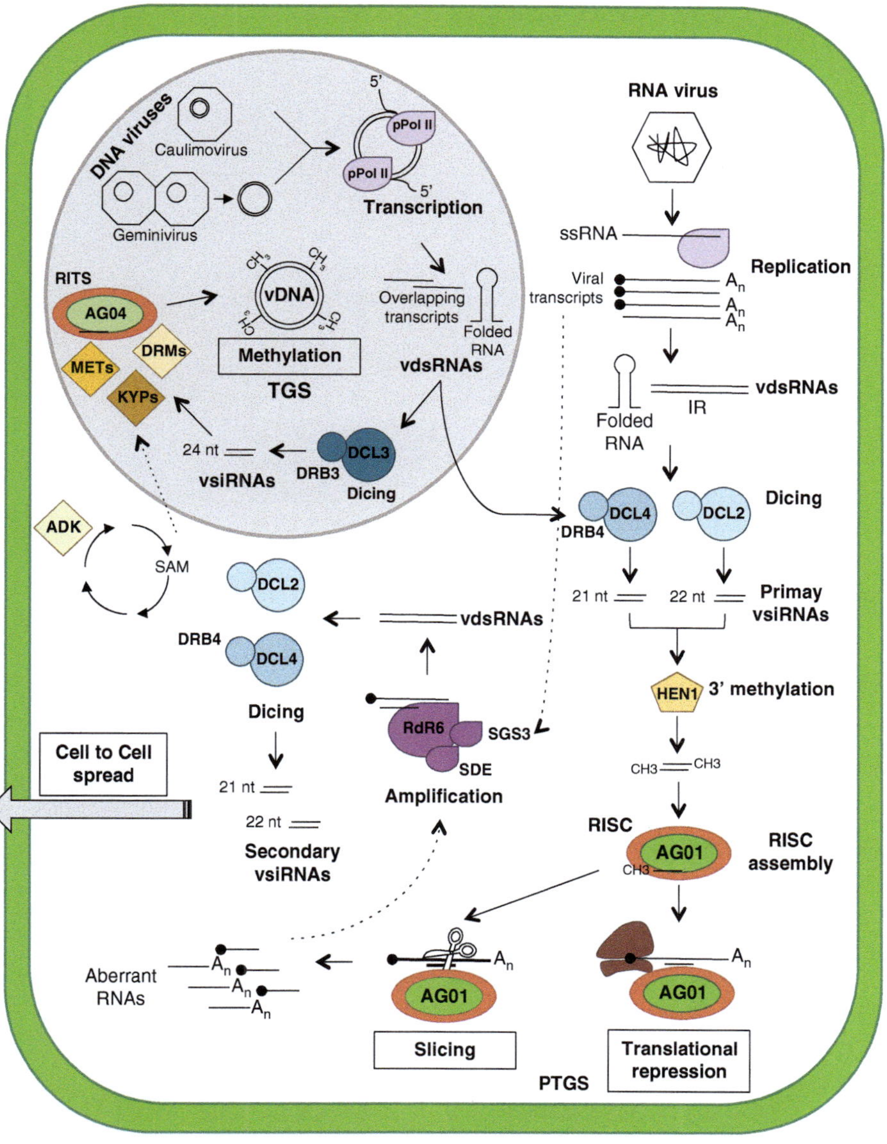


**Fig. 2.1.** Schematic representation of the antiviral silencing pathway in plant cells. The main components involved in the different steps of the general pathway are represented, as well as some of their predicted functions against RNA and DNA plant viruses.

The core components of the silencing machinery are DCL, RDR and AGO proteins. In the model plant *Arabidopsis thaliana* there are four DCLs, ten AGOs and six RDRs, and they play roles in the different silencing pathways in a specific or redundant manner (Csorba *et al.*, 2015). The roles of these factors in the diverse phases of antiviral silencing pathways are described below.

### 2.2.1 vsiRNAs biogenesis

The biogenesis of vsiRNAs is analogous to that of endogenous plant siRNAs, consisting in two well-differentiated phases. The initiation phase gives rise to the production of primary vsiRNAs through the action of DCLs. After viral replication the amplification phase, mediated by RDRs and DCLs enzymes, results in a secondary production of vsiRNAs (Zhang *et al.*, 2015).

#### *Initiation phase (DCLs, recognition and processing)*

Upon virus infection, viral dsRNA molecules are generated in the infected cell and can be used as substrate for DCLs to produce primary vsiRNAs. In the case of positive single-stranded RNA (+ssRNA) viruses, the origin of dsRNAs can rely on replicative intermediates (Ahlquist, 2002) and on highly specific folded regions of the viral RNA (Molnár *et al.*, 2005; Donaire *et al.*, 2008; Donaire *et al.*, 2009; Szittya *et al.*, 2010; Xu *et al.*, 2012), although recent results addressing the relative amounts of + or − small RNAs suggest that technical issues might be masking the true origin (Harris *et al.*, 2015) . For DNA viruses, dsRNAs might have their origin in products of overlapping bidirectional transcription (Chellappan *et al.*, 2004; Aregger *et al.*, 2012), or in highly structured RNA regions, as in the case of the 5´ terminal region of the 35S major RNA transcript of *Cauliflower mosaic virus* (Blevins *et al.*, 2011).

Studies with the model plant *A. thaliana* showed that DCLs, helped by dsRNA binding proteins (DRBs), recognize viral dsRNAs in a hierarchical manner and process them into vsiRNAs of different lengths (DCL4 mainly produces siRNAs of 21 nt, DCL2 of 22 nt and DCL3 of 24 nt) (Zhang *et al.*, 2015). During infections of +ssRNA viruses, DCL4 appears to be the main producer of vsiRNAs, whereas DCL2 can replace DCL4 when its activity is reduced or suppressed (Bouché *et al.*, 2006; Deleris *et al.*, 2006; Fusaro *et al.*, 2006; Du *et al.*, 2007; Qu *et al.*, 2008; Garcia-Ruiz *et al.*, 2010). Furthermore, DCL2 seems to be responsible for vsiRNA generation from specific viral regions (Donaire *et al.*, 2008). Although DCL3 acts in some particular cases, its role is much less relevant in the defence against these viruses (Diaz-Pendon *et al.*, 2007; Donaire *et al.*, 2008; Qu *et al.*, 2008; Azevedo *et al.*, 2010). Interestingly, DCL1 can act as negative regulator of DCL3 and DCL4 expression, repressing the antiviral silencing response (Qu *et al.*, 2008). Regarding viruses with DNA genomes, all four DCLs are involved in the vsiRNA production. DCL2, DCL3 and DCL4 are direct generators of siRNAs, whereas DCL1 processes viral dsRNAs only in some particular cases. Moreover, DCL1 can play an additional role by facilitating the tasks of the other DCLs (Akbergenov *et al.*, 2006; Blevins *et al.*, 2006; Aregger *et al.*, 2012).

Other players are involved in the initiation phase, such as the case of DRBs, whose interactions with DCLs are necessary to reach the optimal ribonuclease activity of the processing complexes. The involvement in defence of the DRB4/DCL4 pair has been reported for infections of both DNA and RNA viruses (Haas *et al.*, 2008; Qu *et al.*, 2008), and the cooperation of DRB3 with DCL3 seems to be relevant for genome methylation of geminiviruses (Raja *et al.*, 2014). Interestingly, a recent report suggests a possible antiviral role of RNase III-like 1 (RTL1), an enzyme that lacks DCL-specific domains, which appears to degrade dsRNA intermediates of viral replication (Shamandi *et al.*, 2015).

Once vsiRNAs are generated, HUA enhancer 1 (HEN1) methylates their 3′ end to avoid U-tailing and further destabilization/degradation (Boutet *et al.*, 2003; Blevins *et al.*, 2006; Lózsa *et al.*, 2008; Zhang *et al.*, 2012). vsiRNAs then trigger diverse silencing responses. It is generally assumed that vsiRNAs produced by the action of DCL2 and DCL4 lead to post-transcriptional gene silencing (PTGS), whereas those derived from DCL3 cleavage are involved in transcriptional gene silencing (TGS) (Rodríguez-Negrete *et al.*, 2009; Raja *et al.*, 2010).

### *Amplification phase (RDRs + DCLs)*

During the amplification phase, plant RDRs trigger the synthesis of dsRNA molecules, which act as substrates of DCLs for the production of secondary vsiRNAs (Zhang *et al.*, 2015).

To generate the dsRNAs precursors of secondary vsiRNAs, RDRs use aberrant viral RNAs lacking certain features, such as the 5′ cap or the polyA tail, as templates (Gazzani *et al.*, 2004; Luo and Chen, 2007; Donaire *et al.*, 2008; Voinnet, 2008; Moreno *et al.*, 2013). Some reports suggest that aberrant RNAs come from cleaved viral fragments, targeted by primary vsiRNAs loaded in AGO proteins (Moissiard *et al.*, 2007; Mallory and Vaucheret, 2009) (see below to know more about the effector phase). On the other hand, primary vsiRNAs could be acting as primers in the amplification event (Moissiard *et al.*, 2007). For RNA viruses, the implication of RDR1, RDR2 and RDR6 has been well demonstrated in the amplification step and, consequently, in the antiviral defence response (Xie *et al.*, 2001; Yu *et al.*, 2003; Schwach *et al.*, 2005; Diaz-Pendon *et al.*, 2007; Donaire *et al.*, 2008; Qu *et al.*, 2008; Wang *et al.*, 2010; Garcia-Ruiz *et al.*, 2010; Lee *et al.*, 2016). However, in infections by DNA viruses it seems that viral transcripts are poor templates for RDRs (Aregger *et al.*, 2012). The efficient action of RDRs requires interaction with cofactors such as the RNA helicase SDE3, trafficking proteins (SDE5, HPR1 and TEX1) and the blocker of RNA degradation SGS3 (Mourrain *et al.*, 2000; Dalmay *et al.*, 2001; Kumakura *et al.*, 2009; Jauvion *et al.*, 2010; Wang *et al.*, 2011). After the synthesis of dsRNA molecules by the action of RDRs, DCL4 and DCL2 process them to give rise to a batch of secondary vsiRNAs (Wang *et al.*, 2010; Parent *et al.*, 2015).

It is thought that the amplification phase not only increases the potency of local antiviral silencing, but might also lead to a systemic response that immunizes non-infected tissues and therefore prevents the spread of the infection to healthy areas of the plant (Qu *et al.*, 2005; Schwach *et al.*, 2005; Ruiz-Ferrer and Voinnet, 2009; Wang *et al.*, 2011).

### 2.2.2 Effector phase (AGOs + RNA-induced silencing complexes (RISC) complex)

The action of DCLs has an impact over the accumulation of viral RNAs, but it is not enough to prevent the spread of viruses through the whole plant (Valli *et al.*, 2009; Carbonell and Carrington, 2015). A robust antiviral silencing relies on the effector phase, in which vsiRNAs are loaded into an Argonaute protein. It seems that the 5′ terminal nucleotide and thermodynamic properties of the siRNA duplex determine the loading of the guide strand into an specific Argonaute (Schwarz *et al.*, 2003; Mi *et al.*, 2008; Schuck *et al.*, 2013). The activated RISCs are then able to target the viral RNA for degradation (slicing) and/or translation repression in a sequence-specific manner (Carbonell and Carrington, 2015).

In plants, slicing of viral RNAs seems to be the main activity of antiviral RISCs, whereas translational repression has been reported only in the AGO1 silencing response against ToRSV, and just suggested for potyviral infections (Ghoshal and Sanfaçon, 2014; Ivanov *et al.*, 2016). The capacity to recruit vsiRNAs has been reported for *Arabidopsis* AGO1, AGO2, AGO5, AGO7 and AGO10 (Qu *et al.*, 2008; Takeda *et al.*, 2008; Carbonell *et al.*, 2012; Garcia-Ruiz *et al.*, 2015) and rice AGO1 and AGO18 (Du *et al.*, 2011a; Wu *et al.*, 2015). AGO1 has been described as the main antiviral slicer against many RNA viruses (Morel *et al.*, 2002; Qu *et al.*, 2008; Wang *et al.*, 2011; Carbonell *et al.*, 2012), being itself controlled by the action of miR168 (Rhoades *et al.*, 2002; Vaucheret *et al.*, 2004). Although AGO2 acts as a second antiviral layer in the absence of AGO1, recent observations have confirmed that AGO2 can adopt a more relevant role as viral slicer in plants where AGO1 activity is somehow compromised (Harvey *et al.*, 2011; Scholthof *et al.*, 2011; Wang *et al.*, 2011; Zhang *et al.*, 2012; Garcia-Ruiz *et al.*, 2015; Ma *et al.*, 2015). The other *A. thaliana* AGOs (AGO5, AGO7 and AGO10) seem to have a limited contribution to the antiviral defence. For example, AGO7 targets less structured RNAs than AGO1; AGO5 can specifically restrict PVX infection; and AGO10 can target viruses in some specific tissues (Qu *et al.*, 2008; Brosseau and Moffett, 2015; Garcia-Ruiz *et al.*, 2015). Taken together, these results suggest that different AGOs work in a coordinated and cooperative manner during viral infection.

As mentioned above, TGS is another demonstrated outcome of RNA silencing activities, and is carried out by RNA-induced transcriptional silencing (RITS) complexes through histone and/or DNA methylation at the genome level (Holoch and Moazed, 2015). In *A. thaliana* infected with DNA viruses, RITS complexes contain AGO4 and vsiRNAs produced by the DCL3/DRB3 processing activity (Raja *et al.*, 2014). Together with other components and enzymes of the methylation pathways (e.g. DRMs, METs, KYPs, ADKs), RITS complexes methylate viral DNA genomes to promote TGS (Raja *et al.*, 2010). Such a defence response at the epigenetic level has been reported during the infection of plants by several geminiviruses (Raja *et al.*, 2008; Rodríguez-Negrete *et al.*, 2009; Raja *et al.*, 2014).

## 2.3 Induction of RNA Silencing to Indirectly Promote Defence Responses

As explained above, plants deploy the silencing pathway to fight directly against viruses by using effector complexes carrying vsiRNAs. Interestingly, these complexes can be also loaded with endogenous sRNAs that downregulate the expression of host genes promoting viral infection, so that RNA silencing can also promote antiviral defence in an indirect manner (Carbonell and Carrington, 2015).

For instance, in the defence against CMV, an antiviral role for endogenous siRNAs associated with AGO4 was proposed (Hamera *et al.*, 2012). More recently, this hypothesis has been supported by the identification of virus-activated small interfering RNAs (vasiRNAs), a new class of sRNAs generated during CMV and TuMV infection, which largely map to genes related with stress responses, among others. These vasiRNAs are active in directing the widespread silencing of certain target genes in order to promote antiviral defence (Cao *et al.*, 2014).

In rice, it has been proposed that viral infection by RSV and RDV can be controlled by induction of AGO18, which sequesters miR168 with the consequent over-accumulation of the antiviral, miR168-targeted, AGO1. This mechanism confers a broad-spectrum viral resistance (Wu *et al.*, 2015).

## 2.4 Viral Counter-defence Strategies against Antiviral RNA Silencing

The simple observation that plant viruses can infect their hosts means that these pathogens have co-evolved strategies to fight back against the plant antiviral silencing pathway and escape from it. It now seems that the most common strategy used by plant viruses to reach this objective is the expression of proteins with anti-silencing activity. These viral factors are termed RSSs. Over the last decade, plant virologists have identified many RSSs and studied their mechanisms of action. It has been found that they can interfere with almost all steps of the silencing cascade, even acting at multiple levels, with AGO proteins and vsiRNAs as the most frequently targeted elements (Fig. 2.2). In this section, we will consider our current knowledge of RSSs and their modes of action. Table 2.1 lists the viruses described here, whereas Table 2.2 summarizes known RSSs and their modes of action.

## 2.5 RNA Silencing Suppressors (RSSs) and Their Mechanism of Action

### 2.5.1 Suppressors targeting the initiation phase of silencing (DCLs)

The production of vsiRNAs during the initiation phase plays a key role during the plant silencing response against viruses. For this reason, it is not surprising that several RSSs are able to block this step. To achieve this objective, RSSs can either prevent the cleavage of dsRNAs into vsiRNA or sequester and/or modify vsiRNAs before their loading into AGO proteins.

Just a few viral suppressors have been reported as inhibitors of DCL proteins. For RNA viruses, some evidence indicates that p27 and p88 proteins from RCNMV disrupt DCL activity (Takeda *et al.*, 2005; Mine *et al.*, 2010). Although the same hypothesis was proposed for TCV p38 (Qu *et al.*, 2003; Deleris *et al.*, 2006), a later study attributed this DCL dysfunction to an indirect effect derived from p38-mediated

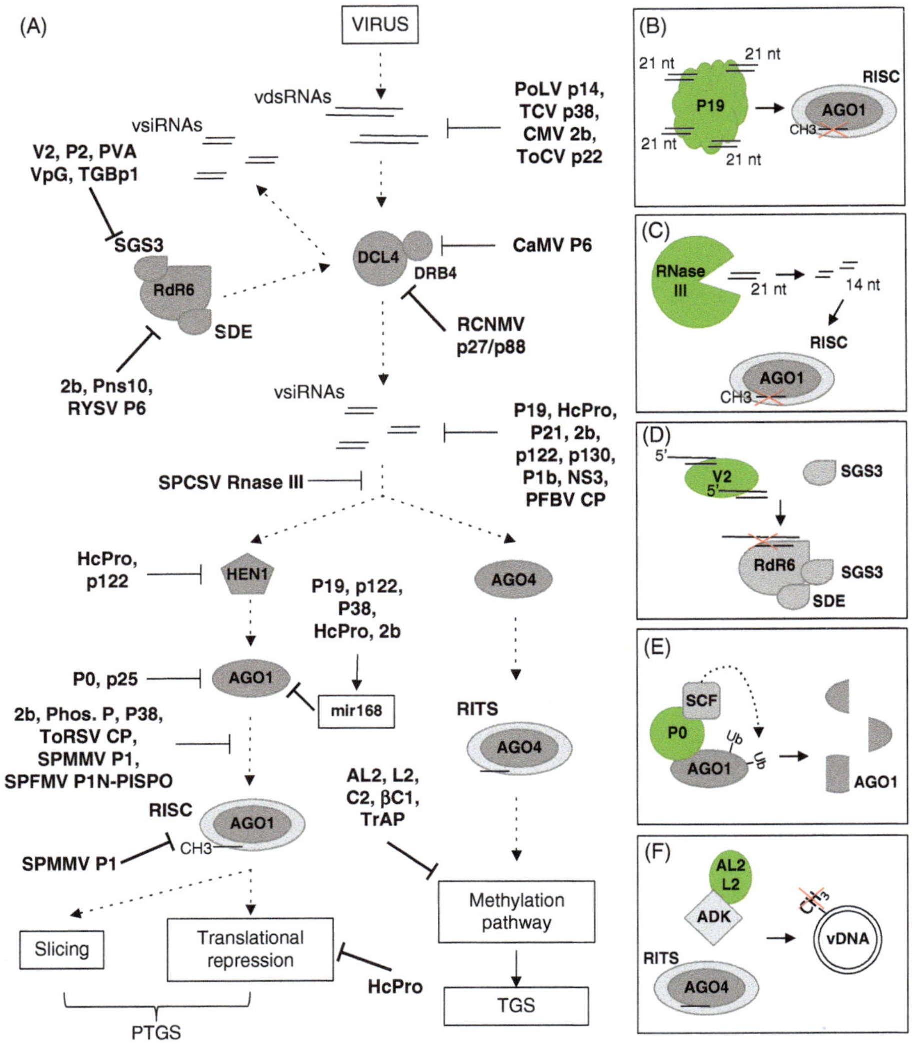


**Fig. 2.2.** Selection of known viral RNA silencing suppressors and their modes of action. (A) Main steps of the silencing pathway are shown, indicating the silencing components targeted by the indicated RSSs. (B–F) Schematic representation of specific RSSs (in pale green) and their modes of action. B: Sequestration of vsiRNAs by P19; C: Degradation of vsiRNAs by RNase III; D: Competition for available dsRNA molecules mediated by V2; E: AGO degradation mediated by P0; F: RITS complex interference mediated by AL2/L2.

AGO blocking (Azevedo *et al.*, 2010). In addition, a different mechanism that affects DCL function has been described for CaMV, where the P6 protein inactivates the DCL4-cofactor DRB4 (Love *et al.*, 2007; Haas *et al.*, 2008; Shivaprasad *et al.*, 2008).

Another group of viral suppressors are able to interact with long dsRNAs precursors. In addition to TCV p38, PoLV p14, CMV 2b and ToCV p22 bind long dsRNAs, preventing their cleavage by DCLs and the consequent production of vsiRNAs (Mérai *et al.*, 2005; Mérai *et al.*, 2006; Goto *et al.*, 2007; Landeo-Ríos *et al.*, 2016).

**Table 2.1.** Alphabetic list of viruses and their abbreviations whose RSS are described in the text.

| Abbreviation | Virus |
|---|---|
| BCTV | *Beet curly top virus* |
| BSCTV | *Beet severe curly top virus* |
| BWYV | *Beet western yellows virus* |
| BYV | *Beet yellows virus* |
| CABYV | *Cucurbit aphid-borne yellows virus* |
| CaLCuV | *Cabbage leaf curl virus* |
| CaMV | *Cauliflower mosaic virus* |
| CIRV | *Carnation Italian ringspot virus* |
| CMV | *Cucumber mosaic virus* |
| CMV Y-sat | *Cucumber mosaic virus Y satellite* |
| CVYV | *Cucumber vein yellowing virus* |
| CymRSV | *Cymbidium ringspot virus* |
| GFkV | *Grapevine fleck virus* |
| GRSPaV | *Grapevine rupestris stem pitting-associated virus* |
| GVA | *Grapevine virus A* |
| LNYV | *Lettuce necrotic yellows virus* |
| PEMV-1 | *Pea enation mosaic virus-1* |
| PFBV | *Pelargonium flower break virus* |
| PlAMV | *Plantago asiatica mosaic virus* |
| PLMVd | *Peach latent mosaic viroid* |
| PLPV | *Pelargonium line pattern virus* |
| PoLV | *Pothos latent virus* |
| PRSV | *Papaya ringspot virus* |
| PSTVd | *Potato spindle tuber viroid* |
| PVA | *Potato virus A* |
| PVX | *Potato virus X* |
| RCNMV | *Red clover necrotic mosaic virus* |
| RDV | *Rice dwarf phytoreovirus* |
| RHBV | *Rice hoja blanca virus* |
| RSV | *Rice stripe virus* |
| RYSV | *Rice yellow stunt rhabdovirus* |
| SCMV | *Sugarcane mosaic virus* |
| SPCSV | *Sweet potato chlorotic stunt virus* |
| SPFMV | *Sweet potato feathery mottle virus* |
| SPMMV | *Sweet potato mild mottle virus* |
| TAV | *Tomato aspermy virus* |
| TBSV | *Tomato bushy stunt virus* |
| TCV | *Turnip crinkle virus* |
| TEV | *Tobacco etch virus* |
| TGMV | *Tomato golden mosaic virus* |
| TMV | *Tobacco mosaic virus* |
| ToCV | *Tomato chlorosis virus* |
| ToMV | *Tomato mosaic virus* |
| ToRSV | *Tomato ringspot virus* |
| TuMV | *Turnip mosaic virus* |
| TYLCCNV | *Tomato yellow leaf curl China virus* |
| TYLCV | *Tomato yellow leaf curl virus* |
| ZYMV | *Zucchini yellow mosaic virus* |

**Table 2.2.** Summary of RSSs described in this work and their mode/s of action. RSSs are classified by viral genome, viral family and genus.

| Genome | Family | Genus | Virus | Abbreviation | RSS | Mode of action | Reference |
|---|---|---|---|---|---|---|---|
| +ssRNA | *Alphaflexiviridae* | *Potexvirus* | *Plantago asiatica mosaic virus* | PIAMV | TGBp1 | RDR6/SGS3 interaction | (Okano *et al.*, 2014) |
| | | | *Potato virus X* | PVX | P25 | AGO1 degradation | (Voinnet *et al.*, 2000; Chiu *et al.*, 2010) |
| | *Betaflexiviridae* | *Vitivirus* | *Grapevine virus A* | GVA | P10 | Single and dúplex siRNA binding | (Zhou *et al.*, 2006) |
| | *Bromoviridae* | *Cucumovirus* | *Cucumber mosaic virus* | CMV | 2b | dsRNA binding, siRNA sequestration<br>AG01 homeostasis<br>AGO1 and AGO4 interaction | (Brigneti *et al.*, 1998; Zhang *et al.*, 2006; Goto *et al.*, 2007; Duan *et al.*, 2012; Hamera *et al.*, 2012; Várallyay and Havelda, 2013; Dong *et al.*, 2016) |
| | | | *Tomato aspermy virus* | TAV | 2b | RDR6 downregulation | (Li *et al.*, 1999; Zhang *et al.*, 2008) |
| | *Closteroviridae* | *Closterovirus* | *Beet yellows virus* | BYV | p21 | siRNA sequestration | (Reed *et al.*, 2003; Chapman *et al.*, 2004) |
| | | *Crinivirus* | *Sweet potato chlorotic stunt virus* | SPCSV | RNase III | siRNA cleavage | (Kreuze *et al.*, 2005; Cuellar *et al.*, 2009; Weinheimer *et al.*, 2014) |
| | | | *Tomato chlorosis virus* | ToCV | P22 | dsRNA binding | (Cañizares *et al.*, 2008; Landeo-Ríos *et al.*, 2016) |
| | *Luteoviridae* | *Enamovirus* | *Pea enation mosaic virus-1* | PEMV-1 | P0 | AG01 degradation | (Fusaro *et al.*, 2012) |
| | | *Polerovirus* | *Beet western yellows virus* | BWYV | P0 | AGO1 degradation | (Pfeffer *et al.*, 2002; Pazhouhandeh *et al.*, 2006; Baumberger *et al.*, 2007; Bortolamiol *et al.*, 2007; Csorba *et al.*, 2010; Derrien *et al.*, 2012) |
| | | | *Cucurbit aphid-borne yellows virus* | CABYV | P0 | AGO1 degradation | (Pfeffer *et al.*, 2002; Pazhouhandeh *et al.*, 2006; Bortolamiol *et al.*, 2007) |
| | *Potyviridae* | *Ipomovirus* | *Cucumber vein yellowing virus* | CVYV | P1b | siRNA sequestration | (Valli *et al.*, 2006; Valli *et al.*, 2008; Valli *et al.*, 2011) |
| | | | *Sweet potato mild mottle virus* | SPMMV | P1 | AGO1 interaction | (Giner *et al.*, 2010) |
| | | *Potyvirus* | *Papaya ringspot virus* | PRSV | HcPro | siRNA sequestration | (Sahana *et al.*, 2014) |
| | | | *Potato virus A* | PVA | HcPro<br>VPg | AGO1 interaction<br>SGS3 interaction | (Rajamäki and Valkonen, 2009; Rajamäki *et al.*, 2014; Ivanov *et al.*, 2016) |
| | | | *Tobacco etch virus* | TEV | HcPro | siRNA sequestration<br>Interfering with the HEN1-mediated methylation<br>AGO1 homeostasis | (Kasschau and Carrington, 1998; Ebhardt *et al.*, 2005; Lakatos *et al.*, 2006; Mérai *et al.*, 2006; Lózsa *et al.*, 2008; Várallyay and Havelda, 2013) |

| | | | | | | | |
|---|---|---|---|---|---|---|---|
| | | | *Turnip mosaic virus* | TuMV | HcPro | siRNA sequestration | (Kasschau *et al.*, 2003; Garcia-Ruiz *et al.*, 2015) |
| | | | *Sugarcane mosaic virus* | SCMV | HcPro | RDR6 downregulation | (Zhang *et al.*, 2008) |
| | | | *Sweet potato feathery mottle virus* | SPFMV | P1N-PISPO | AGO1 interaction? | (Szabó *et al.*, 2012; Rodamilans *et al.*, 2015; Mingot *et al.*, 2016; Untiveros *et al.*, 2016) |
| | | | *Zucchini yellow mosaic virus* | ZYMV | HcPro | Interfering with the HEN1-mediated methylation | (Jamous *et al.*, 2011) |
| | *Secoviridae* | *Nepovirus* | *Tomato ringspot virus* | ToRSV | CP | AGO1 interaction | (Karran and Sanfaçon, 2014) |
| | *Tombusviridae* | *Aureusvirus* | *Pothos latent virus* | PoLV | p14 | dsRNA binding | (Mérai *et al.*, 2005) |
| | | *Carmovirus* | *Pelargonium flower break virus* | PFBV | CP | siRNA sequestration | (Martínez-Turiño and Hernández, 2009) |
| | | | *Pelargonium line pattern virus* | PLPV | P37 (CP) | siRNA sequestration | (Pérez-Cañamás and Hernández, 2015) |
| | | | *Turnip crinkle virus* | TCV | P38 (CP) | dsRNA binding<br>AGO1 and AGO2 interaction<br>AGO1 homeostasis | (Qu *et al.*, 2003; Deleris *et al.*, 2006; Mérai *et al.*, 2006; Azevedo *et al.*, 2010; Zhang *et al.*, 2012; Várallyay and Havelda, 2013) |
| | | *Dianthovirus* | *Red clover necrotic mosaic virus* | RCNMV | p27<br>p88 | Unknown | (Takeda *et al.*, 2005; Mine *et al.*, 2010) |
| | | *Tombusvirus* | *Carnation Italian ringspot virus* | *CIRV* | P19 | siRNA sequestration | (Vargason *et al.*, 2003) |
| | | | *Cymbidium ringspot virus* | *CymRSV* | P19 | siRNA sequestration<br>AGO1 homeostasis | (Lakatos *et al.*, 2004; Lakatos *et al.*, 2006; Várallyay *et al.*, 2010) |
| | | | *Tomato bushy stunt virus* | *TBSV* | P19 | siRNA sequestration | (Voinnet *et al.*, 1999; Silhavy *et al.*, 2002; Ye *et al.*, 2003) |
| | *Virgaviridae* | *Tobamovirus* | *Tobacco mosaic virus* | TMV | p126=p122 | siRNA sequestration<br>Interfering with HEN1-mediated methylation<br>AGO1 homeostasis | (Ding *et al.*, 2004; Csorba *et al.*, 2007; Kurihara *et al.*, 2007; Vogler *et al.*, 2007; Várallyay and Havelda, 2013) |
| | | | *Tomato mosaic virus* | ToMV | p130 | siRNA sequestration | (Kubota *et al.*, 2003; Kurihara *et al.*, 2007) |
| -ssRNA | *Rhabdoviridae* | *Cytorhabdovirus* | *Lettuce necrotic yellows virus* | LNYV | Phosphoprotein P | AGO1, AGO2 and AGO4 interaction<br>RDR6 and SGS3 interaction | (Mann *et al.*, 2015; Mann *et al.*, 2016) |
| | | *Nucleorhabdovirus* | *Rice yellow stunt rhabdovirus* | RYSV | P6 | RDR6 downregulation | (Guo *et al.*, 2013) |

*Continued*

**Table 2.2.** Continued.

| Genome | Family | Genus | Virus | Abbreviation | RSS | Mode of action | Reference |
|---|---|---|---|---|---|---|---|
| | Unassigned | *Tenuivirus* | *Rice hoja blanca virus* | RHBV | NS3 | siRNA sequestration | (Bucher *et al.*, 2003; Hemmes *et al.*, 2007) |
| | | | *Rice stripe virus* | RSV | NS3<br>P2 | Single and duplex siRNA binding<br>SGS3 interaction | (Xiong *et al.*, 2009; Du *et al.*, 2011b) |
| dsRNA | *Reoviridae* | *Phytoreovirus* | *Rice dwarf phytoreovirus* | RDV | Pns10 | RDR6 downregulation | (Cao *et al.*, 2005; Ren *et al.*, 2010) |
| dsDNA | *Caulimoviridae* | *Caulimovirus* | *Cauliflower mosaic virus* | CaMV | P6 | DRB4 inactivation | (Love *et al.*, 2007; Haas *et al.*, 2008; Shivaprasad *et al.*, 2008) |
| ssDNA | *Geminiviridae* | *Begomovirus* | *Tomato yellow leaf curl virus* | TYLCV | V2 | SGS3 interaction (competition by dsRNA) | (Zrachya *et al.*, 2007; Glick *et al.*, 2008; Fukunaga and Doudna, 2009) |
| | | | *Tomato yellow leaf curl China virus* | TYLCCNV | βC1 | SAHH inhibition | (Yang *et al.*, 2011) |
| | | | *Beet severe curly top virus* | BSCTV | C2 | SAMDC1 degradation | (Zhang *et al.*, 2011) |
| | | | *Cabbage leaf curl virus* | CaLCuV | TrAP | KYP inactivation | (Castillo-González *et al.*, 2015) |
| | | | *Tomato golden mosaic virus* | TGMV | AL2=TrAP | ADK inactivation<br>KYP inactivation | (Buchmann *et al.*, 2009; Castillo-González *et al.*, 2015) |
| | | *Curtovirus* | *Beet curly top virus* | BCTV | L2 | ADK inactivation | (Buchmann *et al.*, 2009) |

The most widespread strategy of RNA silencing suppression is based on the interaction of RSSs with vsiRNAs. The capacity to sequester vsiRNAs and, therefore, prevent their incorporation into AGO proteins has been reported for a quite large number of RSSs, such as the tombusviral P19, potyviral HCPro, tobamoviral p122 (TMV) and p130 (ToMV), BYV p21, CMV 2b, CVYV P1b, RHBV NS3 and PFBV CP (Silhavy *et al.*, 2002; Chapman *et al.*, 2004; Lakatos *et al.*, 2006; Mérai *et al.*, 2006; Csorba *et al.*, 2007; Goto *et al.*, 2007; Hemmes *et al.*, 2007; Kurihara *et al.*, 2007; Valli *et al.*, 2008; Martínez-Turiño and Hernández, 2009). Although all these viral suppressors share the ability to bind vsiRNAs, they present some peculiarities. For example, some of them show preferences for vsiRNA of a given size, which seems to be determined for the protein structure, as demonstrated for the tombusviral P19 (Vargason *et al.*, 2003; Ye *et al.*, 2003). In addition, they can also display higher affinity for those sRNA duplexes with 2 nt-long 3´ overhangs, a typical feature of DCL-derived sRNAs (Lakatos *et al.*, 2006; Csorba *et al.*, 2007; Hemmes *et al.*, 2007; Valli *et al.*, 2011). Even in certain cases, such as CVYV P1b, TEV HCPro and CMV 2b, other specific structural features in vsiRNA molecules have been shown to be required for a proper RSS-mediated recognition (Valli *et al.*, 2011; Sahana *et al.*, 2014; Dong *et al.*, 2016). Although no conserved motif has been identified among different RSSs with siRNA binding capacity, the well-known GW/WG motifs, which usually work as AGO hooks (see below) have recently been implicated in the interaction between PLPV p37 and vsiRNAs (Pérez-Cañamás and Hernández, 2015).

Two additional mechanisms have also been found to work by blocking the action of vsiRNAs. The first one corresponds to SPCSV RNase III, an endonuclease able to inactivate vsiRNAs through their cleavage into non-functional 14 bp duplexes (Kreuze *et al.*, 2005; Cuellar *et al.*, 2009; Weinheimer *et al.*, 2014). The second one is used by viral suppressors such as potyviral HCPro and TMV p122, and blocks vsiRNA maturation by inhibition of their HEN1-mediated 3´-end methylation (Ebhardt *et al.*, 2005; Csorba *et al.*, 2007; Vogler *et al.*, 2007; Lózsa *et al.*, 2008; Jamous *et al.*, 2011).

### 2.5.2 Suppressors targeting the amplification phase of silencing (RDRs)

During the amplification phase, secondary vsiRNA synthesis depends on the availability of factors involved in the production of primary vsiRNAs and subsequent effector reactions. Therefore, the inactivation of these components by some RSSs has an indirect inhibitory effect on the amplification steps (Csorba *et al.*, 2007). The fact that some RSSs are able to suppress RNA silencing induced by sense RNAs (which need to be converted to dsRNA to cause silencing) but not that induced by dsRNA, and can downregulate secondary vsiRNA accumulation without altering the amount of primary vsiRNAs, suggests that silencing suppression strategies can also target the silencing amplification phase (Voinnet *et al.*, 2000; Takeda *et al.*, 2002; Chen *et al.*, 2004; Diaz-Pendon *et al.*, 2007; Moissiard *et al.*, 2007; Mlotshwa *et al.*, 2008).

Plant RDR6 and its cofactor SGS3, key elements of the amplification step of RNA silencing, have been reported as targets of several RSSs. For instance, SCMV

HCPro, TAV 2b, RDV Pns10 and RYSV P6 are able to negatively affect the RDR6 mRNA levels (Zhang *et al.*, 2008; Ren *et al.*, 2010; Guo *et al.*, 2013). Different studies have demonstrated that SGS3 interacts with TYLCV V2, RSV P2 and PVA VPg RSSs, supporting the idea that these viral proteins target the SGS3 cofactor to block the RNA silencing amplification step (Glick *et al.*, 2008; Du *et al.*, 2011b; Rajamäki *et al.*, 2014). Indeed, a later study with TYLCV V2 showed that this RSS in particular outcompetes SGS3 for dsRNA recognition (Fukunaga and Doudna, 2009). Finally, another RSS that acts at this level is PlAMV TGBp1, which co-aggregates in the cytoplasm with both RDR6 and SGS3, inhibiting their activities (Okano *et al.*, 2014).

### 2.5.3 Suppressors targeting the effector phase of silencing (AGOs + RISC)

The action of RSSs at the effector phase level has been studied widely in recent years. These viral proteins operate in different ways: (i) interfering with the activity of AGO proteins; (ii) interacting with the guide strand RNA loaded in RISC/RITS complexes; or (iii) even blocking activities downstream of the activation of these complexes.

Several viral suppressors interfere with the normal action of AGO proteins by affecting either their expression or protein stability. As mentioned above, AGO1 homeostasis depends on the miR168 activity (Rhoades *et al.*, 2002; Vaucheret *et al.*, 2004); hence, an increased production of miR168 will correlate with an enhanced downregulation of AGO1. This strategy has been demonstrated for diverse RSSs, such as p19, p122, p38, HCPro and 2b (Várallyay *et al.*, 2010; Várallyay and Havelda, 2013). Interestingly, these RSSs are normally associated with other anti-silencing mechanisms, suggesting that additional suppression strategies at diverse phases might be required under certain conditions to consistently alleviate the antiviral action of RNA silencing.

The F-box motif-containing P0 protein from poleroviruses and enamoviruses interacts with both AGO1 and the SCF ubiquitin ligase and, consequently, promotes the degradation of AGO1 through the autophagy pathway, thus preventing the *de novo* formation of RISC complexes (Pazhouhandeh *et al.*, 2006; Baumberger *et al.*, 2007; Bortolamiol *et al.*, 2007; Csorba *et al.*, 2010; Derrien *et al.*, 2012; Fusaro *et al.*, 2012). PVX p25 also interacts with and promotes the degradation of AGOs, but in this particular case the protein degradation occurs via the proteasome pathway (Chiu *et al.*, 2010).

Another group of RSSs seem to inhibit the activity of effector complexes simply through physical interaction with AGO proteins, such as the case of CMV 2b and LNYV Phosphoprotein P (Zhang *et al.*, 2006; Duan *et al.*, 2012; Hamera *et al.*, 2012; Mann *et al.*, 2016). An equivalent mechanism is displayed by RSSs with WG/GW motifs. The WG/GW motifs, so-called AGO hooks, were first identified in different host proteins where they mediate the interaction of these factors with AGOs to promote RISC assembly and activity (El-Shami *et al.*, 2007; Eulalio *et al.*, 2009). The TCV P38, ToRSV CP and SPMMV P1 RSSs mimic this strategy to directly interact with AGOs, interfering with RISC activity and consequently suppressing the RNA silencing (Azevedo *et al.*, 2010; Giner *et al.*, 2010; Karran and Sanfaçon, 2014). In particular, the binding of SPMMV P1 to AGO1 through its three WG/GW motifs prevents the *de novo* RISC formation, thus it inhibits AGO1 function once RISC is assembled (Giner *et al.*, 2010). The relevance of WG/GW motifs in the anti-silencing capacity of viral proteins was further evidenced with the modification of SPFMV P1, a protein without clear silencing suppressor activity, which was transformed into an active RSS just by the introduction of additional WG/GW motifs (Szabó *et al.*, 2012). In line with this result, recent studies demonstrated that there is an active RSS hidden in the SPFMV P1 cistron: P1N-PISPO, a trans-framed protein produced from a polymerase slippage event at the P1 cistron which gives rise to a different P1 C-terminal half, providing additional WG/GW motifs (Rodamilans *et al.*, 2015; Mingot *et al.*, 2016; Untiveros *et al.*, 2016).

The PVA HCPro also interacts with AGO1 when they are both associated with ribosomes *in planta*, suggesting a role for HCPro in counteracting the translational repression mediated by RISC (Ivanov *et al.*, 2016).

The blocking of RISC/RITS complexes can be also achieved by targeting the guide strand RNA. This assumption is supported by the capacity to bind single-strand siRNAs of GVA P10 and RSV NS3 (Zhou *et al.*, 2006; Xiong *et al.*, 2009).

Some RSSs from DNA viruses inactivate the antiviral silencing pathway after the formation of RITS (Bisaro, 2006). This is the case of the TGMV AL2, BCTV L2, TYLCCNV βC1, BSCTV C2 and CaLCuV TrAP, which interfere with the methylation process by diverse mechanisms (Buchmann *et al.*, 2009; Yang *et al.*, 2011; Zhang *et al.*, 2011). RSSs from RNA viruses such as PVA HCPro and ToCV CP resemble geminivirus RSSs in interfering with the methyl cycle and *S*-adenosylmethionine-dependent methyltransferase activity (Cañizares *et al.*, 2013; Ivanov *et al.*, 2016). Whether this interference is relevant for siRNA methylation (see above), RITS activity or both is still unknown.

Finally, viruses can counteract antiviral RNA silencing not only through protein factors. For instance, massive production of small RNAs from the 5′ leader of the CaMV 35S mRNA have been suggested to serve as a decoy diverting the silencing machinery from relevant viral RNA targets (Blevins *et al.*, 2011).

Summarizing the previous sections: viral silencing suppressors can block almost all the steps of the silencing pathway through diverse mechanisms. Moreover, some RSSs, such as the potyviral HCPro, cucumoviral 2b and tombusviral P19, display multiple modes of action. The capacity of a single RSS to counteract the RNA silencing at different levels may lead to a stronger suppression of this plant defence; but it is important to keep in mind that a disproportioned suppression can be detrimental for viruses. The activity of RSSs needs to be tightly regulated to allow viral replication and dissemination with no or minimal interference of crucial plant physiological processes that might compromise the host and, subsequently, the viral infection. To reach this equilibrium, viruses most likely have evolved strategies aiming to modulate RSS activity, such as the control of RSS sub-cellular localization, or the accumulation of mutated protein forms with different silencing suppression strength (Haas *et al.*, 2008; Torres-Barceló *et al.*, 2008; Du *et al.*, 2014). Considering all these situations, the induction and suppression of the antiviral silencing response needs to be considered as a precisely regulated spatiotemporal mechanism that occurs during plant virus infection.

## 2.6 Effects of Antiviral RNA Silencing on Plant Gene Expression

Potentially, when there is enough nucleotide complementarity, specific vsiRNAs may target host transcripts and the resulting gene downregulation can cause specific disease symptoms. Such an effect has been demonstrated in a few cases (Table 2.3). A particular siRNA that comes from the CMV Y satellite targets a chlorophyll biosynthetic gene (CHLI), leading to a yellow mosaic phenotype in host plants infected with CMV (Shimura *et al.*, 2011; Smith *et al.*, 2011). Many genes in grapevine hosts were identified as putative targets of vsiRNAs, although specific mRNA cleavage was only validated for just a small fraction of them (Miozzi *et al.*, 2013).

Viroids are infectious plant pathogens with circular RNA genomes without protein-coding capacity (Gago-Zachert, 2016). Like viruses, viroids are sources of siRNAs (vdsiRNAS), which also have the potential to target host genes. For instance, two PLMVd-derived siRNAs trigger the cleavage of the chloroplastic heat-shock protein 90 (cHSP90) mRNA. The lack of cHSP90 causes an albino phenotype in leaves infected with PLMVd (Navarro *et al.*, 2012). Other examples have been reported for host genes in relation to pathogen infection, such as the case of some callose synthase mRNAs that are degraded by the action of a PSTVd-derived siRNA during tomato infection (Adkar-Purushothama *et al.*, 2015).

According to these results, regulation of plant gene expression by vsiRNAs or vdsiRNAs might happen, but whether this only occurs on limited occasions, or has a more significant contribution, is still unknown.

## 2.7 Concluding Remarks

The discovery of RNA silencing more than 20 years ago was the starting signal in changing our simplistic view of the RNA molecule. It is not only an intermediate messenger between the hereditary information and protein synthesis, but also has crucial regulatory activities in trans that control gene expression. Many biological meanings of RNA silencing in plants have been discovered, and we have briefly described its key contribution to host defence against viruses. Remarkably, recent studies point to a wider role of RNA silencing as a protective barrier against diverse non-viral plant pathogens. Indeed, current investigations are focusing on understanding the actual mechanism by which RNA silencing fights bacteria, fungi, insects and nematodes.

To avoid their elimination from the host, viruses in turn developed the capacity to circumvent RNA silencing by deploying, in most cases, RSSs. Interestingly, very different viruses use similar strategies to counteract RNA silencing, and it is also surprising that similar viruses – even members of the same genus – may use different ways to escape from silencing-based defences. If RSSs were extremely potent, they would have deleterious effects on the host since they would also interfere with other partially overlapping modules – non-antiviral related – of the host RNA silencing machinery. Hence, there must be some sort of control over the anti-silencing activity (e.g. natural selection of mild RRS variants, or specificity of the RSS to target only the antiviral RNA silencing module), which we are just starting to investigate. On the other hand, and given the above-mentioned role of RNA silencing against non-viral

**Table 2.3.** Summary of vsiRNAs with demonstrated activity to target plant genes.

| Genome | Family | Genus | *Virus* | Abbreviation | Plant gene targeted by vsiRNAs | Reference |
|---|---|---|---|---|---|---|
| satRNA | | | *Cucumber mosaic virus Y satellite* | CMV Y-sat | Chlorophyll biosynthetic gene (CHLI) | (Shimura *et al.*, 2011; Smith *et al.*, 2011) |
| Circular ssRNA | *Avsunviroidae* | *Pelamoviroid* | *Peach latent mosaic viroid* | PLMVd | Chloroplastic heat-shock protein 90 (cHSP90) | (Navarro *et al.*, 2012) |
| Circular ssRNA | *Pospiviroidae* | *Pospiviroid* | *Potato spindle tuber viroid* | PSTVd | Callose synthase genes | (Adkar-Purushothama *et al.*, 2015) |
| +ssRNA | *Tymoviridae* | *Maculavirus* | *Grapevine fleck virus* | GFkV | Ribosomal biosynthetic and stress-related genes | (Miozzi *et al.*, 2013) |
| +ssRNA | *Betaflexiviridae* | *Foveavirus* | *Grapevine rupestris stem pitting-associated virus* | GRSPaV | Ribosomal biosynthetic and stress-related genes | (Miozzi *et al.*, 2013) |

pathogens, it would be expected that bacteria, fungi, insects and nematodes also express suppressors and/or modulators of the RNA silencing-based defences.

Recent studies show that plants also develop mechanisms to counter-counteract the action of RSSs. The most striking example is the recognition that many RSSs are specific elicitors of R-gene-mediated resistance. It has recently been discovered that RSSs are also involved in a more general induction of this defence mechanism. R-genes are constitutively subject to downregulation by miRNAs, which are just released after viral infections due to the capacity of RSSs to block miRNA activity. As a result, R-genes are expressed in infected plants to create an extra layer of defence, providing yet another example of never-ending host/pathogen competition. It is important here to highlight that studies on RSSs are helping to better understand not only the counteracting and counter-counteracting roles of both viral and host proteins, but also more general aspects of the endogenous RNA silencing pathways, as well as peculiarities of complex infection diseases triggered by both single and mixed pathogen infections.

Evidence of more general crosstalk between RNA silencing and other well-described regulatory networks, not only those involved in defence against pathogens, have started to emerge. It will be exciting to elucidate how these networks are distorted in plants suffering attacks. Future studies will help to provide insights that assist scientists in designing resistant varieties of plants, and reduce economic losses caused by diseases.

## References


Adkar-Purushothama, C.R., Brosseau, C., Giguère, T., Sano, T., Moffett, P. and Perreault, J.-P. (2015) Small RNA derived from the virulence modulating region of the potato spindle tuber viroid silences callose synthase genes of tomato plants. *The Plant Cell* 27(8), 2178–2194.

Ahlquist, P. (2002) RNA-dependent RNA polymerases, viruses, and RNA silencing. *Science* 296(5571), 1270–1273.

Akbergenov, R., Si-Ammour, A., Blevins, T., Amin, I., Kutter, C., Vanderschuren, H., Zhang, P., Gruisem, W., Meins, F., Hohn, T. and Pooggin, M.M. (2006) Molecular characterization of geminivirus-derived small RNAs in different plant species. *Nucleic Acids Research* 34(2), 462–471.

Aregger, M., Borah, B.K., Seguin, J., Rajeswaran, R., Gubaeva, E.G., Zvereva, A.S., Windels, D., Vazquez, F., Blevins, T., Farinelli, L., Pooggin, M.M. (2012) Primary and secondary siRNAs in geminivirus-induced gene silencing. *PLoS Pathogens* 8(9), e1002941.

Azevedo, J., Garcia, D., Pontier, D., Ohnesorge, S., Yu, A., Garcia, S., Braun, L., Bergdoll, M., Hakimi, M.A., Lagrange, T., Voinnet, O. (2010) Argonaute quenching and global changes in Dicer homeostasis caused by a pathogen-encoded GW repeat protein. *Genes and Development* 24(9), 904–915.

Baumberger, N., Tsai, C.-H., Lie, M., Havecker, E. and Baulcombe, D.C. (2007) The Polerovirus silencing suppressor P0 targets ARGONAUTE proteins for degradation. *Current Biology* 17(18), 1609–1614.

Bisaro, D.M. (2006) Silencing suppression by geminivirus proteins. *Virology* 344(1), 158–168.

Blevins, T., Rajeswaran, R., Shivaprasad, P.V., Beknazariants, D., Si-Ammour, A., Park, H.-S., Vazquez, F., Robertson, D., Meins, F. Jr, Hohn, T., Pooggin, M.M. (2006) Four plant Dicers mediate viral small RNA biogenesis and DNA virus induced silencing. *Nucleic Acids Research* 34(21), 6233–6246.

Blevins, T., Rajeswaran, R., Aregger, M., Borah, B.K., Schepetilnikov, M., Baerlocher, L., Farinelli, L., Meins, F. Jr, Hohn, T., Pooggin, M.M. (2011) Massive production of small RNAs from a non-coding region of *Cauliflower mosaic virus* in plant defense and viral counter-defense. *Nucleic Acids Research* 39(12), 5003–5014.

Bologna, N.G. and Voinnet, O. (2014) The diversity, biogenesis, and activities of endogenous silencing small RNAs in *Arabidopsis*. *Annual Review of Plant Biology* 65, 473–503.

Bortolamiol, D., Pazhouhandeh, M., Marrocco, K., Genschik, P., Ziegler-Graff, V. (2007) The Polerovirus F box protein P0 targets ARGONAUTE1 to suppress RNA silencing. *Current Biology* 17(18), 1615–1621.

Bouché, N., Lauressergues, D., Gasciolli, V. and Vaucheret, H. (2006) An antagonistic function for *Arabidopsis* DCL2 in development and a new function for DCL4 in generating viral siRNAs. *The EMBO Journal* 25(14), 3347–3356.

Boutet, S., Vazquez, F., Liu, J., Béclin, C., Fagard, M., Gratias, A., Morel, J.B., Crété, P., Chen, X., Vaucheret, H. (2003) *Arabidopsis* HEN1: a genetic link between endogenous miRNA controlling development and siRNA controlling transgene silencing and virus resistance. *Current Biology* 13(10), 843–848.

Brigneti, G., Voinnet, O., Li, W.X., Ji, L.H., Ding, S.W. and Baulcombe, D.C. (1998) Viral pathogenicity determinants are suppressors of transgene silencing in Nicotiana benthamiana. *The EMBO Journal* 17(22), 6739–6746.

Brosseau, C. and Moffett, P. (2015) Functional and genetic analysis identify a role for *Arabidopsis* ARGONAUTE5 in antiviral RNA silencing. *The Plant Cell* 27(6), 1742–1754.

Bucher, E., Sijen, T., De Haan, P., Goldbach, R. and Prins, M. (2003) Negative-strand tospoviruses and tenuiviruses carry a gene for a suppressor of gene silencing at analogous genomic positions. *Journal of Virology* 77(2), 1329–1336.

Buchmann, R.C., Asad, S., Wolf, J.N., Mohannath, G. and Bisaro, D.M. (2009) Geminivirus AL2 and L2 proteins suppress transcriptional gene silencing and cause genome-wide reductions in cytosine methylation. *Journal of Virology* 83(10), 5005–5013.

Cañizares, M.C., Navas-Castillo, J. and Moriones, E. (2008) Multiple suppressors of RNA silencing encoded by both genomic RNAs of the crinivirus, tomato chlorosis virus. *Virology* 379(1), 168–174.

Cañizares, M.C., Lozano-Durán, R., Canto, T., Bejarano, E.R., Bisaro, D.M., Navas-Castillo, J. and Moriones, E. (2013) Effects of the crinivirus coat protein-interacting plant protein SAHH on post-transcriptional RNA silencing and its suppression. *Molecular Plant-Microbe Interactions* 26(9), 1004–1015.

Cao, X., Zhou, P., Zhang, X., Zhu, S., Zhong, X., Xiao, Q., Ding, B., Li, Y. (2005) Identification of an RNA silencing suppressor from a plant double-stranded RNA virus. *Journal of Virology* 79(20), 13018–13027.

Cao, M., Du, P., Wang, X., Yu, Y.-Q., Qiu, Y.-H., Li, W., Gal-On, A., Zhou, C., Li, Y., Ding, S.-W. (2014) Virus infection triggers widespread silencing of host genes by a distinct class of endogenous siRNAs in *Arabidopsis*. *Proceedings of the National Academy of Sciences of the United States of America* 111(40), 14613–14618.

Carbonell, A. and Carrington, J.C. (2015) Antiviral roles of plant ARGONAUTES. *Current Opinion in Plant Biology* 27, 111–117.

Carbonell, A., Fahlgren, N., Garcia-Ruiz, H., Gilbert, K.B., Montgomery, T.A., Nguyen, T., Cuperus, J.T., Carrington, J.C. (2012) Functional analysis of three *Arabidopsis* ARGONAUTES using slicer-defective mutants. *The Plant Cell* 24(9), 3613–3629.

Castillo-González, C., Liu, X., Huang, C., Zhao, C., Ma, Z., Hu, T., Sun, F., Zhou, Y., Zhou, X., Wang, X.J., Zhang, X. (2015) Geminivirus-encoded TrAP suppressor inhibits the histone methyltransferase SUVH4/KYP to counter host defense. *eLife* 4, e06671.

Chapman, E.J., Prokhnevsky, A.I., Gopinath, K., Dolja, V.V. and Carrington, J.C. (2004) Viral RNA silencing suppressors inhibit the microRNA pathway at an intermediate step. *Genes and Development* 18(10), 1179–1186.

Chellappan, P., Vanitharani, R., Pita, J. and Fauquet, C.M. (2004) Short interfering RNA accumulation correlates with host recovery in DNA virus-infected hosts, and gene silencing targets specific viral sequences. *Journal of Virology* 78(14), 7465–7477.

Chen, J., Li, W.X., Xie, D., Peng, J.R. and Ding, S.W. (2004) Viral virulence protein suppresses RNA silencing-mediated defense but upregulates the role of microrna in host gene expression. *The Plant Cell* 16(5), 1302–1313.

Chiu, M.-H., Chen, I.-H., Baulcombe, D.C. and Tsai, C.-H. (2010) The silencing suppressor P25 of potato virus X interacts with Argonaute1 and mediates its degradation through the proteasome pathway. *Molecular Plant Pathology* 11(5), 641–649.

Csorba, T., Bovi, A., Dalmay, T. and Burgyán, J. (2007) The p122 subunit of tobacco mosaic virus replicase is a potent silencing suppressor and compromises both small interfering RNA- and microRNA-mediated pathways. *Journal of Virology* 81(21), 11768–11780.

Csorba, T., Lózsa, R., Hutvágner, G. and Burgyán, J. (2010) Polerovirus protein P0 prevents the assembly of small RNA-containing RISC complexes and leads to degradation of ARGONAUTE1. *The Plant Journal* 62(3), 463–472.

Csorba, T., Kontra, L. and Burgyán, J. (2015) Viral silencing suppressors: tools forged to fine-tune host-pathogen coexistence. *Virology* 479–480, 85–103.

Cuellar, W.J., Kreuze, J.F., Rajamäki, M.-L., Cruzado, K.R., Untiveros, M. and Valkonen, J.P.T. (2009) Elimination of antiviral defense by viral RNase III. *Proceedings of the National Academy of Sciences of the United States of America* 106(25), 10354–10358.

Dalmay, T., Horsefield, R., Braunstein, T.H. and Baulcombe, D.C. (2001) SDE3 encodes an RNA helicase required for post-transcriptional gene silencing in *Arabidopsis*. *The EMBO Journal* 20(8), 2069–2078.

Deleris, A., Gallego-Bartolome, J., Bao, J., Kasschau, K.D., Carrington, J.C. and Voinnet, O. (2006) Hierarchical action and inhibition of plant Dicer-like proteins in antiviral defense. *Science* 313(5783), 68–71.

Derrien, B., Baumberger, N., Schepetilnikov, M., Viotti, C., De Cillia, J., Ziegler-Graff, V., Isono, E., Schumacher, K., Genschik, P. (2012) Degradation of the antiviral component ARGONAUTE1 by the autophagy pathway. *Proceedings of the National Academy of Sciences of the United States of America* 109(39), 15942–15946.

Diaz-Pendon, J.A., Li, F., Li, W.-X. and Ding, S.-W. (2007) Suppression of antiviral silencing by cucumber mosaic virus 2b protein in *Arabidopsis* is associated with drastically reduced accumulation of three classes of viral small interfering RNAs. *The Plant Cell* 19(6), 2053–2063.

Ding, X.S., Liu, J., Cheng, N.-H., Folimonov, A., Hou, Y.-M., Bao, Y., Katagi, C., Carter, S.A., Nelson, R.S. (2004) The Tobacco mosaic virus 126-kDa protein associated with virus replication and movement suppresses RNA silencing. *Molecular Plant–Microbe Interactions* 17(6), 583–592.

Donaire, L., Barajas, D., Martínez-García, B., Martínez-Priego, L., Pagán, I. and Llave, C. (2008) Structural and genetic requirements for the biogenesis of tobacco rattle virus-derived small interfering RNAs. *Journal of Virology* 82(11), 5167–5177.

Donaire, L., Wang, Y., Gonzalez-Ibeas, D., Mayer, K.F., Aranda, M.A. and Llave, C. (2009) Deep-sequencing of plant viral small RNAs reveals effective and widespread targeting of viral genomes. *Virology* 392(2), 203–214.

Dong, K., Wang, Y., Zhang, Z., Chai, L.-X., Tong, X., Xu, J., Li, D., Wang, X.-B. (2016) Two amino acids near the N-terminus of Cucumber mosaic virus 2b play critical roles in the suppression of RNA silencing and viral infectivity. *Molecular Plant Pathology* 17(2), 173–183.

Du, Q.-S., Duan, C.-G., Zhang, Z.-H., Fang, Y.-Y., Fang, R.-X., Xie, Q. and Guo, H.-S. (2007) DCL4 targets cucumber mosaic virus satellite RNA at novel secondary structures. *Journal of Virology* 81(17), 9142–9151.

Du, P., Wu, J., Zhang, J., Zhao, S., Zheng, H., Gao, G., Wei, L., Li, Y. (2011a) Viral infection induces expression of novel phased microRNAs from conserved cellular microRNA precursors. *PLoS Pathogens* 7(8), e1002176.

Du, Z., Xiao, D., Wu, J., Jia, D., Yuan, Z., Liu, Y., Hu, L., Han, Z., Wei, T., Lin, Q., Wu, Z., Xie, L. (2011b) p2 of rice stripe virus (RSV) interacts with OsSGS3 and is a silencing suppressor. *Molecular Plant Pathology* 12(8), 808–814.

Du, Z., Chen, A., Chen, W., Liao, Q., Zhang, H., Bao, Y., Roossinck, M.J., Carr, J.P. (2014) Nuclear-cytoplasmic partitioning of cucumber mosaic virus protein 2b determines the balance between its roles as a virulence determinant and an RNA-silencing suppressor. *Journal of Virology* 88(10), 5228–5241.

Duan, C.-G., Fang, Y.-Y., Zhou, B.-J., Zhao, J.-H., Hou, W.-N., Zhu, H., Ding, S.W., Guo, H.-S. (2012) Suppression of *Arabidopsis* ARGONAUTE1-mediated slicing, transgene-induced RNA silencing, and DNA methylation by distinct domains of the *Cucumber mosaic virus* 2b protein. *The Plant Cell* 24(1), 259–274.

Ebhardt, H.A., Thi, E.P., Wang, M.-B. and Unrau, P.J. (2005) Extensive 3′ modification of plant small RNAs is modulated by helper component-proteinase expression. *Proceedings of the National Academy of Sciences of the United States of America* 102(38), 13398–13403.

El-Shami, M., Pontier, D., Lahmy, S., Braun, L., Picart, C., Vega, D., Hakimi, M.A., Jacobsen, S.E., Cooke, R., Lagrange, T. (2007) Reiterated WG/GW motifs form functionally and evolutionarily conserved ARGONAUTE-binding platforms in RNAi-related components. *Genes and Development* 21(20), 2539–2544.

Eulalio, A., Tritschler, F. and Izaurralde, E. (2009) The GW182 protein family in animal cells: new insights into domains required for miRNA-mediated gene silencing. *RNA* 15(8), 1433–1442.

Fukunaga, R. and Doudna, J.A. (2009) dsRNA with 5′ overhangs contributes to endogenous and antiviral RNA silencing pathways in plants. *The EMBO Journal* 28(5), 545–555.

Fusaro, A.F., Matthew, L., Smith, N.A., Curtin, S.J., Dedic-Hagan, J., Ellacott, G.A., Watson, J.M., Wang, M.B., Brosnan, C., Carroll, B.J., Waterhouse, P.M. (2006) RNA interference-inducing hairpin RNAs in plants act through the viral defence pathway. *EMBO Reports* 7(11), 1168–1175.

Fusaro, A.F., Correa, R.L., Nakasugi, K., Jackson, C., Kawchuk, L., Vaslin, M.F.S. and Waterhouse, P.M. (2012) The Enamovirus P0 protein is a silencing suppressor which inhibits local and systemic RNA silencing through AGO1 degradation. *Virology* 426(2), 178–187.

Gago-Zachert, S. (2016) Viroids, infectious long non-coding RNAs with autonomous replication. *Virus Research* 212, 12–24.

Garcia-Ruiz, H., Takeda, A., Chapman, E.J., Sullivan, C.M., Fahlgren, N., Brempelis, K.J. and Carrington, J.C. (2010) *Arabidopsis* RNA-dependent RNA polymerases and dicer-like proteins in antiviral defense and small interfering RNA biogenesis during turnip mosaic virus infection. *The Plant Cell* 22(2), 481–496.

Garcia-Ruiz, H., Carbonell, A., Hoyer, J.S., Fahlgren, N., Gilbert, K.B., Takeda, A., Giampetruzzi, A., Garcia Ruiz, M.T., McGinn, M.G., Lowery, N., Martinez Baladejo, M.T., Carrington, J.C. (2015) Roles and programming of *Arabidopsis* ARGONAUTE proteins during turnip mosaic virus infection. *PLoS Pathogens* 11(3), e1004755.

Gazzani, S., Lawrenson, T., Woodward, C., Headon, D. and Sablowski, R. (2004) A link between mRNA turnover and RNA interference in *Arabidopsis*. *Science* 306(5698), 1046–1048.

Ghoshal, B. and Sanfaçon, H. (2014) Temperature-dependent symptom recovery in *Nicotiana benthamiana* plants infected with tomato ringspot virus is associated with reduced translation of viral RNA2 and requires ARGONAUTE 1. *Virology* 456–457, 188–197.

Giner, A., Lakatos, L., García-Chapa, M., López-Moya, J.J. and Burgyán, J. (2010) Viral protein inhibits RISC activity by argonaute binding through conserved WG/GW motifs. *PLoS Pathogens* 6(7), e1000996.

Glick, E., Zrachya, A., Levy, Y., Mett, A., Gidoni, D., Belausov, E., Citovsky, V., Gafni, Y. (2008) Interaction with host SGS3 is required for suppression of RNA silencing by tomato yellow leaf curl virus V2 protein. *Proceedings of the National Academy of Sciences of the United States of America* 105(1), 157–161.

Goto, K., Kobori, T., Kosaka, Y., Natsuaki, T. and Masuta, C. (2007) Characterization of silencing suppressor 2b of cucumber mosaic virus based on examination of its small RNA-binding abilities. *Plant and Cell Physiology* 48(7), 1050–1060.

Guo, H., Song, X., Xie, C., Huo, Y., Zhang, F., Chen, X., Geng, Y., Fang, R. (2013) Rice yellow stunt rhabdovirus protein 6 suppresses systemic RNA silencing by blocking RDR6-mediated secondary siRNA synthesis. *Molecular Plant-Microbe Interactions* 26(8), 927–936.

Haas, G., Azevedo, J., Moissiard, G., Geldreich, A., Himber, C., Bureau, M., Fukuhara, T., Keller, M., Voinnet, O. (2008) Nuclear import of CaMV P6 is required for infection and suppression of the RNA silencing factor DRB4. *The EMBO Journal* 27(15), 2102–2112.

Hamera, S., Song, X., Su, L., Chen, X. and Fang, R. (2012) Cucumber mosaic virus suppressor 2b binds to AGO4-related small RNAs and impairs AGO4 activities. *The Plant Journal* 69(1), 104–115.

Hamilton, A.J. and Baulcombe, D.C. (1999) A species of small antisense RNA in posttranscriptional gene silencing in plants. *Science* 286(5441), 950–952.

Harris, C.J., Molnar, A., Muller, S.Y. and Baulcombe, D.C. (2015) FDF-PAGE: a powerful technique revealing previously undetected small RNAs sequestered by complementary transcripts. *Nucleic Acids Research* 43, 7590–7599.

Harvey, J.J.W., Lewsey, M.G., Patel, K., Westwood, J., Heimstädt, S., Carr, J.P. and Baulcombe, D.C. (2011) An antiviral defense role of AGO2 in plants. *PloS One* 6(1), e14639.

Hemmes, H., Lakatos, L., Goldbach, R., Burgyán, J. and Prins, M. (2007) The NS3 protein of *Rice hoja blanca tenuivirus* suppresses RNA silencing in plant and insect hosts by efficiently binding both siRNAs and miRNAs. *RNA* 13(7), 1079–1089.

Holoch, D. and Moazed, D. (2015) RNA-mediated epigenetic regulation of gene expression. *Nature Reviews Genetics* 16(2), 71–84.

Ivanov, K.I., Eskelin, K., Bašić, M., De, S., Lõhmus, A., Varjosalo, M. and Mäkinen, K. (2016) Molecular insights into the function of the viral RNA silencing suppressor HCPro. *The Plant Journal* 85(1), 30–45.

Jamous, R.M., Boonrod, K., Fuellgrabe, M.W., Ali-Shtayeh, M.S., Krczal, G. and Wassenegger, M. (2011) The helper component-proteinase of the *Zucchini yellow mosaic virus* inhibits the Hua Enhancer 1 methyltransferase activity in vitro. *The Journal of General Virology* 92(Pt 9), 2222–2226.

Jauvion, V., Elmayan, T. and Vaucheret, H. (2010) The conserved RNA trafficking proteins HPR1 and TEX1 are involved in the production of endogenous and exogenous small interfering RNA in *Arabidopsis*. *The Plant Cell* 22(8), 2697–2709.

Karran, R.A. and Sanfaçon, H. (2014) Tomato ringspot virus coat protein binds to ARGONAUTE 1 and suppresses the translation repression of a reporter gene. *Molecular Plant-Microbe Interactions* 27(9), 933–943.

Kasschau, K.D. and Carrington, J.C. (1998) A counterdefensive strategy of plant viruses: suppression of posttranscriptional gene silencing. *Cell* 95(4), 461–470.

Kasschau, K.D., Xie, Z., Allen, E., Llave, C., Chapman, E.J., Krizan, K.A. and Carrington, J.C. (2003) P1/HC-Pro, a viral suppressor of RNA silencing, interferes with *Arabidopsis* development and miRNA unction. *Developmental Cell* 4(2), 205–217.

Kreuze, J.F., Savenkov, E.I., Cuellar, W., Li, X. and Valkonen, J.P.T. (2005) Viral class 1 RNase III involved in suppression of RNA silencing. *Journal of Virology* 79(11), 7227–7238.

Kubota, K., Tsuda, S., Tamai, A. and Meshi, T. (2003) *Tomato mosaic virus* replication protein suppresses virus-targeted posttranscriptional gene silencing. *Journal of Virology* 77(20), 11016–11026.

Kumakura, N., Takeda, A., Fujioka, Y., Motose, H., Takano, R. and Watanabe, Y. (2009) SGS3 and RDR6 interact and colocalize in cytoplasmic SGS3/RDR6-bodies. *FEBS Letters* 583(8), 1261–1266.

Kurihara, Y., Inaba, N., Kutsuna, N., Takeda, A., Tagami, Y. and Watanabe, Y. (2007) Binding of tobamovirus replication protein with small RNA duplexes. *The Journal of General Virology* 88(Pt 8), 2347–2352.

Lakatos, L., Szittya, G., Silhavy, D. and Burgyán, J. (2004) Molecular mechanism of RNA silencing suppression mediated by p19 protein of tombusviruses. *The EMBO Journal* 23(4), 876–884.

Lakatos, L., Csorba, T., Pantaleo, V., Chapman, E.J., Carrington, J.C., Liu, Y.-P., Dolja, V.V., Calvino, L.F., López-Moya, J.J., Burgyán, J. (2006) Small RNA binding is a common strategy to suppress RNA silencing by several viral suppressors. *The EMBO Journal* 25(12), 2768–2780.

Landeo-Ríos, Y., Navas-Castillo, J., Moriones, E. and Cañizares, M.C. (2016) The p22 RNA silencing suppressor of the crinivirus *tomato chlorosis virus* preferentially binds long dsRNAs preventing them from cleavage. *Virology* 488, 129–136.

Lee, W.-S., Fu, S.-F., Li, Z., Murphy, A.M., Dobson, E.A., Garland, L., Chaluvadi, S.R., Lewsey, M.G., Nelson, R.S., Carr, J.P. (2016) Salicylic acid treatment and expression of an RNA-dependent RNA polymerase 1 transgene inhibit lethal symptoms and meristem invasion during *tobacco mosaic virus* infection in *Nicotiana benthamiana*. *BMC Plant Biology* 16(1), 15.

Li, H., Li, W.X. and Ding, S.W. (2002) Induction and suppression of RNA silencing by an animal virus. *Science* 296(5571), 1319–1321.

Li, H.W., Lucy, A.P., Guo, H.S., Li, W.X., Ji, L.H., Wong, S.M. and Ding, S.W. (1999) Strong host resistance targeted against a viral suppressor of the plant gene silencing defence mechanism. *The EMBO Journal* 18(10), 2683–2691.

Li, Y., Lu, J., Han, Y., Fan, X. and Ding, S.-W. (2013) RNA interference functions as an antiviral immunity mechanism in mammals. *Science* 342(6155), 231–234.

Love, A.J., Laird, J., Holt, J., Hamilton, A.J., Sadanandom, A. and Milner, J.J. (2007) *Cauliflower mosaic virus* protein P6 is a suppressor of RNA silencing. *The Journal of General Virology* 88(Pt 12), 3439–3444.

Lózsa, R., Csorba, T., Lakatos, L. and Burgyán, J. (2008) Inhibition of 3′ modification of small RNAs in virus-infected plants require spatial and temporal co-expression of small RNAs and viral silencing-suppressor proteins. *Nucleic Acids Research* 36(12), 4099–4107.

Lu, R., Maduro, M., Li, F., Li, H.W., Broitman-Maduro, G., Li, W.X. and Ding, S.W. (2005) Animal virus replication and RNAi-mediated antiviral silencing in *Caenorhabditis elegans*. *Nature* 436(7053), 1040–1043.

Luo, Z. and Chen, Z. (2007) Improperly terminated, unpolyadenylated mRNA of sense transgenes is targeted by RDR6-mediated RNA silencing in *Arabidopsis*. *The Plant Cell* 19(3), 943–958.

Ma, X., Nicole, M.-C., Meteignier, L.-V., Hong, N., Wang, G. and Moffett, P. (2015) Different roles for RNA silencing and RNA processing components in virus recovery and virus-induced gene silencing in plants. *Journal of Experimental Botany* 66(3), 919–932.

Maillard, P.V., Ciaudo, C., Marchais, A., Li, Y., Jay, F., Ding, S.W. and Voinnet, O. (2013) Antiviral RNA interference in mammalian cells. *Science* 342(6155), 235–238.

Mallory, A.C. and Vaucheret, H. (2009) ARGONAUTE 1 homeostasis invokes the coordinate action of the microRNA and siRNA pathways. *EMBO Reports* 10(5), 521–526.

Mann, K.S., Johnson, K.N. and Dietzgen, R.G. (2015) Cytorhabdovirus phosphoprotein shows RNA silencing suppressor activity in plants, but not in insect cells. *Virology* 476, 413–418.

Mann, K.S., Johnson, K.N., Carroll, B.J. and Dietzgen, R.G. (2016) Cytorhabdovirus P protein suppresses RISC-mediated cleavage and RNA silencing amplification in planta. *Virology* 490, 27–40.

Martínez-Turiño, S. and Hernández, C. (2009) Inhibition of RNA silencing by the coat protein of *Pelargonium flower break virus*: distinctions from closely related suppressors. *The Journal of General Virology* 90(Pt 2), 519–525.

Mérai, Z., Kerényi, Z., Molnár, A., Barta, E., Válóczi, A., Bisztray, G., Havelda, Z., Burgyán, J. Silhavy, D. (2005) Aureusvirus P14 is an efficient RNA silencing suppressor that binds double-stranded RNAs without size specificity. *Journal of Virology* 79(11), 7217–7226.

Mérai, Z., Kerényi, Z., Kertész, S., Magna, M., Lakatos, L. and Silhavy, D. (2006) Double-stranded RNA binding may be a general plant RNA viral strategy to suppress RNA silencing. *Journal of Virology* 80(12), 5747–5756.

Mi, S., Cai, T., Hu, Y., Chen, Y., Hodges, E., Ni, F., Wu, L., Li, S., Zhou, H., Long, C., Chen, S., Hannon, G.J. and Qi, Y. (2008) Sorting of small RNAs into *Arabidopsis* argonaute complexes is directed by the 5′ terminal nucleotide. *Cell* 133(1), 116–127.

Mine, A., Hyodo, K., Takeda, A., Kaido, M., Mise, K. and Okuno, T. (2010) Interactions between p27 and p88 replicase proteins of *red clover necrotic mosaic virus* play an essential role in viral RNA replication and suppression of RNA silencing via the 480-kDa viral replicase complex assembly. *Virology* 407(2), 213–224.

Mingot, A., Valli, A., Rodamilans, B., San León, D., Baulcombe, D.C., García, J.A. and López-Moya, J.J. (2016) The P1N-PISPO trans-frame gene of *sweet potato feathery mottle potyvirus* is produced during virus infection and functions as RNA silencing suppressor. *Journal of Virology* 90, 3543–3557.

Miozzi, L., Gambino, G., Burgyan, J. and Pantaleo, V. (2013) Genome-wide identification of viral and host transcripts targeted by viral siRNAs in *Vitis vinifera*. *Molecular Plant Pathology* 14(1), 30–43.

Mlotshwa, S., Pruss, G.J., Peragine, A., Endres, M.W., Li, J., Chen, X., Poethig, R.S., Bowman, L.H. and Vance, V. (2008) DICER-LIKE2 plays a primary role in transitive silencing of transgenes in *Arabidopsis*. *PloS One* 3(3), e1755. DOI: 10.1371/journal.pone.0001755

Moissiard, G., Parizotto, E.A., Himber, C. and Voinnet, O. (2007) Transitivity in *Arabidopsis* can be primed, requires the redundant action of the antiviral Dicer-like 4 and Dicer-like 2, and is compromised by viral-encoded suppressor proteins. *RNA* 13(8), 1268–1278.

Molnár, A., Csorba, T., Lakatos, L., Várallyay, E., Lacomme, C. and Burgyán, J. (2005) Plant virus-derived small interfering RNAs originate predominantly from highly structured single-stranded viral RNAs. *Journal of Virology* 79(12), 7812–7818.

Morel, J.-B., Godon, C., Mourrain, P., Béclin, C., Boutet, S., Feuerbach, F., Proux, F., Vaucheret, H. (2002) Fertile hypomorphic ARGONAUTE (ago1) mutants impaired in post-transcriptional gene silencing and virus resistance. *The Plant Cell* 14(3), 629–639.

Moreno, A.B., Martínez de Alba, A.E., Bardou, F., Crespi, M.D., Vaucheret, H., Maizel, A. and Mallory, A.C. (2013) Cytoplasmic and nuclear quality control and turnover of single-stranded RNA modulate post-transcriptional gene silencing in plants. *Nucleic Acids Research* 41(8), 4699–4708.

Mourrain, P., Béclin, C., Elmayan, T., Feuerbach, F., Godon, C., Morel, J.B., Jouette, D., Lacombe, A.M., Nikic, S., Picault, N., Rémoué, K., Sanial, M., Vo, T.A. and Vaucheret, H. (2000) *Arabidopsis* SGS2 and SGS3 genes are required for posttranscriptional gene silencing and natural virus resistance. *Cell* 101(5), 533–542.

Navarro, B., Gisel, A., Rodio, M.E., Delgado, S., Flores, R. and Di Serio, F. (2012) Small RNAs containing the pathogenic determinant of a chloroplast-replicating viroid guide the degradation of a host mRNA as predicted by RNA silencing. *The Plant Journal* 70(6), 991–1003.

Okano, Y., Senshu, H., Hashimoto, M., Neriya, Y., Netsu, O., Minato, N., Yoshida, T., Maejima, K., Oshima, K., Komatsu, K., Yamaji, Y., Namba, S. (2014) In planta recognition of a double-stranded RNA synthesis protein complex by a potexviral RNA silencing suppressor. *The Plant Cell* 26(5), 2168–2183.

Parent, J.-S., Bouteiller, N., Elmayan, T. and Vaucheret, H. (2015) Respective contributions of *Arabidopsis* DCL2 and DCL4 to RNA silencing. *The Plant Journal* 81(2), 223–232.

Pazhouhandeh, M., Dieterle, M., Marrocco, K., Lechner, E., Berry, B., Brault, V., Hemmer, O., Kretsch, T., Richards, K.E., Genschik, P. and Ziegler-Graff, V. (2006) F-box-like domain in the polerovirus protein P0 is required for silencing suppressor function. *Proceedings of the National Academy of Sciences of the United States of America* 103(6), 1994–1999.

Pérez-Cañamás, M. and Hernández, C. (2015) Key importance of small RNA binding for the activity of a glycine-tryptophan (GW) motif-containing viral suppressor of RNA silencing. *The Journal of Biological Chemistry* 290(5), 3106–3120.

Pfeffer, S., Dunoyer, P., Heim, F., Richards, K.E., Jonard, G. and Ziegler-Graff, V. (2002) P0 of *beet Western yellows virus* is a suppressor of posttranscriptional gene silencing. *Journal of Virology* 76(13), 6815–6824.

Qu, F., Ren, T. and Morris, T.J. (2003) The coat protein of turnip crinkle virus suppresses post-transcriptional gene silencing at an early initiation step. *Journal of Virology* 77(1), 511–522.

Qu, F., Ye, X., Hou, G., Sato, S., Clemente, T.E. and Morris, T.J. (2005) RDR6 has a broad-spectrum but temperature-dependent antiviral defense role in *Nicotiana benthamiana*. *Journal of Virology* 79(24), 15209–15217.

Qu, F., Ye, X. and Morris, T.J. (2008) *Arabidopsis* DRB4, AGO1, AGO7, and RDR6 participate in a DCL4-initiated antiviral RNA silencing pathway negatively regulated by DCL1. *Proceedings of the National Academy of Sciences of the United States of America* 105(38), 14732–14737.

Raja, P., Sanville, B.C., Buchmann, R.C. and Bisaro, D.M. (2008) Viral genome methylation as an epigenetic defense against geminiviruses. *Journal of Virology* 82(18), 8997–9007.

Raja, P., Wolf, J.N. and Bisaro, D.M. (2010) RNA silencing directed against geminiviruses: post-transcriptional and epigenetic components. *Biochimica et Biophysica Acta* 1799(3–4), 337–351.

Raja, P., Jackel, J.N., Li, S., Heard, I.M. and Bisaro, D.M. (2014) *Arabidopsis* double-stranded RNA binding protein DRB3 participates in methylation-mediated defense against geminiviruses. *Journal of Virology* 88(5), 2611–2622.

Rajamäki, M.-L. and Valkonen, J.P.T. (2009) Control of nuclear and nucleolar localization of nuclear inclusion protein a of picorna-like *potato virus A* in *Nicotiana* species. *The Plant Cell* 21(8), 2485–2502.

Rajamäki, M.-L., Streng, J. and Valkonen, J.P.T. (2014) Silencing suppressor protein VPg of a potyvirus interacts with the plant silencing-related protein SGS3. *Molecular Plant-Microbe Interactions* 27(11), 1199–1210.

Reed, J.C., Kasschau, K.D., Prokhnevsky, A.I., Gopinath, K., Pogue, G.P., Carrington, J.C. and Dolja, V.V. (2003) Suppressor of RNA silencing encoded by *beet yellows virus*. *Virology* 306(2), 203–209.

Ren, B., Guo, Y., Gao, F., Zhou, P., Wu, F., Meng, Z., Wei, C. and Li, Y. (2010) Multiple functions of *Rice dwarf phytoreovirus* Pns10 in suppressing systemic RNA silencing. *Journal of Virology* 84(24), 12914–12923.

Rhoades, M.W., Reinhart, B.J., Lim, L.P., Burge, C.B., Bartel, B. and Bartel, D.P. (2002) Prediction of plant microRNA targets. *Cell* 110(4), 513–520.

Rodamilans, B., Valli, A., Mingot, A., San León, D., Baulcombe, D., López-Moya, J.J. and García, J.A. (2015) RNA polymerase slippage as a mechanism for the production of frameshift gene products in plant viruses of the *Potyviridae* family. *Journal of Virology* 89(13), 6965–6967.

Rodríguez-Negrete, E.A., Carrillo-Tripp, J. and Rivera-Bustamante, R.F. (2009) RNA silencing against geminivirus: complementary action of posttranscriptional gene silencing and transcriptional gene silencing in host recovery. *Journal of Virology* 83(3), 1332–1340.

Ruiz-Ferrer, V. and Voinnet, O. (2009) Roles of plant small RNAs in biotic stress responses. *Annual Review of Plant Biology* 60, 485–510.

Sahana, N., Kaur, H., Jain, R.K., Palukaitis, P., Canto, T. and Praveen, S. (2014) The asparagine residue in the FRNK box of potyviral helper-component protease is critical for its small RNA binding and subcellular localization. *The Journal of General Virology* 95(Pt 5), 1167–1177.

Scholthof, H.B., Alvarado, V.Y., Vega-Arreguin, J.C., Ciomperlik, J., Odokonyero, D., Brosseau, C., Jaubert, M., Zamora, A. and Moffett, P. (2011) Identification of an ARGONAUTE for antiviral RNA silencing in *Nicotiana benthamiana*. *Plant Physiology* 156(3), 1548–1555.

Schuck, J., Gursinsky, T., Pantaleo, V., Burgyán, J. and Behrens, S.-E. (2013) AGO/RISC-mediated antiviral RNA silencing in a plant in vitro system. *Nucleic Acids Research* 41(9), 5090–5103.

Schwach, F., Vaistij, F.E., Jones, L. and Baulcombe, D.C. (2005) An RNA-dependent RNA polymerase prevents meristem invasion by *potato virus X* and is required for the activity but not the production of a systemic silencing signal. *Plant Physiology* 138(4), 1842–1852.

Schwarz, D.S., Hutvágner, G., Du, T., Xu, Z., Aronin, N. and Zamore, P.D. (2003) Asymmetry in the assembly of the RNAi enzyme complex. *Cell* 115(2), 199–208.

Segers, G.C., Zhang, X., Deng, F., Sun, Q. and Nuss, D.L. (2007) Evidence that RNA silencing functions as an antiviral defense mechanism in fungi. *Proceedings of the National Academy of Sciences of the United States of America* 104(31), 12902–12906.

Shamandi, N., Zytnicki, M., Charbonnel, C., Elvira-Matelot, E., Bochnakian, A., Comella, P., Mallory, A.C., Lepère, G., Sáez-Vásquez, J. and Vaucheret, H. (2015) Plants encode a general siRNA suppressor that is induced and suppressed by viruses. *PLOS Biology* 13(12), e1002326.

Shimura, H., Pantaleo, V., Ishihara, T., Myojo, N., Inaba, J., Sueda, K., Burgyán, J. and Masuta, C. (2011) A viral satellite RNA induces yellow symptoms on tobacco by targeting a gene involved in chlorophyll biosynthesis using the RNA silencing machinery. *PLoS Pathogens* 7(5), e1002021.

Shivaprasad, P.V., Rajeswaran, R., Blevins, T., Schoelz, J., Meins, F., Hohn, T. and Pooggin, M.M. (2008) The CaMV transactivator/viroplasmin interferes with RDR6-dependent trans-acting and secondary siRNA pathways in *Arabidopsis*. *Nucleic Acids Research* 36(18), 5896–5909.

Silhavy, D., Molnár, A., Lucioli, A., Szittya, G., Hornyik, C., Tavazza, M. and Burgyán, J. (2002) A viral protein suppresses RNA silencing and binds silencing-generated *21*- to 25-nucleotide double-stranded RNAs. *The EMBO Journal* 21(12), 3070–3080.

Siomi, H. and Siomi, M.C. (2009) On the road to reading the RNA-interference code. *Nature* 457(7228), 396–404.

Smith, N.A., Eamens, A.L. and Wang, M.-B. (2011) Viral small interfering RNAs target host genes to mediate disease symptoms in plants. *PLoS Pathogens* 7(5), e1002022.

Szabó, E.Z., Manczinger, M., Göblös, A., Kemény, L. and Lakatos, L. (2012) Switching on RNA silencing suppressor activity by restoring argonaute binding to a viral protein. *Journal of Virology* 86(15), 8324–8327.

Szittya, G., Moxon, S., Pantaleo, V., Toth, G., Rusholme Pilcher, R.L., Moulton, V., Burgyan, J. and Dalmay, T. (2010) Structural and functional analysis of viral siRNAs. *PLoS Pathogens* 6(4), e1000838.

Takeda, A., Sugiyama, K., Nagano, H., Mori, M., Kaido, M., Mise, K., Tsuda, S. and Okuno, T. (2002) Identification of a novel RNA silencing suppressor, NSs protein of *tomato spotted wilt virus*. *FEBS Letters* 532(1–2), 75–79.

Takeda, A., Tsukuda, M., Mizumoto, H., Okamoto, K., Kaido, M., Mise, K. and Okuno, T. (2005) A plant RNA virus suppresses RNA silencing through viral RNA replication. *The EMBO Journal* 24(17), 3147–3157.

Takeda, A., Iwasaki, S., Watanabe, T., Utsumi, M. and Watanabe, Y. (2008) The mechanism selecting the guide strand from small RNA duplexes is different among argonaute proteins. *Plant and Cell Physiology* 49(4), 493–500.

Torres-Barceló, C., Martín, S., Daròs J.A., Elena, S.F. (2008) From hypo- to hypersuppression: Effect of amino acid substitutions on the RNA-silencing suppressor activity of the *Tobacco etch potyvirus* HC-Pro. *Genetics* 180, 1039–1049.

Untiveros, M., Olspert, A., Artola, K., Firth, A.E., Kreuze, J.F. and Valkonen, J.P.T. (2016) A novel *sweet potato potyvirus* ORF is expressed via polymerase slippage and suppresses RNA silencing. *Molecular Plant Pathology* 17, 1111–1123.

Valli, A., Martín-Hernández, A.M., López-Moya, J.J. and García, J.A. (2006) RNA silencing suppression by a second copy of the P1 serine protease of *cucumber vein yellowing ipomovirus*, a member of the family *Potyviridae* that lacks the cysteine protease HCPro. *Journal of Virology* 80(20), 10055–10063.

Valli, A., Dujovny, G. and García, J.A. (2008) Protease activity, self interaction, and small interfering RNA binding of the silencing suppressor p1b from *cucumber vein yellowing ipomovirus*. *Journal of Virology* 82(2), 974–986.

Valli, A., López-Moya, J.J. and García, J.A. (2009) RNA silencing and its suppressors in the plant-virus interplay. In: *Encyclopedia of Life Sciences (ELS)*. John Wiley & Sons, Ltd, Chichester, UK. DOI: 10.1002/9780470015902.a0021261

Valli, A., Oliveros, J.C., Molnar, A., Baulcombe, D. and García, J.A. (2011) The specific binding to 21-nt double-stranded RNAs is crucial for the anti-silencing activity of *Cucumber vein yellowing virus* P1b and perturbs endogenous small RNA populations. *RNA* 17(6), 1148–1158.

Várallyay, E. and Havelda, Z. (2013) Unrelated viral suppressors of RNA silencing mediate the control of ARGONAUTE1 level. *Molecular Plant Pathology* 14(6), 567–575.

Várallyay, E., Válóczi, A., Agyi, A., Burgyán, J. and Havelda, Z. (2010) Plant virus-mediated induction of miR168 is associated with repression of ARGONAUTE1 accumulation. *The EMBO Journal* 29(20), 3507–3519.

Vargason, J.M., Szittya, G., Burgyán, J. and Hall, T.M.T. (2003) Size selective recognition of siRNA by an RNA silencing suppressor. *Cell* 115(7), 799–811.

Vaucheret, H., Vazquez, F., Crété, P. and Bartel, D.P. (2004) The action of ARGONAUTE1 in the miRNA pathway and its regulation by the miRNA pathway are crucial for plant development. *Genes and Development* 18(10), 1187–1197.

Vogler, H., Akbergenov, R., Shivaprasad, P.V., Dang, V., Fasler, M., Kwon, M.-O., Zhanybekova, S., Hohn, T. and Heinlein, M. (2007) Modification of small RNAs associated with suppression of RNA silencing by tobamovirus replicase protein. *Journal of Virology* 81(19), 10379–10388.

Voinnet, O. (2008) Use, tolerance and avoidance of amplified RNA silencing by plants. *Trends in Plant Science* 13(7), 317–328.

Voinnet, O., Pinto, Y.M. and Baulcombe, D.C. (1999) Suppression of gene silencing: a general strategy used by diverse DNA and RNA viruses of plants. *Proceedings of the National Academy of Sciences of the United States of America* 96(24), 14147–14152.

Voinnet, O., Lederer, C. and Baulcombe, D.C. (2000) A viral movement protein prevents spread of the gene silencing signal in *Nicotiana benthamiana*. *Cell* 103(1), 157–167.

Wang, X.-B., Wu, Q., Ito, T., Cillo, F., Li, W.-X., Chen, X., Yu, J.L. and Ding, S.-W. (2010) RNAi-mediated viral immunity requires amplification of virus-derived siRNAs in *Arabidopsis thaliana*. *Proceedings of the National Academy of Sciences of the United States of America* 107(1), 484–489.

Wang, X.-B., Jovel, J., Udomporn, P., Wang, Y., Wu, Q., Li, W.-X., Gasciolli, V., Vaucheret, H. and Ding, S.-W. (2011) The 21-nucleotide, but not 22-nucleotide, viral secondary small interfering RNAs direct potent antiviral defense by two cooperative argonautes in *Arabidopsis thaliana*. *The Plant Cell* 23(4), 1625–1638.

Weinheimer, I., Boonrod, K., Moser, M., Wassenegger, M., Krczal, G., Butcher, S.J. and Valkonen, J.P.T. (2014) Binding and processing of small dsRNA molecules by the class 1 RNase III protein encoded by *sweet potato chlorotic stunt virus*. *The Journal of General Virology* 95(Pt 2), 486–495.

Wilkins, C., Dishongh, R., Moore, S.C., Whitt, M.A., Chow, M. and Machaca, K. (2005) RNA interference is an antiviral defence mechanism in *Caenorhabditis elegans*. *Nature* 436(7053), 1044–1047.

Wu, J., Yang, Z., Wang, Y., Zheng, L., Ye, R., Ji, Y., Zhao, S., Ji, S., Liu, R., Xu, L., Zheng, H., Zhou, Y., Zhang, X., Cao, X., Xie, L., Wu, Z., Qi, Y. and Li, Y. (2015) Viral-inducible Argonaute18 confers broad-spectrum virus resistance in rice by sequestering a host microRNA. *eLife* 4, e05733. DOI: 10.7554/eLife.05733

Xie, Z., Fan, B., Chen, C. and Chen, Z. (2001) An important role of an inducible RNA-dependent RNA polymerase in plant antiviral defense. *Proceedings of the National Academy of Sciences of the United States of America* 98(11), 6516–6521.

Xiong, R., Wu, J., Zhou, Y. and Zhou, X. (2009) Characterization and subcellular localization of an RNA silencing suppressor encoded by *rice stripe tenuivirus*. *Virology* 387(1), 29–40.

Xu, Y., Huang, L., Fu, S., Wu, J. and Zhou, X. (2012) Population diversity of *rice stripe virus*-derived siRNAs in three different hosts and RNAi-based antiviral immunity in *Laodelphax striatellus*. *PloS One* 7(9), e46238.

Yang, X., Xie, Y., Raja, P., Li, S., Wolf, J.N., Shen, Q., Bisaro, D.M. and Zhou, X. (2011) Suppression of methylation-mediated transcriptional gene silencing by βC1-SAHH protein interaction during geminivirus-betasatellite infection. *PLoS Pathogens* 7(10), e1002329.

Ye, K., Malinina, L. and Patel, D.J. (2003) Recognition of small interfering RNA by a viral suppressor of RNA silencing. *Nature* 426(6968), 874–878.

Yu, D., Fan, B., MacFarlane, S.A. and Chen, Z. (2003) Analysis of the involvement of an inducible *Arabidopsis* RNA-dependent RNA polymerase in antiviral defense. *Molecular Plant-Microbe Interactions* 16(3), 206–216.

Zhang, C., Wu, Z., Li, Y. and Wu, J. (2015) Biogenesis, function, and applications of virus-derived small RNAs in plants. *Frontiers in Microbiology* 6, 1237.

Zhang, X., Yuan, Y.-R., Pei, Y., Lin, S.-S., Tuschl, T., Patel, D.J. and Chua, N.-H. (2006) *Cucumber mosaic virus*-encoded 2b suppressor inhibits *Arabidopsis* Argonaute1 cleavage activity to counter plant defense. *Genes and Development* 20(23), 3255–3268.

Zhang, X., Du, P., Lu, L., Xiao, Q., Wang, W., Cao, X., Ren, B., Wei, C. and Li, Y. (2008) Contrasting effects of HC-Pro and 2b viral suppressors from *sugarcane mosaic virus* and *tomato aspermy cucumovirus* on the accumulation of siRNAs. *Virology* 374(2), 351–360.

Zhang, X., Zhang, X., Singh, J., Li, D. and Qu, F. (2012) Temperature-dependent survival of *Turnip crinkle virus*-infected *Arabidopsis* plants relies on an RNA silencing-based defense that requires dcl2, AGO2, and HEN1. *Journal of Virology* 86(12), 6847–6854.

Zhang, Z., Chen, H., Huang, X., Xia, R., Zhao, Q., Lai, J., Teng, K., Li, Y., Liang, L., Du, Q., Zhou, X., Guo, H. and Xie, Q. (2011) BSCTV C2 attenuates the degradation of SAMDC1 to suppress DNA methylation-mediated gene silencing in *Arabidopsis*. *The Plant Cell* 23(1), 273–288.

Zhou, Z.S., Dell'Orco, M., Saldarelli, P., Turturo, C., Minafra, A. and Martelli, G.P. (2006) Identification of an RNA-silencing suppressor in the genome of *Grapevine virus A*. *The Journal of General Virology* 87(Pt 8), 2387–2395.

Zrachya, A., Glick, E., Levy, Y., Arazi, T., Citovsky, V. and Gafni, Y. (2007) Suppressor of RNA silencing encoded by *tomato yellow leaf curl virus-Israel*. *Virology* 358(1), 159–165.

# 3 Artificial Induction and Maintenance of Epigenetic Variations in Plants

**Soumita Das, Rahul Raj Singh and P.V. Shivaprasad***
*National Centre for Biological Sciences, Bangalore, India*

## 3.1 Introduction

The term 'epigenetic' was first coined in 1942 by Waddington, a British developmental biologist, referring to differentiation of cells from their totipotent state. The term expresses the external manifestation of the genetic activity in organisms. Epigenetics has now emerged as a broad field of science that investigates a range of biological phenomena with novel molecular mechanisms. Epigenetics works on two levels – at the DNA level, and at the level of histones. Methylation of cytosines is context independent and reversible, and plays a crucial role especially if the modification is in the promoter regions of transcription units. Cytosine methylation controls transposons and unruly elements in genomes. The second level of epigenetics is at the level of modifications of DNA-bound histone proteins. Many of the histone-modifying proteins have been characterized, although still many more remain to be discovered, and the sequence/priority order of the histone modifications are being elucidated. Among plants, small RNAs play an essential role in establishment and maintenance of DNA methylation, through a novel process termed 'RNA directed DNA methylation' (RdDM). In animals, the developmental programme is planned in the embryo; but in plants, it is designed post-embryonically through a poorly understood mechanism. Early indications are that epigenetic variations play a significant role in such programming. Thus, a case for epigenetic studies has been set, while the stepwise molecular events of such a phenomenon are largely unknown and currently an intense area of research.

Nonetheless, what has been clearly demonstrated is a case for artificial induction and maintenance of epigenetic marks in plants. Some of the studies initiated a decade ago to understand the behaviour of transgenes have provided the first clear evidence for transgene-mediated induction and subsequent inheritance of epigenetic marks. Now there is enough proof to show that it is possible to induce methylation of DNA and histones and also to maintain them during subsequent generations without the need for the inducer. This review brings together recent advances in

* Corresponding author e-mail: shivaprasad@ncbs.res.in

such artificial induction and suggests ways forward to make this technology an equivalent – if not better – strategy than genetic modification, which is clouded with never-ending controversies.

## 3.2 Cytosine Methylation, an Important Epigenetic Landscape among Plants

Cytosine methylation is one of the most extensively studied reversible epigenetic marks among all eukaryotes. Biochemically, it involves addition of a methyl group at the fifth position of the pyrimidine ring of cytosines. Methylation of DNA at the cytosine residue serves as an origin of genomic imprinting in successive generations, regulating gene expression and maintaining the silenced status of the transposable element (TE) as well as other repeat loci. Genome-wide methylome studies in different flowering plants (*Arabidopsis*, maize, rice and poplar; Vaughn and Martienssen, 2005; Gehring and Henikoff, 2007; Matzke *et al.*, 2007) revealed that more than 80–90% of cytosine methylation is concentrated in transposons and repeat loci whereas only 10–20% is distributed in the gene-rich regions. Most sites with cytosine methylation are enclosed within or close to DNA elements which have the ability to transpose from one location in the genome to another. This observation reflects a primary role of DNA methylation in genome defence. However, transposon-associated DNA methylation also affects transcription of adjacent genes so that the silencing of TEs can influence patterns of gene expression and has the potential to alter phenotypes of the plant.

Among plants, cytosine residue in either of the two alternate sequence contexts CHG or CHH (where H can be A, T or G) is also methylated. However, CG methylation is usually predominant among those three different types of methylation. In the *Arabidopsis thaliana* genome, distribution of CG, CHG and CHH methylation was found to be 24%, 6.7% and 1.7%, respectively. Intriguingly, CHG and CHH classes of cytosine methylation patterns are evolved and restricted only within the plant genome. DNA METHYLTRANSFERASE 1 (MET1), a homologue of DNMT1 among animals, maintains CG methylation in plants, copying the methylation onto the opposite cytosine residue of the daughter strand (Aufsatz *et al.*, 2004). The methyl binding proteins VARIANT IN METHYLATION 1 (VIM1), VIM2 and VIM3 interact with hemi-methylated CG sites in the parent DNA and recruit MET1 to carry out the methylation (Woo *et al.*, 2008). Plant-specific symmetric CHG and asymmetric CHH methylation is carried out by two different methyltransferases, CHROMOMETHYLASE 3 (CMT3) and CHROMOMETHYLASE 2 (CMT2), respectively (Zemach *et al.*, 2013). Both CMT2 and CMT3 recognize dimethylated histone 3 tails at lysine 9 (H3K9m2) on the nucleosomes and maintain the methylation in the newly synthesized DNA strand (Cao and Jacobsen, 2002).

Among plants, components and mechanisms for the establishment of methylation in the unmethylated cytosine residue (*de novo* methylation) is guided by small RNAs. Specifically, small interfering (si) RNAs of 24nt are the key components in establishment and maintenance of RdDM. Studies of the biogenesis of these siRNAs and their effect on genomes have unearthed key components of the machinery that makes these siRNAs. The most surprising aspect is the presence of two additional DNA-dependent RNA polymerases, PolIV and PolV. These polymerases transcribe non-coding transcripts from numerous regions in the genomes. The SAWADEE HOMEODOMAIN HOMOLOG (SHH)1 binds and recruits PolIV to transcribe methylated DNA (Law *et al.*, 2013). Transcripts from PolIV are converted into dsRNA through the action of one of the plant-specific groups of RNA-dependent RNA polymerases (RdRP), named RDR2. These transcripts, the nature of which has been intensely debated recently, are processed in the nucleus through an RNAseIII-type endonuclease called DICER-LIKE 3 (DCL3) into siRNAs of 24 nt in size. These siRNAs are exported to the cytoplasm and associate with specific Argonautes of the AGO4/6/9 family. Recent studies indicate that AGO6 has a redundant function and AGO9 has tissue-specific roles, while AGO4 appears to be the workhorse. Following loading of these siRNAs onto AGO4, the ribonucleoprotein complexes are re-imported to the nucleus, where the siRNA guides the targeting of nascent scaffold transcripts from PolV by sequence complementarity. This interaction recruits the *de novo* methyltransferase DRM2 that modifies C residues of the DNA strand, acting as the template for PolV (Henderson *et al.*, 2010; Zhong *et al.*, 2014). Methylation of cytosines in all classes of sequence contexts (that is, CG, CHG and CHH, where H represents A, T or G) is achieved subsequently. This results in transcriptional silencing at the genomic loci which are transcribed by PolV, particularly transposons and other repetitive DNA.

There is yet another, newly recognized, pathway of RdDM. Here, neither DCL3 (which generates 24 nt siRNAs) nor PolIV is required for the establishment of RdDM (Pontier *et al.*, 2012). This RdDM pathway depends on RDR6 to generate substrates; these cleave into 21 or 22 nt siRNAs which either re-establish or *de novo* methylate cognate DNAs (Nuthikattu *et al.*, 2013). The pathway requires PolV but depends on the PTGS trigger through 21 and 22 nt siRNAs processed by DCL4 and DCL2, respectively (Bond and Baulcombe, 2015).

The methylated cytosine residue (mC) interacts with different methyl domain-binding proteins and chromosome remodeller complexes to initiate and sustain condensed chromatin architecture which results in repression of associated transcription. Hence, cytosine methylation is generally considered a basic repressive epigenetic signature. However, the specific region where methylated cytosine is present is also important. By and large, promoter DNA methylation is repressive in nature and associated with reduced, if not non-existent, transcription of the gene driven by this promoter. Some well-studied examples include methylation in the promoter of *FLOWERING WAGENINGEN* (FWA) and SDC (*Suppressor of drm1 drm2 cmt3*) genes, which result in suppression of the transcription necessary for plant growth and development (Soppe *et al.*, 2002; Hogenhout *et al.*, 2009). However, the nature and distribution of the mC within the coding region serves as an exception. Interestingly, modestly expressed genes contain a high degree of cytosine methylation within the coding region when compared to low or highly expressed genes (Zilberman and Henikoff, 2007; Cokus *et al.*, 2008; Verdel *et al.*, 2009; Zhang and Zhu, 2011). The mC-mediated repression of aberrant transcription from the cryptic promoter positioned within the coding region of modestly expressed genes results in increased levels of transcription (Tran *et al.*, 2005a; Zilberman and Henikoff, 2007). In the coding region cytosine methylation occurs exclusively at the CG site.

## 3.3 Histone Modifications

The amino-terminal tails of the core histones are subjected to various posttranslational modifications, including acetylation, methylation, ubiquitination, phosphorylation, glycosylation, ADP-ribosylation and sumoylation. The histone code hypothesis predicts that these covalent modifications might provide specificity for effector proteins which bind the modification marks and interpret the code into functional outcomes (Ooi and Henikoff, 2007; Liu *et al.*, 2010; Chen *et al.*, 2011; Simon and Meyers, 2011). Although most of the histone modifications are conserved across different kingdoms, the establishment and maintenance of these modifications in plants are related but not identical to fungi and animals (Pfluger and Wagner, 2007; Numa *et al.*, 2010). These differences are likely to be due to the presence of unique processes in plant development and the requirement to pass on environmental cues to the next generation for a better survival. For example, the double fertilization of flowering plants, in which the egg cell is fertilized by one sperm to form the embryo while a second sperm fertilizes the two polar nuclei to form the endosperm, allows for unique mechanisms of epigenetic inheritance. Also, the floral transition makes permanent changes in shoot apical meristem identity and provides an excellent example of an epigenetically regulated event that is determined by developmental and environmental cues (e.g., vernalization). Similarly, transgenerational inheritance of environmental cues has been proposed (and validated at least in *A. thaliana*) to involve DNA and histone modifications.

Histone methylation plays an essential role in diverse biological processes ranging from transcriptional regulation to heterochromatin formation. As one of the most complex modifications, it not only occurs at different residues (lysine and arginine) and distinct sites but also differs in the number of methyl groups added. In *Arabidopsis*, histone lysine methylation occurs mainly at Lys4 (K4), Lys9 (K9), Lys27 (K27) and Lys36 (K36) of histone H3. These modifications are written by different histone lysine methyltransferases (HKMTs). In contrast to mammals and yeast, in which Lys20 (K20) of histone H4 (H4K20) is methylated, H4K20 is acetylated in *Arabidopsis*, though mono-methylated H4K20 (H4K20me1) has been reported to be detected by immunostaining (Simon and Meyers, 2011; Saze *et al.*, 2012). *Arabidopsis* H3 lysine 79 (H3K79) methyltransferase, called DOT1, is required for telomeric silencing (Frederiks *et al.*, 2008). In addition, *Arabidopsis* and rice were shown to have much higher levels of H3K4 di-methylation (H3K4me2) than among animals.

Lysine methylation of histones is an important and complex epigenetic mark that decorates both transcriptionally silenced and active chromatin domains, depending on which lysine residues are methylated and the degree of methylation. Lysine methylation does not affect the net charge of the modified residues, but it elevates the hydrophobicity and may alter intra- or intermolecular interactions or create new binding surfaces for reader proteins that bind preferentially to the methylated domain. Generally, histone H3K9 and H3K27 methylation are associated with silenced regions, whereas H3K4 and H3K36 methylation are associated with active genes (Rando, 2007; Ahmad *et al.*, 2010).

Characterization of the interplay between histone H3K9 methylation and DNA methylation revealed that H3K9 methylation is critical for maintenance of genome-wide transcriptional gene silencing and genome stability (Deleris *et al.*, 2012). In the kyp/suvh4 mutant, reduced H3K9me2 leads to loss of non-CG DNA methylation catalyzed by CMT3, thereby derepressing the silenced SUP and PAI and endogenous TA3 transposon (Tran *et al.*, 2005b; Fischer *et al.*, 2006). KYP/SUVH4 also acts at the SINE retroelement AtSN1(*Arabidopsis thaliana* short interspersed element 1), a region with high levels of CHH methylation and intermediate levels of CHG methylation, but only low levels of CG methylation (Tran *et al.*, 2005b; Pouch-Pélissier *et al.*, 2008). DNA methylation profiling revealed that CMT3 targets genome-wide transposons and retrotransposons (Cao *et al.*, 2003), and expression profiling of kyp/suvh4 and cmt3 identified hundreds of reactivated TEs as common targets of both KYP/SUVH4 and CMT3. High resolution genome-wide ChIP-chip analysis further revealed a very high coincidence between H3K9me2 and CHG methylation (Schubert *et al.*, 2006; Deleris *et al.*, 2012). These results demonstrated that maintenance of non-CG DNA methylation requires histone H3K9 methylation, and that these modifications are critical for repression of TEs, arrays of ribosomal RNA genes (rDNA) and other repetitive sequences.

Other lines of evidence suggest that DNA methylation also reinforces histone methylation in a positive feedback loop. For example, impaired CG DNA methylation in met1 mutants is associated with decreased H3K9me2 at 180 bp centromeric repeats, endogenous Ta2 transposon and other TEs (Deleris *et al.*, 2012; Zubko *et al.*, 2012). Histone H3K9 methylation patterns regulated by CG and non-CG methylation are locus specific and do not overlap. Two critical features for such locus specificity include the density of CG, CHG and CHH sequences at each locus, and the presence of either inverted repeats or tandem repeats. The mechanisms that tie histone methylation to DNA methylation have been partially elucidated by the identification of methylated DNA-binding proteins. For example, YDG/SRA domains of KYP/SUVH4 and SUVH6 preferentially bind methylated CHG DNA, suggesting a role of DNA methylation in recruiting H3K9 methyltransferases (Fischer *et al.*, 2006).

It can be suggested that the plant genome is more stringently regulated under a complex and exceedingly evolved epigenetic mechanism compared to the mammalian genome. Expression of any endogenous plant gene is not only driven by its native promoter but also controlled by epigenomics. In other words, epigenetic modifiers or components can be used as targets to introduce new characters in the plant or to manipulate gene expression.

## 3.4 Two Additional DNA-dependent RNA Polymerases

RdDM in plants is exceptional among related pathways of chromatin modifications in other model systems. It depends on a specialized transcriptional machinery that is centred around two plant-specific RNA polymerases, PolIV and PolV, both of which evolved from PolII (Matzke and Mosher, 2014). PolV has been characterized only from angiosperms whereas PolIV exists in all plants. Interestingly, algae lack both these additional polymerases (Matzke and Mosher, 2014). Biochemically, PolII, PolIV and PolV all have 12 subunits, many of which are shared by the three polymerases; but each also has specialized subunits (Ream *et al.*, 2009; Tucker *et al.*, 2010). Subunits are named nuclear RNA polymerase B (NRPB) for PolII subunits, NRPD for PolIV subunits and NRPE for PolV subunits. *In vitro* experiments indicate that PolIV and PolV can carry out RNA-primed transcription of DNA and transcribe from bipartite RNA–DNA templates. PolIV can also transcribe bipartite RNA–RNA templates (Haag *et al.*, 2012), the significance of which is not known. Similar to PolII, PolIV and PolV may require factors that assist entry into the nucleus from the cytoplasm, in which the subunits are synthesized and assembled. DNA-binding proteins, co-factors and methyltransferases associated with these two plant-specific RNA polymerases have been identified, but the exact mechanism is still unclear (Kanno and Habu, 2011).

## 3.5 Artificial Induction of Epigenetic Variations

A surprising link with siRNAs of 24 nt and DNA methylation coincided with the observation that artificially produced siRNAs could induce DNA methylation. This ability to introduce modifications through artificial means turned out to be an amazing part of the plant epigenome. Since siRNAs could be designed and produced in a given plant relatively easily, attempts were made to target promoters of genes (both transgenes and endogenes) and then observe if promoters of target genes attracted DNA methylation. Further, the heritability of DNA methylation was checked. Two such possibilities have been tried and tested with moderate success (Fig. 3.1).

### 3.5.1 Inverted repeat mediated RdDM induction

It has been established that the introduction of transcribed sense transgenes could subdue the expression of homologous endogenous genes mediated through a process called cosuppression (Jorgensen *et al.*, 2006). Thus, it was a logical step to use *Agrobacterium*-mediated T-DNA transfer and integration as a method to introduce transgenes that could initiate TGS, as the method was very successful in inducing silencing. However, in PTGS extensively used in functional genomics, the sequences used as transgenes included regions derived from protein-coding regions. These sequences were usually in antisense orientation, to make double-stranded (ds) RNA trigger DCL-mediated cleavage of target transcripts. This technique was modified to target promoters of genes to induce silencing. In such strategies, either a full-length promoter (Mette *et al.*, 1999; Jones *et al.*, 2001) or a part of the promoter was used either in sense or antisense orientation. Although the importance of 24 nt siRNAs was elucidated much later, these experiments provided proof of concept that transgene promoters can be methylated and silenced (Fig. 3.1).

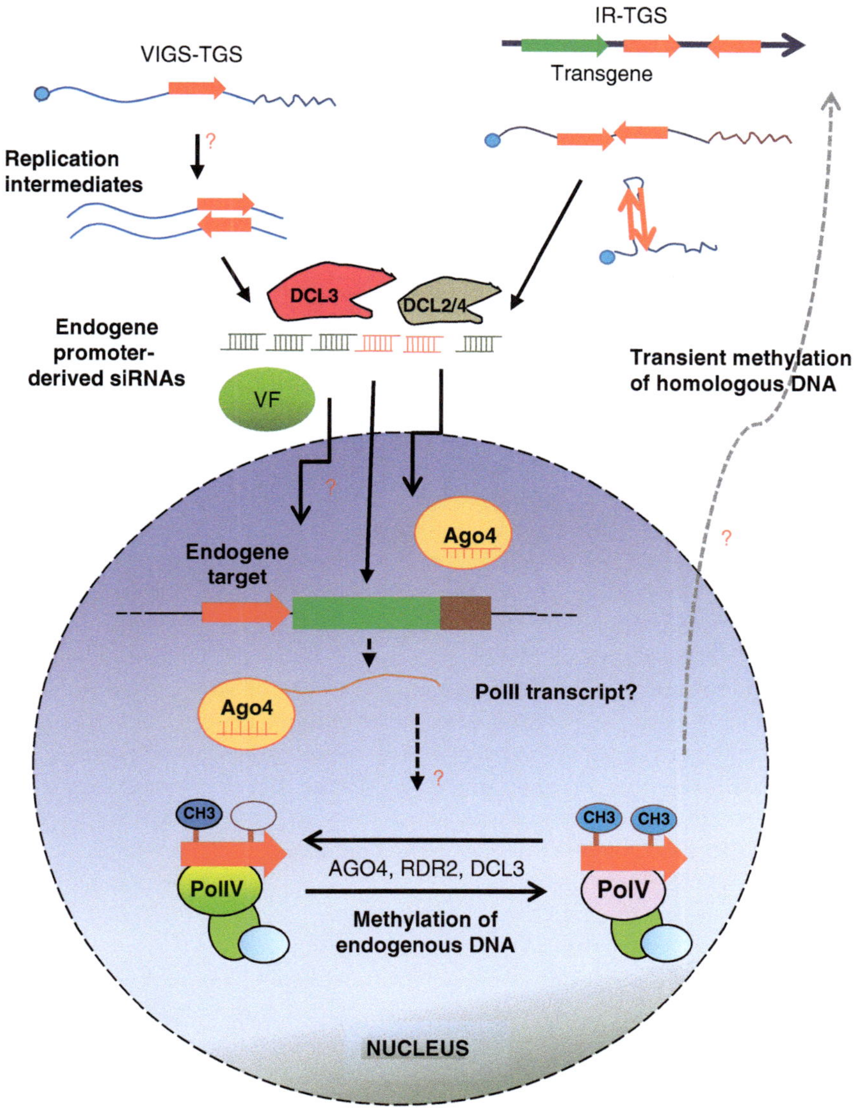


**Fig. 3.1.** Schematic representation of artificial transcriptional silencing technology. It is possible to induce heritable TGS by at least two methods: (i) using viruses as vectors that generate dsRNA intermediates during replication, resulting in promoter-derived siRNAs which are transported to the nucleus through viral proteins (VF); and (ii) using transgenic introduction of inverted repeats derived from endogenous promoters.

The theoretical understanding of the mechanism was not clear at that time. It was also demonstrated (Matzke *et al*., 1994) that assimilation of multiple copies of a transgene in a particular fashion may lead to methylation and subsequently transcriptional gene silencing and downregulation of homologous genes. This strategy served as an outstanding tool in functional genomics.

A lucid projection of this technique was demonstrated by Mette *et al.* (1999), where promoter sequence of nopaline synthase gene was expressed under 35S promoter. The expression instigated the trans-TGS and methylation of homologous NOS promoter. This tactic incorporated the use of hairpin conformation owing to inverted repeats, with target sequences in the middle. This form of construct was reported to be more effective as compared to the sense and anti-sense silencers (Matzke and Matzke, 1998). In contrast to the transgenics that carried one locus having a direct tandem repeat of T-DNA leading to post-transcriptional gene silencing, transgenics having more than two copies of T-DNA exhibited transcriptional gene silencing (Vaucheret and Fagard, 2001).

One of the most successful efforts was the targeting of endogenous gene FWA, a homeodomain transcription factor that represses flowering time in *Arabidopsis*. Here the target was a pair of direct repeats in the promoter of FWA. If these direct repeats are not methylated, FWA is expressed and the plants exhibit a late-flowering phenotype. However, *Agrobacterium*-mediated transformation with an inverted repeat transgene producing small RNAs complementary to the direct repeats triggered RdDM at the promoter (Kinoshita *et al.*, 2007). The silent state was inherited into the next generation, even if the inverted repeat transgene segregated away from the silenced locus. In segregated lines, CG methylation was observed in the FWA promoter.

Despite several advantages associated with the strategy, it is arduous to employ inverted repeats to establish the inverted repeats as a foundation for TGS of target genes. A significant setback is the problem of heritability. No *de novo* methylation was observed in *Nicotiana benthamiana* even when sequence-specific degradation of RNA was induced (Jones *et al.*, 1999). Moreover, successful reports of gene silencing through promoter-targeted siRNA was reported in *Arabidopsis* (Melquist and Bender, 2003) and maize (Mark Cigan *et al.*, 2005), but parallel positive results were not clearly projected in rice (Okano *et al.*, 2008). Single-handedly, RdDM may not be sufficient to suppress endogenous rice genes, so alteration in histone signatures may also have vital roles in silencing.

Another major disadvantage of this system is the need to have tedious transformation of inverted repeats. Usually such repeats are difficult to integrate stably (Sunitha *et al.*, 2012). In addition, demand for a thorough molecular analysis is quite high (picking single copy transgenics, making sure that both arms of inverted repeats are present in the transgenes, absence of recombination and deletion of repeats in *Agrobacterium*, etc.). We predict that this may not be the method of choice for artificial induction and inheritance.

### 3.5.2 Virus induced gene silencing (VIGS), an artificial tool to introduce epigenetic modification among plants

The term 'virus induced gene silencing' (VIGS) was first applied to describe the process of host recovery after viral infection (van Kammen, 1997). Later, it was modified and developed as a tool for functional genomics in plants (Ruiz *et al.*, 1998; Baulcombe, 1999). It has been used to perform both forward and reverse genetics to identify plant genes involved in several plant processes. However, this technology has not yet been used to its full potential as it has limitations, such as the inability to silence genes during seed germination and the non-stable nature of silencing.

*Tobacco mosaic virus* (TMV) was the first virus selected for creating a recombinant virus suitable for PTGS (Kumagai *et al.*, 1995). Phytoene desaturase (PDS) was the target gene used for the study and the host was the tobacco species *N. benthamiana*. Currently, more than 20 different virus vectors are available for VIGS, most commonly *Cucumber mosaic virus* (CMV), PVX, TGMV, TRV and BGMV (Kjemtrup *et al.*, 1998; Ruiz *et al.*, 1998; Dalmay *et al.*, 2000; Ratcliff *et al.*, 2001; Holzberg *et al.*, 2002; Kanazawa, 2008; Kanazawa *et al.*, 2011). VIGS vectors differ in their host range, tissue specificity, efficiency and level of producing symptoms followed by infection. Most of the viruses used to design vectors contain RNA as genomic material. RNA viruses produce dsRNA intermediate during viral replication.

The principle behind this technology is that viral RNA is targeted by the host defence system and is finally degraded by an siRNA-mediated gene-silencing pathway. Therefore, a recombinant virus carrying the host's endogenous gene triggers an siRNA pathway against the homologous sequence and degrades the transcript, leading to post-transcriptional gene silencing or PTGS (Fig. 3.1). In another method of VIGS action, a recombinant virus containing a host endogenous promoter element leads to promoter methylation and TGS. The best example elucidating the potential of small RNA-mediated transgenerational epigenetic inheritance came from the Baulcombe group (Jones *et al.*, 2001). Here, a plant expressing GFP under *Cauliflower mosaic virus* 35S promoter was infected with *Tobacco rattle virus* containing a portion of DNA sequence corresponding to 35S promoter of the reporter gene. On infection with the recombinant virus, the plant's antiviral defence mechanism produced small RNAs against the virus and the promoter of GFP through VIGS (Lu *et al.*, 2003). The reporter gene GFP was silenced in the infected plants due to cytosine methylation at the promoter in all DNA sequence contexts (CG, CHG and CHH). Extraordinarily, the 35S promoter remained methylated and GFP was silent in later generations where the virus was no longer present. However, the maintenance of silencing in these progeny plants was not associated with methylation of CHH motifs and, unlike the infected plants, MET1 was required for continued silencing. This explained why small RNAs are not required for maintenance of promoter methylation. It is possible that the PolIV/PolV loop that is active at the 35S promoter region of the transgene now operates on its own.

However, application of TGS in epigenetic modification of the endogenous plant gene promoter was attempted only very recently. Kanazawa *et al.* (2011) first showed that the CMV vector carrying a promoter sequence of endogenous CHS A gene induces methylation in the endogenous promoter of CHS A gene, followed by heterochromatin formation and CHS A silencing in the petunia plant (Kanazawa *et al.*, 2011; Kanazawa and Kasai, 2015). These 24nt siRNAs are capable of passing through the germ line and induce transgenerational gene silencing. CMV-based vectors are gaining importance in TGS due to the presence of nuclear signalling protein 2b, which selectively binds to dsRNA and transports it to the nucleus to establish TGS (Kanazawa *et al.*, 2011). Hence, TGS can be applied for incorporation of a new stable character into a plant system. However, the functions of all the components involved in this RdDM mechanism, as well as in TGS, are not well characterized to date. Suitable vectors are also needed to increase the efficiency and efficacy of this technology.

## 3.6 VIGS-mediated TGS Mechanism: Gaps in Our Understanding

In a VIGS pathway, the siRNAs derived from the virus must target a transcribed region from the endogene or a transgene to initiate TGS. In such a scenario, the density with which the 24 nt siRNAs generate should be directly proportional to the DNA methylation effect. However, there are successful examples (Kanazawa *et al.*, 2011) where VIGS-induced TGS involved a cytoplasmic virus making almost negligible levels of 24 nt siRNAs which were then transported to the nucleus through the action of virus-encoded suppressors (Fig. 3.1). If this is the case, then why do some other viruses – which have nuclear-targeted suppressors that also can bind to siRNAs of multiple sizes including 24 nt siRNAs – not effectively induce TGS? In an endogenous native TGS pathway, PolIV usually transcribes a region that is ultimately methylated. PolIV transcription of a viral region or a transgene has not been reported even for DNA viruses that spend a substantial part of their life cycle in the nucleus. In all likelihood, PolIV transcription must start from the endogenous promoter region at some point and begin making substrates that will ultimately shift the normal siRNA pathway to the PolIV–PolV cycle. Another important question is the role of the CG/non-CG context in the promoters. It is well known that CG methylation can be maintained without the inducer by MET1, while non-CG methylation requires additional information such as siRNAs to maintain methylation. Does this mean that induction of TGS needs a target region of the promoter with a high CG context? Some recent experiments have indicated a higher frequency of inheritance of DNA methylation if the target region is rich in GC (Kanazawa and Kasai, 2015).

The most effective promoter to induce TGS has been the 35S viral promoter. Besides being a viral promoter exogenous in nature, it is one of the strongest promoters known in plant systems. This poses two questions. Is there an inherent bias for exogenous promoters being more sensitive towards RdDM? Is there a precondition that a promoter must be constitutive and strong, thereby inducing aberrant RNAs from this region that ultimately end up as substrates for DCL3 to generate the all-important 24 nt siRNAs? We do not have data to answer these questions at the moment, but there is a likelihood that all these are prerequisites for setting up heritable RdDM.

## 3.7 Recipe for High Efficiency for Inheritance of Methylation

While it is possible to induce and maintain heritable epigenetic variations, a perfect recipe is lacking. To modify the epigenome (or to undo the centuries-old modifications that took place during crop domestication), and offer a possible mechanism for altering plant phenotypes, the following should be tried:

**1.** Improved vector systems. The viruses that cause the least disruptive symptoms and those that can get into the meristems should be preferred. PVX has been shown to invade meristems, but this vector system has not been studied extensively. Similarly monocots, a group to which major crops belong, are yet to be artificially induced to have TGS. Obviously viruses that are disarmed, or those that do not result in strong phenotypes should be preferred, as they are not likely to result in reduced seed set. Kanazawa *et al.* (2011) bypassed this by using vector RNAs from multiple strains of the viruses. A comprehensive understanding of the nature and combination of pseudo-recombinants is required for better induction of TGS is needed.

**2.** Vectors with viral suppressors that bind/protect 24 nt siRNAs and preferably target the nucleus. Examples can be TRV vectors having 2b protein from CMV, or any other vector that is also additionally supplied with such suppressors.

**3.** Co-silencing of RdDM antagonistic genes such as demethylases, similar to that reported by Otagaki *et al.* (2013). This group reported higher DNA methylation of targets upon silencing ROS1 demethylase. Few DNA viral suppressors have the ability to demethylate their promoters as well as endogenous promoters. Designing vectors without such suppressors can be another attractive strategy.

**4.** The vector should be a preferred substrate for DCL3. In such a scenario, ssDNA viruses such as geminiviruses and nanoviruses that replicate in the nucleus and are predominantly sources of 24 nt siRNAs should be tested for their ability to induce heritable changes. Caulimoviruses, which spend most of their life cycle in the nucleus, are to be preferred.

**5.** The promoter should be enriched for GC sequences, as shown by Kanazawa *et al.* (2011). The ability of the 35S promoter to be more amenable than other tested promoters could be due to its high percentage of CGs. Endogenous promoters that also have these features are likely to be more suitable for TGS.

**6.** The target region should have some PolI transcription activity, for example as in FWA. One way to design a construct for inducing heritable TGS is by incorporating 5′UTR sequences.

**7.** Repeats inside the promoter sequences. As in the FWA example, there is a short repeat inside the promoter region and FWA is one of the earliest genes where epiallele formation was first demonstrated. It should be noted that the CaMV 35S promoter, which appears to be easily amenable for epigenetic inheritance, has some short repeat regions inside the promoter. Could these need PolIV or PolV for their transcription? Unfortunately, PolIV and PolV transcript characters are not well known. It is possible that they have a special liking for repeats to induce transcription. Indeed, they show increased genome-wide activity in repeat-rich regions such as TEs. It would be useful to have such structures embedded inside the promoter sequence.

## 3.8 Conclusions

RdDM in plants is an amazing pathway with unparalleled implications. Many speculate that RdDM plays a major role in the way that plants, even after being sedentary, are able to tolerate external stimuli and stresses by virtue of remembering what they or their ancestors went through. With the above-mentioned ability to induce heritable epigenetic variations artificially, plants can be altered without the need for genome modifications, and consequently the controversies surrounding transgenic crops are unlikely to affect epigenetic modifications of crop plants (Kasai and Kanazawa, 2013). Unlike many competing technologies, the effects on the genome can be specified. With VIGS, we have a method that lacks a selection marker requirement or a requirement for relatively cumbersome crop transformation. Along with genome-editing technologies, we can anticipate artificial induction of RdDM technology playing a major role in new crop improvement strategies.

## Acknowledgements

PVS acknowledges support from the Ramanujan Fellowship (SR/S2/RJN-109/2012; Department of Science and Technology, Government of India). The Principal Investigator's laboratory is supported by NCBS-TIFR core funding and a grant (BT/PR12394/AGIII/103/891/2014) from the Department of Biotechnology, Government of India. We thank N.D. Sunitha for comments.

## References

Ahmad, A., Zhang, Y. and Cao, X.F. (2010) Decoding the epigenetic language of plant development. *Molecular Plant* 3, 719–728.

Aufsatz, W., Mette, M.F., Matzke, A.J.M. and Matzke, M. (2004) The role of MET1 in RNA-directed *de novo* and maintenance methylation of CG dinucleotides. *Plant Molecular Biology* 54, 793–804.

Baulcombe, D.C. (1999) Gene silencing: RNA makes RNA makes no protein. *Current Biology* 9, 599–601.

Bond, D.M. and Baulcombe, D.C. (2015) Epigenetic transitions leading to heritable, RNA-mediated *de novo* silencing in *Arabidopsis thaliana*. *Proceedings of the National Academy of the Sciences of the United States of America* 112, 917–922.

Cao, X. and Jacobsen, S.E. (2002) Role of the *Arabidopsis* DRM methyltransferases in *de novo* DNA methylation and gene silencing. *Current Biology* 12, 1138–1144.

Cao, X., Aufsatz, W., Zilberman, D., Mette, M.F., Huang, M.S., Matzke, M. and Jacobsen, S.E. (2003) Role of the DRM and CMT3 methyltransferases in RNA-directed DNA methylation. *Current Biology* 13, 2212–2217.

Chen, X., Hu, Y. and Zhou, D.X. (2011) Epigenetic gene regulation by plant Jumonji group of histone demethylase. *Biochimica et Biophysica Acta - Gene Regulatory Mechanisms* 1809, 421–426.

Cokus, S.J., Feng, S., Zhang, X., Chen, Z., Merriman, B., Haudenschild, C.D., Pradhan, S., Nelson, S.F., Pellegrini, M. and Jacobsen, S.E. (2008) Shotgun bisulphite sequencing of the *Arabidopsis* genome reveals DNA methylation patterning. *Nature* 452, 215–219.

Dalmay, T., Hamilton, A., Mueller, E. and Baulcombe, D.C. (2000) *Potato virus X* amplicons in *Arabidopsis* mediate genetic and epigenetic gene silencing. *Plant Cell* 12, 369–379.

Deleris, A., Stroud, H., Bernatavichute, Y., Johnson, E., Klein, G., Schubert, D. and Jacobsen, S.E. (2012) Loss of the DNA methyltransferase MET1 induces H3K9 hypermethylation at PcG target genes and redistribution of H3K27 trimethylation to transposons in *Arabidopsis thaliana*. *PLoS Genetics* 8, e1003062.

Fischer, A., Hofmann, I., Naumann, K. and Reuter, G. (2006) Heterochromatin proteins and the control of heterochromatic gene silencing in *Arabidopsis*. *Journal of Plant Physiology* 163, 358–368.

Frederiks, F., Tzouros, M., Oudgenoeg, G., van Welsem, T., Fornerod, M., Krijgsveld, J. and van Leeuwen, F. (2008) Nonprocessive methylation by Dot1 leads to functional redundancy of histone H3K79 methylation states. *Nature Structural and Molecular Biology* 15, 550–557.

Gehring, M. and Henikoff, S. (2007) DNA methylation dynamics in plant genomes. *Biochimica et Biophysica Acta* 1769, 276–286.

Haag, J.R., Ream, T.S., Marasco, M., Nicora, C.D., Norbeck, A.D., Pasa-Tolic, L. and Pikaard, C.S. (2012) In vitro transcription activities of Pol IV, Pol V, and RDR2 reveal coupling of Pol IV and RDR2 for dsRNA synthesis in plant RNA silencing. *Molecular Cell* 48, 811–818.

Henderson, I.R., Deleris, A., Wong, W., Zhong, X., Chin, H.G., Horwitz, G.A., Kelly, K.A., Pradhan, S. and Jacobsen, S.E. (2010) The *de novo* cytosine methyltransferase DRM2 requires intact UBA domains and a catalytically mutated paralog DRM3 during RNA-directed DNA methylation in *Arabidopsis thaliana*. *PLoS Genetics* 6, 1–11.

Hogenhout, S.A., Van der Hoorn, R.A.L., Terauchi, R. and Kamoun, S. (2009) Emerging concepts in effector biology of plant-associated organisms. *Molecular Plant Microbe Interactions* 22, 115–122.

Holzberg, S., Brosio, P., Gross, C. and Pogue, G.P. (2002) Barley stripe mosaic virus-induced gene silencing in a monocot plant. *The Plant Journal* 30, 315–327.

Jones, L., Hamilton, A.J., Voinnet, O., Thomas, C.L., Maule, A.J. and Baulcombe, D.C. (1999) RNA-DNA interactions and DNA methylation in post-transcriptional gene silencing. *Plant Cell* 11, 2291–2301.

Jones, L., Ratcliff, F. and Baulcombe, D.C. (2001) RNA-directed transcriptional gene silencing in plants can be inherited independently of the RNA trigger and requires Met1 for maintenance. *Current Biology* 11, 747–757.

Jorgensen, R.A., Doetsch, N., Müller, A., Que, Q., Gendler, K. and Napoli, C.A. (2006) A paragenetic perspective on integration of RNA silencing into the epigenome and its role in the biology of higher plants. *Cold Spring Harbor Symposia on Quantitative Biology* 71, 481–485.

Kanazawa, A. (2008) RNA silencing manifested as visibly altered phenotypes in plants. *Plant Biotechnology* 25, 423–435.

Kanazawa, A. and Kasai, M. (2015) Induction of stable epigenetic gene silencing in plants using a virus vector. In: Mysore, K.S. and Senthil-Kumar, S. (eds) *Plant Gene Silencing Methods and Protocols*. Springer, New York, pp. 129–137.

Kanazawa, A., Inaba, J.I., Shimura, H., Otagaki, S., Tsukahara, S., Matsuzawa, A., Kim, B.M., Goto, K. and Masuta, C. (2011) Virus-mediated efficient induction of epigenetic modifications of endogenous genes with phenotypic changes in plants. *The Plant Journal* 65, 156–168.

Kanno, T. and Habu, Y. (2011) SiRNA-mediated chromatin maintenance and its function in *Arabidopsis thaliana*. *Biochimica et Biophysica Acta Gene Regulatory Mechanisms* 1809, 444–451.

Kasai, M. and Kanazawa, A. (2013) Induction of RNA-directed DNA methylation and heritable transcriptional gene silencing as a tool to engineer novel traits in plants. *Plant Biotechnology* 30, 233–241.

Kinoshita, Y., Saze, H., Kinoshita, T., Miura, A., Soppe, W.J.J., Koornneef, M. and Kakutani, T. (2007) Control of FWA gene silencing in *Arabidopsis thaliana* by SINE-related direct repeats. *The Plant Journal* 49, 38–45.

Kjemtrup, S., Sampson, K.S., Peele, C.G., Nguyen, L.V., Conkling, M.A., Thompson, W.F. and Robertson, D. (1998) Gene silencing from plant DNA carried by a *Geminivirus*. *The Plant Journal* 14, 91–100.

Kumagai, M.H., Donson, J., Della-Cioppa, G., Harvey, D., Hanley, K. and Grill, L.K. (1995) Cytoplasmic inhibition of carotenoid biosynthesis with virus-derived RNA. *Proceedings of the National Academy of the Sciences of the United States of America* 92, 1679–1683.

Law, J.A., Du, J., Hale, C.J., Feng, S., Krajewski, K., Palanca, A.M.S., Strahl, B.D., Patel, D.J. and Jacobsen, S.E. (2013) Polymerase IV occupancy at RNA-directed DNA methylation sites requires SHH1. *Nature* 498, 385–389.

Liu, C., Lu, F., Cui, X. and Cao, X. (2010) Histone methylation in higher plants. *Annual Review of Plant Biology* 61, 395–420.

Lu, R., Martin-Hernandez, A.M., Peart, J.R., Malcuit, I. and Baulcombe, D.C. (2003) Virus-induced gene silencing in plants. *Methods* 30, 296–303.

Mark Cigan, A., Unger-Wallace, E. and Haug-Collet, K. (2005) Transcriptional gene silencing as a tool for uncovering gene function in maize. *The Plant Journal* 43, 929–940.

Matzke, M.A. and Matzke, A.J.M. (1998) Epigenetic silencing of plant transgenes as a consequence of diverse cellular defence responses. *Cellular and Molecular Life Sciences* 54, 94–103.

Matzke, M.A. and Mosher, R.A. (2014) RNA-directed DNA methylation: an epigenetic pathway of increasing complexity. *Nature Reviews Genetics* 15, 394–408.

Matzke, A., Neuhubar, F., Park, Y.-D., Ambros, P. and Matzke, M. (1994) Homology-dependent gene silencing in transgenic plants: epistatic silencing loci contain multiple copies of methylated transgenes. *Molecular Genetics and Genomics* 244, 21.

Matzke, M., Kanno, T., Huettel, B., Daxinger, L. and Matzke, A.J.M. (2007) Targets of RNA-directed DNA methylation. *Current Opinion in Plant Biology* 10, 512–519.

Melquist, S. and Bender, J. (2003) Transcription from an upstream promoter controls methylation signaling from an inverted repeat of endogenous genes in *Arabidopsis*. *Genes and Development* 17, 2036–2047.

Mette, M.F., Van Der Winden, J., Matzke, M.A. and Matzke, A.J.M. (1999) Production of aberrant promoter transcripts contributes to methylation and silencing of unlinked homologous promoters in trans. *The EMBO Journal* 18, 241–248.

Numa, H., Kim, J.-M., Matsui, A., Kurihara, Y., Morosawa, T., Ishida, J., Mochizuki, Y., Kimura, H., Shinozaki, K., Toyoda, T., Seki, M., Yoshikawa, M. and Habu, Y. (2010) Transduction of RNA-directed DNA methylation signals to repressive histone marks in *Arabidopsis thaliana*. *The EMBO Journal* 29, 352–362.

Nuthikattu, S., McCue, A.D., Panda, K., Fultz, D., DeFraia, C., Thomas, E.N. and Slotkin, R.K. (2013) The initiation of epigenetic silencing of active transposable elements is triggered by RDR6 and 21-22 nucleotide small interfering RNAs. *Plant Physiology* 162, 116–131.

Okano, Y., Miki, D. and Shimamoto, K. (2008) Small interfering RNA (siRNA) targeting of endogenous promoters induces DNA methylation, but not necessarily gene silencing, in rice. *The Plant Journal* 53, 65–77.

Ooi, S.L. and Henikoff, S. (2007) Germline histone dynamics and epigenetics. *Current Opinion in Cell Biology* 19, 257–265.

Otagaki, S., Kasai, M., Masuta, C. and Kanazawa, A. (2013) Enhancement of RNA-directed DNA methylation of a transgene by simultaneously downregulating a ROS1 ortholog using a virus vector in *Nicotiana benthamiana*. *Frontiers in Genetics* 4, 1–10.

Pfluger, J. and Wagner, D. (2007) Histone modifications and dynamic regulation of genome accessibility in plants. *Current Opinion in Plant Biology* 10, 645–652.

Pontier, D., Picart, C., Roudier, F., Garcia, D., Lahmy, S., Azevedo, J., Alart, E., Laudié, M., Karlowski, W.M., Cooke, R., Colot, V., Voinnet, O. and Lagrange, T. (2012) NERD, a plant-specific GW protein, defines an additional RNAi-dependent chromatin-based pathway in *Arabidopsis*. *Molecular Cell* 48, 121–132.

Pouch-Pélissier, M.N., Pélissier, T., Elmayan, T., Vaucheret, H., Boko, D., Jantsch, M.F. and Deragon, J.M. (2008) SINE RNA induces severe developmental defects in *Arabidopsis thaliana* and interacts with HYL1 (DRB1), a key member of the DCL1 complex. *PLoS Genetics* 4, e1000096.

Rando, O.J. (2007) Global patterns of histone modifications. *Current Opinion in Genetics and Development* 17, 94–99.

Ratcliff, F., Martin-Hernandez, A.M. and Baulcombe, D.C. (2001) Technical advance. *Tobacco rattle virus* as a vector for analysis of gene function by silencing. *The Plant Journal* 25, 237–245.

Ream, T.S., Haag, J.R., Wierzbicki, A.T., Nicora, C.D., Norbeck, A.D., Zhu, J.K., Hagen, G., Guilfoyle, T.J., Pasa-Tolić, L. and Pikaard, C.S. (2009) Subunit compositions of the RNA-silencing enzymes Pol IV and Pol V reveal their origins as specialized forms of RNA Polymerase II. *Molecular Cell* 33, 192–203.

Ruiz, M.T., Voinnet, O. and Baulcombe, D.C. (1998) Initiation and maintenance of virus-induced gene silencing. *Plant Cell* 10, 937–946.

Saze, H., Tsugane, K., Kanno, T. and Nishimura, T. (2012) DNA methylation in plants: relationship to small rnas and histone modifications, and functions in transposon inactivation. *Plant and Cell Physiology* 53, 766–784.

Schubert, D., Primavesi, L., Bishopp, A., Roberts, G., Doonan, J., Jenuwein, T. and Goodrich, J. (2006) Silencing by plant Polycomb-group genes requires dispersed trimethylation of histone H3 at lysine 27. *The EMBO Journal* 25, 4638–4649.

Simon, S.A. and Meyers, B.C. (2011) Small RNA-mediated epigenetic modifications in plants. *Current Opinion in Plant Biology* 14, 148–155.

Soppe, W.J.J., Jasencakova, Z., Houben, A., Kakutani, T., Meister, A., Huang, M.S., Jacobsen, S.E., Schubert, I. and Fransz, P.F. (2002) DNA methylation controls histone H3 lysine 9 methylation and heterochromatin assembly in *Arabidopsis*. *The EMBO Journal* 21, 6549–6559.

Sunitha, S., Shivaprasad, P.V., Sujata, K. and Veluthambi, K. (2012) High frequency of T-DNA deletions in transgenic plants transformed with intron-containing hairpin RNA genes. *Plant Molecular Biology Report* 30, 158–167.

Tran, R.K., Henikoff, J.G., Zilberman, D., Ditt, R.F., Jacobsen, S.E. and Henikoff, S. (2005a) DNA methylation profiling identifies CG methylation clusters in *Arabidopsis* genes. *Current Biology* 15, 154–159.

Tran, R.K., Zilberman, D., de Bustos, C., Ditt, R.F., Henikoff, J.G., Lindroth, A.M., Delrow, J., Boyle, T., Kwong, S., Bryson, T.D., Jacobsen, S.E. and Henikoff, S. (2005b) Chromatin and siRNA pathways cooperate to maintain DNA methylation of small transposable elements in *Arabidopsis*. *Genome Biology* 6, R90.

Tucker, S.L., Reece, J., Ream, T.S. and Pikaard, C.S. (2010) Evolutionary history of plant multisubunit RNA polymerases IV and V: subunit origins via genome-wide and segmental gene duplications, retrotransposition, and lineage-specific subfunctionalization. Cold Spring Harbor Symposia on *Quantitative Biology* 75, 285–297.

van Kammen, A. (1997) Virus-induced gene silencing in infected and transgenic plants. *Trends in Plant Science* 2, 409–411.

Vaucheret, H. and Fagard, M. (2001) Transcriptional gene silencing in plants: targets, inducers and regulators. *Trends in Genetics* 17, 19–35.

Vaughn, M.W. and Martienssen, R.A. (2005) Finding the right template: RNA pol IV, a plant-specific RNA polymerase. *Molecular Cell* 17, 754–756.

Verdel, A., Vavasseur, A., Le Gorrec, M. and Touat-Todeschini, L. (2009) Common themes in siRNA-mediated epigenetic silencing pathways. *The International Journal of Developmental Biology* 53, 245–257.

Woo, H.R., Dittmer, T.A. and Richards, E.J. (2008) Three SRA-domain methylcytosine-binding proteins cooperate to maintain global CpG methylation and epigenetic silencing in *Arabidopsis*. *PLoS Genetics* 4, e1000156.

Zemach, A., Kim, M.Y., Hsieh, P.H., Coleman-Derr, D., Eshed-Williams, L., Thao, K., Harmer, S.L. and Zilberman, D. (2013) The *Arabidopsis* nucleosome remodeler DDM1 allows DNA methyltransferases to access H1-containing heterochromatin. *Cell* 153, 193–205.

Zhang, H. and Zhu, J.K. (2011) RNA-directed DNA methylation. *Current Opinion in Plant Biology* 14, 142–147.

Zhong, X., Du, J., Hale, C.J., Gallego-Bartolome, J., Feng, S., Vashisht, A.A., Chory, J., Wohlschlegel, J.A., Patel, D.J. and Jacobsen, S.E. (2014) Molecular mechanism of action of plant DRM *de novo* DNA methyltransferases. *Cell* 157, 1050–1060.

Zilberman, D. and Henikoff, S. (2007) Genome-wide analysis of DNA methylation patterns. *Development* 134, 3959–3965.

Zubko, E., Gentry, M., Kunova, A. and Meyer, P. (2012) *De novo* DNA methylation activity of METHYLTRANSFERASE 1 (MET1) partially restores body methylation in *Arabidopsis thaliana*. *The Plant Journal* 71, 1029–1037.

# 4 Gene Silencing in Archaeplastida Algae

**Xinrong Ma, Eun-Jeong Kim and Heriberto Cerutti***
*School of Biological Sciences and Center for Plant Science Innovation, University of Nebraska-Lincoln, Lincoln, USA*

## 4.1 Introduction

Algae are a phylogenetically diverse group of aquatic eukaryotes, commonly comprising photosynthetic organisms found in marine, freshwater and even terrestrial systems. They play important roles in global carbon cycling and other ecosystem functions (Worden and Allen, 2010; Tirichine and Bowler, 2011). Recently, they have attracted interest due to their great potential as feedstock for the production of biofuels and biomaterials. In particular, microalgae are capable of harnessing sunlight and $CO_2$ to synthesize many useful chemical compounds (Gimpel *et al*., 2015; Guarnieri and Pienkos, 2015). However, despite the recent sequencing of several algal genomes, our knowledge of gene function and regulation in most algae is rather limited, constraining biotechnological advances and genetic engineering (Gimpel *et al*., 2015; Kim *et al*., 2015). In this context, the study of gene silencing in algae may provide insights into gene expression mechanisms as well as facilitate the development of tools for biotechnological applications (Cerutti *et al*., 2011).

Gene regulation in eukaryotes is highly complex and involves, as in prokaryotes, mechanisms operating at the transcriptional and post-transcriptional levels. However, eukaryotic DNA is tightly packaged around a core of structural proteins, the histones, to generate a chromatin nucleosome array (Bannister and Kouzarides, 2011). Since chromatin represents a physical barrier to transcription, factors able to modify chromatin structure and/or the accessibility of regulators can influence gene activation or silencing (Venkatesh and Workman, 2015). Mechanisms that modulate chromatin organization/accessibility include: histone variant replacement, post-translational histone modifications, nucleosome remodelling, DNA methylation and RNA-mediated processes that contribute to establishing and/or maintaining specific chromatin states (Bannister and Kouzarides, 2011; Saze and Kakutani, 2011; Breiling and Lyko, 2015; Holoch and Moazed, 2015; Venkatesh and Workman, 2015). For instance, genetic and biochemical studies have indicated that di- or trimethylation of histone H3 on lysine 27 (H3K27) is often associated with silenced chromatin (Shaver *et al*., 2010; Bannister and Kouzarides, 2011;

* Corresponding author e-mail: hcerutti1@unl.edu

Holoch and Moazed, 2015). When present at or near gene regulatory regions, DNA cytosine methylation also plays a role in repression and, at least in some organisms, there appears to be a complex crosstalk between certain post-translational histone modifications and DNA methylation in order to establish a silenced chromatin state (Saze and Kakutani, 2011; Breiling and Lyko, 2015; Du *et al.*, 2015; Holoch and Moazed, 2015).

RNA-mediated mechanisms also participate in the regulation of gene expression. The role of long double-stranded RNAs (dsRNAs) as precursors of small RNAs, in triggering gene silencing, was first reported in *Caenorhabditis elegans* and termed RNA interference (RNAi) (Fire *et al.*, 1998). Since then it has become clear that RNA-mediated silencing is an evolutionarily conserved process in eukaryotes by which small RNAs (~20–30 nt in length) induce the inactivation of cognate sequences through a variety of mechanisms including translation repression, RNA degradation, transcriptional inhibition or, in a few organisms, DNA elimination (Baulcombe, 2004; Cerutti *et al.*, 2011; Borges and Martienssen, 2015; Holoch and Moazed, 2015; Iwakawa and Tomari, 2015). Other RNAs, particularly long noncoding RNAs, have also been implicated in regulatory circuitries (Gardini and Shiekhattar, 2015; Holoch and Moazed, 2015). Moreover, RNA-mediated silencing and the mechanisms that modulate chromatin organization/accessibility affect many cellular processes besides the regulation of gene expression, such as DNA repair and recombination, chromosome condensation/stability and defence responses against transposons and virus infection (Cerutti and Casas-Mollano, 2006; Cerutti *et al.*, 2011; Swarts *et al.*, 2014; Borges and Martienssen, 2015; Gardini and Shiekhattar, 2015).

Epigenetics, although an evolving concept, habitually refers to (meta)stable, heritable variations in gene expression that occur without alterations in DNA sequence (Saze and Kakutani, 2011; Pikaard and Mittelsten Scheid, 2014). Epigenetic mechanisms in eukaryotes allow for the stable regulation of gene expression through multiple rounds of cell division, and play an important role in developmental processes including cell differentiation and genome reprogramming associated with sexual reproduction, as well as in defence responses against parasitic genetic elements such as viruses and transposons (Saze and Kakutani, 2011; Pikaard and Mittelsten Scheid, 2014; Borges and Martienssen, 2015; Holoch and Moazed, 2015). Epigenetic mechanisms are also widely used by organisms, including unicellular ones, that require plastic but (meta)stable acclimation to their physical environment (e.g. periodic changes in temperature, light, nutrients) (Pikaard and Mittelsten Scheid, 2014; Vriet *et al.*, 2015). Key processes in epigenetic control in eukaryotes include those modulating chromatin structure and transcriptional activity as well as certain post-transcriptional (largely RNA-mediated) mechanisms (Saze and Kakutani, 2011; Pikaard and Mittelsten Scheid, 2014; Borges and Martienssen, 2015; Holoch and Moazed, 2015).

Some gene silencing mechanisms (like DNA cytosine methylation and H3K27 trimethylation) are epigenetic in nature and provide reversible cellular memory, whereas others (like the small RNA-mediated repression of certain transcripts) participate in more dynamic, transient modulation of gene expression. All these processes are important for the integration of environmental and intrinsic stimuli in organismal acclimation to abiotic and biotic stresses (Pikaard and Mittelsten Scheid, 2014; Vriet *et al.*, 2015). Patterns of gene expression can be altered in response to environmental or physiological signals, repressing genes when they are not needed and reactivating their expression when internal or external conditions change again. Yet, despite their acknowledged significance, gene silencing mechanisms remain largely uncharacterized in algae (Cerutti *et al.*, 2011; Kim *et al.*, 2015). This review will examine the existence of DNA cytosine methyltransferases and core components of the RNAi machinery, as well as briefly discuss the known or inferred biological role(s) of these mechanisms in microalgae. However, an added level of complexity is the evolutionary divergence of algal species, belonging to multiple eukaryotic supergroups (Worden and Allen, 2010; Tirichine and Bowler, 2011), which complicates the recognition of shared mechanisms. For this reason, our analysis will be limited to microalgae in the Archaeplastida eukaryotic supergroup which includes the red algae (Rhodophyta), the green algae (Chlorophyta and Charophyta), a small group of unicellular microalgae called glaucophytes (Glaucophyta) and the land plants (Embryophyta) (Table 4.1). For information related to additional chromatin modulation processes in microalgae, in particular histone post-translational modifications, the reader is referred to several recent articles (Shaver *et al.*, 2010; Kim *et al.*, 2015; Ngan *et al.*, 2015).

## 4.2 DNA Cytosine Methyltransferases in Microalgae

### 4.2.1 Phylogenetic analysis and domain organization of DNA cytosine methyltransferases

The modified base 5-methylcytosine (5mC) has been identified in a wide array of prokaryotes and eukaryotes (Feng *et al.*, 2010; Zemach *et al.*, 2010; Huff and Zilberman, 2014; Breiling and Lyko, 2015). A large family of DNA methyltransferases catalyses methylation at the C5 position of cytosine, including six subfamilies characterized by conserved active domains associated with distinct N-terminal or C-terminal extensions (Goll and Bestor, 2005; Ponger and Li, 2005; Huff and Zilberman, 2014). DNA cytosine methylation can occur in three sequence contexts: CG, CHG and CHH (where H = A, T, or C) and, in land plants, three DNA methyltransferase subfamilies have been implicated in the establishment and/or maintenance of methylation at these sequences (Goll and Bestor, 2005; Du *et al.*, 2015). CG methylation is maintained by the Dnmt1/MET1 subfamily (with strong preference for hemimethylated DNA), CHG methylation is mediated by the plant-specific CMT3 chromomethylase, and CHH methylation is mainly dependent on the Dnmt3/DRM (Domains Rearranged Methyltransferase) enzymes (Goll and Bestor, 2005; Du *et al.*, 2015). Symmetrical CG methylation in certain eukaryotes is carried out by a divergent cytosine methyltransferase, Dnmt5 (Ponger and Li, 2005; Huff and Zilberman, 2014). To begin characterizing the occurrence and the role(s) of DNA cytosine methyltransferases in microalgae, we surveyed 12 complete or near-complete Archaeplastida algal genomes for the presence of genes encoding the corresponding enzymes (Table 4.1).

Table 4.1. Distribution of core gene silencing components in Archaeplastida.

| Species | Genome size (Mb) | RNA Interference | | | DNA Cytosine Methyltransferases | | | | | References[b] |
|---|---|---|---|---|---|---|---|---|---|---|
| | | AGO-PIWI | Dicer | RDR | Dnmt1/ MET1 | CMTs/ CMT3 | Other DMTase[a] | Dnmt3/ DRM2 | Dnmt5 | |
| **Chlorophyta** | | | | | | | | | | |
| **Chlorophyceae** | | | | | | | | | | |
| *Chlamydomonas reinhardtii* | 120.0 | 3[c] | 3 | 0 | 3[d] | 0 | 1 | 0 | 0 | 1 |
| *Volvox carteri* | 138.0 | 2 | 1 | 0 | 1 | 0 | 0 | 0 | 0 | 2 |
| **Trebouxiophyceae** | | | | | | | | | | |
| *Chlorella sorokiniana* | 56.8 | 1 | 1 | 0 | 1 | 1 | 1 | 0 | 0 | Draft |
| *Chlorella variabilis* NC64A | 46.0 | 1 | 1 | 0 | 1 | 1 | 1 | 0 | 0 | 3 |
| *Coccomyxa subellipsoidea* | 48.8 | 2 | 1[e] | 1 | 0 | 0 | 2 | 0 | 0 | 4 |
| **Mamiellophyceae** | | | | | | | | | | |
| *Bathycoccus prasinos* | 15.1 | 0 | 0 | 0 | 0 | 0 | 0 | 0 | 1 | 5 |
| *Micromonas pusilla* CCMP1545 | 21.9 | 0 | 0 | 0 | 0 | 0 | 0 | 0 | 1 | 6 |
| *Ostreococcus lucimarinus* | 13.2 | 0 | 0 | 0 | 0 | 0 | 0 | 0 | 1 | 7 |
| **Charophyta** | | | | | | | | | | |
| *Klebsormidium flaccidum* | 104.0 | 6 | 2 | 4 | 1 | 1 | 1 | 2/2[f] | 0 | 8 |
| **Embryophyta** | | | | | | | | | | |
| *Physcomitrella patens* | 480.0 | 7 | 4 | 2 | 1 | 1 | 0 | 0/2[f] | 0 | 9 |
| *Arabidopsis thaliana* | 125.0 | 10 | 4 | 6 | 4 | 3 | 0 | 2/1[f] | 0 | 10 |
| **Rhodophyta** | | | | | | | | | | |
| *Porphyridium purpureum* | 19.7 | 3 | 1 | 3 | 0 | 0 | 0 | 0 | 0 | 11 |
| *Cyanidioschyzon merolae* | 16.5 | 0 | 0 | 0 | 0 | 0 | 0 | 1 | 0 | 12 |
| **Glaucophyta** | | | | | | | | | | |
| *Cyanophora paradoxa* | 70.0 | 1[g] | 0 | 1 | 0 | 0 | 1 | 0 | 0 | 13 |

[a]DNA methyltransferases that cannot be clearly categorized (see text for details).
[b]References: 1, Merchant *et al.*, 2007; 2, Prochnik *et al.*, 2010; 3, Blanc *et al.*, 2010; 4, Blanc *et al.*, 2012; 5, Moreau *et al.*, 2012; 6, Worden *et al.*, 2009; 7, Palenik *et al.*, 2007; 8, Hori *et al.*, 2014; 9, Rensing *et al.*, 2008; 10, The Arabidopsis Genome Initiative, 2000; 11, Bhattacharya *et al.*, 2013; 12, Matsuzaki *et al.*, 2004; 13, Price *et al.*, 2012.
[c]Total number of genes in the genome encoding a certain factor.
[d]Includes the chloroplast-targeted DMT1a and DMT1b DNA methyltransferases (see text for details).
[e]Predicted protein fairly divergent from canonical model.
[f]Predicted proteins indicated after the slash have only a Dnmt3-like catalytic domain without any associated motifs.
[g]Draft protein model with only a PIWI domain.

Polypeptides with conserved DNA methyltransferase catalytic domains were identified by either BLAST or PSI-BLAST searches of protein and/or translated genomic DNA databases, using as queries well defined *Arabidopsis thaliana*, *Homo sapiens* or *Micromonas pusilla* sequences corresponding to Dnmt3/DRM2, CMT3, Dnmt1/MET1 or Dnmt5 enzymes (Table 4.1). Homologues of Dnmt3/DRM proteins, implicated in *de novo* DNA methylation (Goll and Bestor, 2005; Ponger and Li, 2005; Du *et al.*, 2015), were found exclusively in the red alga *Cyanidioschyzon merolae* and the charophyte *Klebsormidium flaccidum*, a terrestrial alga related to the ancestor(s) of land plants (Hori *et al.*, 2014) (Table 4.1). Intriguingly, the *C. merolae* predicted polypeptide showed a structural organization similar to vertebrate Dnmt3s, whereas the *K. flaccidum* proteins appeared more closely related to land plant DRMs (Fig. 4.1). Chromomethylase-like methyltransferases seemed to be restricted to the chlorophytes *Chlorella sorokiniana* and *C. variabilis* and to the charophyte *K. flaccidum* (Table 4.1) and they were structurally similar to land plant CMT3, except for the lack of the chromodomain in the *Chlorella* enzymes (Fig. 4.1).

Dnmt1/MET1 homologues displayed a slightly broader distribution, encompassing species in the Trebouxiophyceae (*C. sorokiniana* and *C. variabilis*) and Chlorophyceae (*Chlamydomonas reinhardtii* and *Volvox carteri*) classes as well as in the division Charophyta (*K. flaccidum*) (Table 4.1). In contrast, Dnmt5-related enzymes appeared limited to the Mamiellophyceae class (*Bathycoccus prasinos*, *M. pusilla* and *Ostreococcus lucimarinus*) (Table 4.1). The glaucophyte *Cyanophora paradoxa* contained a single DNA methyltrasferase that cannot be unequivocally categorized (Table 4.1), with the caveat that its draft genome is incomplete and proteins may be missing in the database. Thus, DNA cytosine methyltransferases show wide distribution among Archaeplastida microalgae, but in a patchy pattern, with specific subfamily enzymes largely limited to subgroups of organisms (Table 4.1 and Fig. 4.1). Indeed, only *K. flaccidum* appears to have a set of DNA cytosine methyltransferases similar to that in land plants (Table 4.1).

Phylogenetic analyses support clustering of the *K. flaccidum* DNA methyltransferases with those from land plants (Fig. 4.1), whereas homologues from other Archaeplastida algae tended to group independently within each subfamily. Thus, a parsimonious interpretation of the data suggests that gene losses and duplications of Dnmt1/MET1, CMT, Dnmt3/DRM or Dnmt5 enzymes may have occurred after the divergence of red and most green algae from the lineage leading to land plants and charophytes (Fig. 4.1). Algal Dnmt1/MET1 proteins show high sequence similarity to the land plant polypeptides in the DNA methyltransferase catalytic domain (Fig. 4.1). Additionally, most of these algal proteins have N-terminal extensions with D-RFD (DNA methyltransferase replication foci domain) and bromo adjacent homology (BAH) motifs (Fig. 4.1), as observed in the canonical enzyme (Goll and Bestor, 2005; Ponger and Li, 2005).

However, *Chlamydomonas reinhardtii* also contains two Dnmt1/MET1-related polypeptides lacking conserved motifs in the N-terminal region (Fig. 4.1). These paralogues have been termed DMT1a and DMT1b and a seemingly hybrid recombinant protein has been characterized as a novel DNA methyltransferase with nonselective *de novo* cytosine methylation activity (Nishiyama *et al.*, 2004; Lopez *et al.*, 2015). Moreover, this enzyme localizes to *Chlamydomonas* chloroplasts and

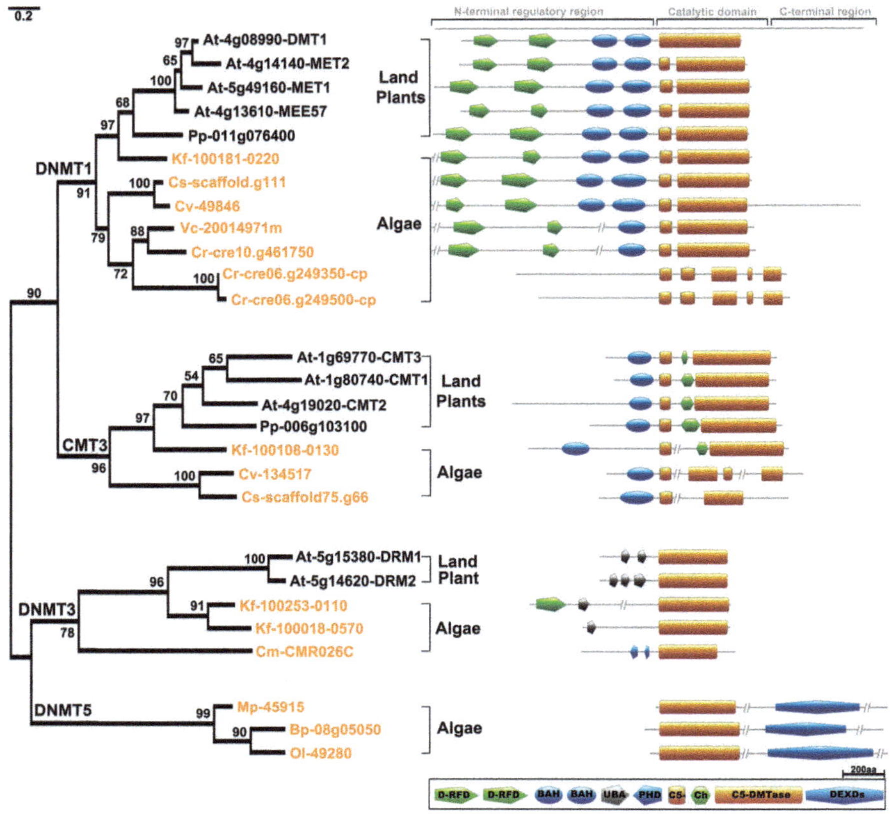


**Fig. 4.1.** Maximum-likelihood tree showing the phylogenetic relationship among DNA cytosine methyltransferases. Numbers on branches indicate bootstrap values (%) based on 1000 pseudoreplicates. The scale bar at the upper left denotes 0.2 amino acid substitutions per site. Sequences corresponding to the DMTase domain from different organisms were aligned using Muscle and the tree was drawn using the MEGA v6.06 program (Tamura *et al*., 2013). Species are designated by a two-letter abbreviation preceding the name of each protein: At, *Arabidopsis thaliana*; Bp, *Bathycoccus prasinos*; Cm, *Cyanidioschyzon merolae*; Cr, *Chlamydomonas reinhardtii*; Cs, *Chlorella sorokiniana*; Cv, *Chlorella variabilis* NC64A; Kf, *Klebsormidium flaccidum*; Mp, *Micromonas pusilla* CCMP1545; Ol, *Ostreococcus lucimarinus*; Pp, *Physcomitrella patens*; and Vc, *Volvox carteri*. Accession numbers of proteins used to draw the tree are: At-4g08990-DMT1, NP_192638.1; At-4g14140-MET2, NP_001190725.1; At-4g13610-MEE57, NP_193097.1; At-5g49160-MET1, NP_199727.1; At-1g80740-CMT1, NP_565245.1; At-4g19020-CMT2, NP_193637.2; At-1g69770-CMT3, NP_177135.1; At-5g15380-DRM1, NP_197042.2; At-5g14620-DRM2, NP_196966.2; Bp-08g05050, XP_007511811.1; Cm-CMR026C, gnl|CMER|CMR026C at http://merolae.biol.s.u-tokyo.ac.jp; Cr-Cre10.g461750, (XP_001698604.1), Cr-Cre06.g249350-cp (BAC99051.1), and Cr-Cre06.g249500-cp (XP_001696666.1) at http://phytozome.jgi.doe.gov; Cs-scaffold75.g66 and Cs-scaffold89.g111 from *Chlorella sorokiniana* draft genomic sequence assembly at the University of Nebraska-Lincoln; Cv-49846, XP_005851422.1; Cv-134517, XP_005847336.1; Kf-I00253-0110 (GAQ85766.1), Kf-I00108-0130 (GAQ82337.1), Kf-I00181-0220 (GAQ84242.1), and Kf-I00018-0570 (GAQ78775.1)

influences plastid DNA methylation (Nishiyama *et al.*, 2004; Lopez *et al.*, 2015). Several microalgae also encode putative DNA cytosine methyltransferases that cannot be clearly categorized (Table 4.1, Other DMTase). These predicted proteins contain catalytic domains somewhat related to those of the Dnmt1/MET1 and/or the CMT subfamilies but they lack N-terminal extensions or conserved domains in the N-terminal extensions. *Chlamydomonas* DMT4 belongs to this group and it is tempting to speculate that some of these enzymes might be responsible for DNA methylation processes unique to microalgae (Lopez *et al.*, 2015).

### 4.2.2 Biological role(s) of DNA cytosine methylation in microalgae

Modified 5mC is not essential for eukaryotic life since it has not been detected in several model organisms such as *Saccharomyces cerevisiae*, *Schizosaccharomyces pombe* and *Caenorhabditis elegans* (Capuano *et al.*, 2014; Breiling and Lyko, 2015). Nonetheless, in land plants or mammals, DNA cytosine methylation is ubiquitous; is often associated with the transcriptional silencing of transposable elements, repetitive DNA sequences and some protein coding genes; and plays a critical role in genomic imprinting, X-chromosome inactivation, and/or chromosome stability (Feng *et al.*, 2010; Saze and Kakutani, 2011; Du *et al.*, 2015). Gene body methylation is also highly conserved although its precise function(s) remains mysterious (Feng *et al.*, 2010; Zemach *et al.*, 2010; Huff and Zilberman, 2014; Breiling and Lyko, 2015). By contrast, the role of DNA cytosine methylation in microalgae is still poorly understood.

Nuclear genome 5mC patterns have been recently profiled in several microalgae, including *Chlorella variabilis*, *Volvox carteri*, *Chlamydomonas reinhardtii* and some Mamiellophyceae members (Feng *et al.*, 2010; Zemach *et al.*, 2010; Huff and Zilberman, 2014; Lopez *et al.*, 2015). In *Chlorella variabilis*, genes are universally CG methylated within their bodies (Zemach *et al.*, 2010). CHG methylation is also substantial but, similarly to land plants, concentrated in repetitive (presumably transposon) sequences and excluded from genes (Zemach *et al.*, 2010). By comparison, the *V. carteri* genome showed much lower methylation and exclusively in the CG context. Transposons and repeats are preferentially methylated but a weak negative correlation between promoter methylation and transcript abundance was also observed (Zemach *et al.*, 2010). In *Volvox*, DNA cytosine methylation had previously been implicated in the transcriptional silencing of transgenes (Babinger *et al.*, 2007).

---

**Fig. 4.1.** Continued.
at http://www.plantmorphogenesis.bio.titech.ac.jp; Mp-45915, XP_003063363.1; Ol-49280, XP_001417393.1; Pp-006g103100, XP_001769709.1; Pp-011g076400, XP_001758167.1; and Vc-20014971m, XP_002949905.1. The domain organization of the proteins is indicated on the right. Conserved protein motifs were identified based on the SMART 7 database (http://smart.embl-heidelberg.de) or InterPro 55.0 database (https://www.ebi.ac.uk/interpro). Domain abbreviations: D-RFD, DNA (cytosine-5)methyltransferase Replication Foci Domain; BAH, Bromo Adjacent Homology domain; UBA, UBiquitin-Associated domain; PHD, Plant HomeoDomain zinc finger domain; C5- or C5-DMTase, C-5 DNA Methyltransferase; Ch, Chromo-like domain; and DEXDs, DEXD-like helicases superfamily.

In *C. reinhardtii*, as in *Volvox*, the nuclear genome is methylated at low levels and CG methylation is preferentially enriched in putative transposon sequences and repeats (Feng *et al.*, 2010; Lopez *et al.*, 2015). Intriguingly, CHG and CHH methylation were also observed uniformly along chromosomes and showed little enhancement in transposons/repeats (Feng *et al.*, 2010). However, *Chlamydomonas* does not contain CMT3 or Dnmt3/DRM homologues (Table 4.1 and Feng *et al.*, 2010) and the methyltransferase (or methyltransferases) responsible for the later modifications has not been characterized. DNA cytosine methylation has also been associated with the transcriptional silencing of transgenes, particularly tandem repeats, in *Chlamydomonas* (Cerutti *et al.*, 1997). In contrast, Mamiellophyceae microalgae such as *O. lucimarinus*, *M. pusilla* and *B. prasinos* possess DNA methyltransferases of the Dnmt5 subfamily and exhibit densely clustered CG methylation in nucleosome linkers, possibly contributing to nucleosome positioning and proper chromatin compaction in very small nuclei (Huff and Zilberman, 2014).

Overall, as in land plants and vertebrates, preferential DNA cytosine methylation of transposable elements, repetitive sequences and gene bodies has been observed in some microalgae (Feng *et al.*, 2010; Zemach *et al.*, 2010; Lopez *et al.*, 2015). This is consistent with a role of 5mC in the repression of transgenes, transposons and, possibly, some protein coding genes (Cerutti *et al.*, 1997; Babinger *et al.*, 2007; Zemach *et al.*, 2010; Du *et al.*, 2015). Moreover, the pattern of nuclear genome 5mC and the complement of DNA methyltransferases in species of the genus *Chlorella*, except for the lack of a Dnmt3/DRM homologue (Table 4.1), are similar to those in land plants. On the other hand, *C. reinhardtii* and *V. carteri* methylate preferentially transposons and repeats in the CG (rather than the CHG/CHH) context, and this is likely to reflect the divergence in their DNA cytosine methyltransferases (and, probably, in specific molecular pathways) from those present in land plants (Table 4.1).

Additionally, certain microalgae show unique nuclear genome patterns of 5mC, like in the Mamiellophyceae class (Huff and Zilberman, 2014), suggestive of novel functions for this modification. In this respect, the chloroplast genome of *Chlamydomonas* is dynamically methylated throughout the life cycle. Chloroplast DNA 5mC is low in vegetative cells; increases during gametogenesis, in a mating-type dependent manner; and reaches a peak during zygote development (Lopez *et al.*, 2015). The purpose of this massive methylation is unknown but could be related to the packaging and protection of chloroplast DNA in zygospores (Lopez *et al.*, 2015). Furthermore some microalgae, such as *C. reinhardtii*, also exhibit DNA adenine methylation in the nuclear genome, catalysed by an undefined methyltransferase. $N^6$-methyladenine mainly locates at AT dinucleotides around transcription start sites, shares little correlation with 5mC and seems to mark active genes (Fu *et al.*, 2015). Thus, the distribution and function(s) of DNA methylation in microalgae appear to be highly varied. In some cases, 5mC seems to be associated with gene silencing – particularly of transposons/repeats, as in higher eukaryotes – but in other instances DNA methylation appears to reflect algal-specific processes that remain to be fully characterized.

## 4.3 The RNA Interference Machinery in Microalgae

### 4.3.1 Taxonomic distribution and phylogenetic analysis of core RNAi components

Despite the mechanistic diversity of RNA-mediated silencing, in the best characterized pathways small RNAs (sRNAs) are incorporated into effector complexes containing at their core Argonaute-PIWI (AGO-PIWI) proteins, which include two major families of polypeptides named after *A. thaliana* ARGONAUTE1 (AGO1) and *Drosophila melanogaster* P-element induced wimpy testis (PIWI) (Cerutti and Casas-Mollano, 2006; Rogers and Chen, 2013; Swarts *et al.*, 2014). Other key components of the RNAi machinery include an RNaseIII-like endonuclease (Dicer), involved in the processing of long dsRNAs into sRNAs, and a RNA-dependent RNA polymerase (RDR), involved in the generation of dsRNA from single-stranded transcripts and/or in the production of secondary sRNAs (Baulcombe, 2004; Cerutti and Casas-Mollano, 2006; Burroughs *et al.*, 2014).

The presence of AGO-PIWI, Dicer and RDR proteins was examined in the 12 microalgae with sequenced genomes belonging to the Archaeplastida supergroup (Table 4.1). As previously reported (Cerutti and Casas-Mollano, 2006; Cerutti *et al.*, 2011), core RNAi machinery components appear to be absent from some algae with small nuclear genomes such as the green algae *B. prasinos*, *M. pusilla* and *O. lucimarinus* and the red alga *Cyanidioschyzon merolae*. However, *Porphyridium purpureum*, that possesses a nuclear genome of comparable size, seems to encode a complete and diversified RNAi machinery (Table 4.1). Currently, the exact reason for the loss of key RNAi factors in certain eukaryotic lineages remains unclear, particularly because of the existence of sister species with similar lifestyles that differ in their presence or absence (Burroughs *et al.*, 2014). Nonetheless, core RNAi components are present in algal species belonging to each of the surveyed Archaeplastida divisions, albeit with irregular distribution (Table 4.1).

AGO-PIWI and Dicer proteins are relatively well conserved and have similar taxonomic distributions among the examined algae (Table 4.1). *Cyanophora paradoxa* (Glaucophyta) appears to be an exception since it lacks a canonical Dicer and contains a polypeptide with only a PIWI motif, instead of the typical domain organization of the AGO-PIWIs (see below). However, this genome is in draft stage and some proteins may be missing from the database whereas others may have errors in their predicted structure (Price *et al.*, 2012). In several algae AGO-PIWI proteins and – to a lower degree – Dicer polypeptides have undergone duplications (Table 4.1; Fig. 4.2). By contrast, consistent with a likely ancillary role in the RNAi pathway (Cerutti and Casas-Mollano, 2006; Cerutti *et al.*, 2011), RDRs show a more limited distribution and appear to be absent from many algal species (Table 4.1).

In the most extensively characterized RNAi pathways, short dsRNA molecules are loaded into AGO-PIWI proteins; one strand (guide) is selectively retained whereas the other strand (passenger) is removed; and then AGO-PIWIs use the retained strand as a guide to bind to complementary RNAs for inactivation (Swarts *et al.*, 2014; Iwakawa and Tomari, 2015). Some AGO-PIWI proteins function as sRNA-guided endonucleases (slicers) that cleave complementary transcripts, whereas

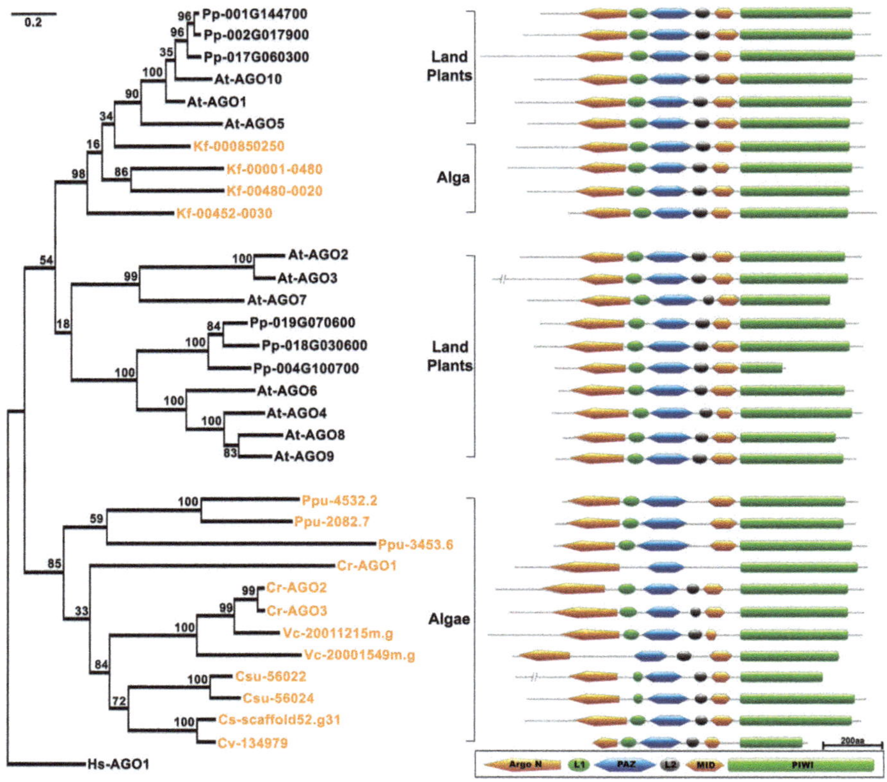


**Fig. 4.2.** Maximum-likelihood tree showing the phylogenetic relationship among AGO-PIWI proteins. Numbers on branches indicate bootstrap values (%) based on 1000 pseudoreplicates. The scale bar at the upper left denotes 0.2 amino acid substitutions per site. Full-length sequences of Argonaute proteins from different organisms were aligned using Muscle and the tree was drawn using the MEGA v6.06 program. Species are designated by a two- or three-letter abbreviation preceding the name of each protein, as described in the legend to Fig. 4.1, except for: Csu, *Coccomyxa subellipsoidea*; Hs, *Homo sapiens*; and Ppu, *Porphyridium purpureum*. Accession numbers of proteins used to draw the tree are: At-AGO1, NP_849784; At-AGO2, NP_174413; At-AGO3, NP_174414; At-AGO4, NP_001189613; At-AGO5, NP_850110; At-AGO6, NP_180853; At-AGO7, NP_177103; At-AGO8, NP_197602; At-AGO9, NP_197613; At-AGO10, NP_199194; Cr-AGO1, XP_001694840.1; Cr-AGO2, XP_001698670.1; Cr-AGO3, XP_001698906.1; Cs-scaffold52.g31 from *Chlorella sorokiniana* draft genomic sequence assembly at the University of Nebraska-Lincoln; Csu-56022, XP_005646873.1; Csu-56024, XP_005646875.1; Cv-134979, XP_005847187.1; Hs-AGO1, 4OLA_A; Kf-00085-0250 (GAQ81609.1), Kf-00001-0480 (GAQ77599.1), Kf-00480-0020 (GAQ88686.1) and Kf-00452-0030 (GAQ89017.1) at http://www.plantmorphogenesis.bio.titech.ac.jp; Pp-001G144700, XP_001757611.1; Pp-002G017900, XP_001752991.1; Pp-017G060300, XP_001774237.1; Pp-019G070600, XP_001784983.1; Pp-018G030600, XP_001755804.1; Pp-004G100700, XP_001769106.1; Ppu-contig-4532.2 (evm.model.contig4532.2), Ppu-contig-2082.8 (evm.model.contig2082.7) and Ppu-contig-3453.6 (evm.model.contig3453.6) at http://cyanophora.rutgers.edu/porphyridium; Vc-20011215m.g, XP_002952894.1; and Vc-20001549m.g, XP_002952090.1. The

others lack endonucleolytic activity and repress their targets through other mechanisms such as translation inhibition (Swarts *et al.*, 2014; Iwakawa and Tomari, 2015). Typical eukaryotic AGO-PIWI polypeptides are characterized by a conserved structure (Fig. 4.2), comprising the following domains: (i) an N (N-terminal) domain, which plays a role in the dissociation of the passenger strand; (ii) the PAZ (PIWI-Argonaute-Zwille) domain, which binds the 3′ end of the guide strand; (iii) the MID (middle) domain, which binds the 5′ end of the guide strand; and (iv) the PIWI domain, an RNaseH-like fold domain, which may have endonucleolytic activity (Burroughs *et al.*, 2014; Swarts *et al.*, 2014). This structure is highly conserved in the majority of the algal AGO-PIWIs (Fig. 4.2).

Phylogenetic analysis of the AGO-PIWI polypeptides was performed with the algal sequences as well as with those from two land plants (*A. thaliana* and *Physcomitrella patens*) and, as an outgroup, one metazoan (*H. sapiens*) (Fig. 4.2). The AGO-PIWI proteins fell into two relatively well-supported groups (Fig. 4.2) that included either exclusively polypeptides from algae or those from land plants and from the charophyte *K. flaccidum*, related to the ancestor(s) of land plants (Hori *et al.*, 2014). Interestingly, AGO-PIWI proteins underwent expansion in several of the lineages studied. However, as with DNA cytosine methyltransferases (Fig. 4.1), gene duplications appear to have occurred after the divergence of red and most green algae from the lineage leading to land plants and charophytes (Fig. 4.2). This expansion of algal AGO-PIWIs is consistent with proposals of functional specialization and binding to distinct sRNA classes, as extensively described in land plants (Casas-Mollano *et al.*, 2008; Rogers and Chen, 2013; Borges and Martienssen, 2015; Voshall *et al.*, 2015; Yamasaki *et al.*, 2016).

Two major classes of sRNAs have been recognized in many eukaryotes: microRNAs (miRNAs) and small interfering RNAs (siRNAs) (Bartel, 2009; Axtell, 2013; Rogers and Chen, 2013; Burroughs *et al.*, 2014). MicroRNAs generally originate from single-stranded noncoding RNAs or introns, which fold into imperfectly paired hairpin structures (Bartel, 2009; Axtell, 2013; Rogers and Chen, 2013). They usually modulate gene expression, affecting many biological processes such as development, metabolism or stress responses. siRNAs are produced from long, near-perfect complementarity dsRNAs of diverse origins, including the products of convergent transcription or RDR activity; viral or transposon RNAs; or dsRNAs experimentally introduced into cells (Baulcombe, 2004; Cerutti and Casas-Mollano, 2006; Axtell, 2013; Burroughs *et al.*, 2014; Borges and Martienssen, 2015). These siRNAs play various roles in suppression of viruses and transposable elements, post-transcriptional regulation of gene expression, DNA double-strand break repair, DNA methylation and/or heterochromatin formation (Baulcombe, 2004; Cerutti and Casas-Mollano, 2006; Axtell, 2013; Burroughs *et al.*, 2014; Borges and Martienssen, 2015).

---

**Fig. 4.2.** Continued.
domain organization of the proteins is indicated on the right. Conserved protein motifs were identified based on the SMART 7 or InterPro 55.0 databases. Domain abbreviations: ArgoN, Argonaute N-terminal domain; L1, Linker 1 domain; PAZ, PAZ domain; L2, Linker 2 domain; MID, MID (middle) domain; and PIWI, PIWI domain (see text for details).

The occurrence of sRNAs has been examined in very few Archaeplastida algae. The chlorophyte *Chlamydomonas reinhardtii* has been shown to contain a complex set of endogenous sRNAs, including miRNAs, phased siRNAs and siRNAs originating from transposons and repeats (Molnár *et al.*, 2007; Zhao *et al.*, 2007; Shu and Hu, 2012; Evers *et al.*, 2015; Voshall *et al.*, 2015). Likewise, endogenous noncoding small RNAs have been identified, by sRNA library sequencing, in the glaucophyte *Cyanophora paradoxa* (Gross *et al.*, 2013), the red algae *Porphyra yezoensis* (Liang *et al.*, 2010) and *Eucheuma denticulatum* (Gao *et al.*, 2016), as well as the green alga *V. carteri* (Li *et al.*, 2014; Evers *et al.*, 2015). Based on the ability of genomic sRNA loci (or of the corresponding transcripts) to fold back into hairpin secondary structures resembling miRNA precursors, putative miRNAs have been predicted in some of these algal species (Liang *et al.*, 2010; Li *et al.*, 2014; Evers *et al.*, 2015; Gao *et al.*, 2016). Interestingly, with the possible exception of candidate miRNAs identified in *P. yezoensis* and *E. denticulatum* (Liang *et al.*, 2010; Gao *et al.*, 2016), there appears to be little conservation of miRNA genes among algae or with those encoded in the genomes of land plants or animals. This suggests, as previously proposed (Cerutti and Casas-Mollano, 2006; Bartel, 2009; Axtell, 2013; Voshall *et al.*, 2015), that (many) miRNA genes and their potential regulatory interaction(s) with target mRNAs may have evolved independently in different eukaryotic lineages.

### 4.3.2 Biological roles of the RNAi machinery in microalgae

In a wide array of eukaryotes, RNA-mediated silencing is involved in various processes including transposon silencing, viral defence, endogenous gene regulation, heterochromatin formation, DNA methylation, DNA repair and maintenance of genome stability (Baulcombe, 2004; Cerutti and Casas-Mollano, 2006; Axtell, 2013; Rogers and Chen, 2013; Borges and Martienssen, 2015; Du *et al.*, 2015; Holoch and Moazed, 2015). In contrast to this wealth of information, little is known about the biological role(s) of RNAi in microalgae. The presence of key components of the RNAi machinery, of endogenous miRNAs/siRNAs and the observation of experimentally triggered RNAi in algae (Cerutti *et al.*, 2011; Kim *et al.*, 2015) suggest that RNA-mediated silencing is functionally relevant in these organisms. However, it is also becoming apparent that RNAi is not essential for unicellular life (Cerutti and Casas-Mollano, 2006; Voshall *et al.*, 2015) and it seems to have been lost in certain algal species.

Core components of the RNAi machinery have undergone duplication in *Chlamydomonas reinhardtii*, which contains three AGOs and three Dicers, suggestive of functional diversification (Casas-Mollano *et al.*, 2008; Voshall *et al.*, 2015; Yamasaki *et al.*, 2016). Indeed, miRNA-mediated post-transcriptional gene regulation appears to rely primarily on AGO3 (Yamasaki *et al.*, 2016). Similarly, one of the Dicer proteins, DCL1, seems to be predominantly required for the post-transcriptional silencing of transposons (Casas-Mollano *et al.*, 2008). However, *Chlamydomonas* also possesses a DCL1-independent mechanism(s), which is chromatin-mediated, for the transcriptional repression of transposable elements (Casas-Mollano *et al.*, 2008; Shaver *et al.*, 2010). Intriguingly, this chromatin-mediated silencing is more effective at 17°C than at 25°C (Cerutti *et al.*, 1997). Conversely,

RNAi-mediated post-transcriptional repression, in both invertebrates and land plants, appears to be more efficient at 25–29°C (Fortier and Belote, 2000; Szittya *et al.*, 2003). Thus, multiple silencing mechanisms may act to suppress transposon mobilization over a wide range of environmental conditions in certain microalgae. Moreover, consistent with a function of RNAi in transposon silencing, sRNAs mapping to transposons and other repeated regions in the genome have been identified in both *C. reinhardtii* and *V. carteri* (Molnár *et al.*, 2007; Zhao *et al.*, 2007; Li *et al.*, 2014). Additionally, in *Chlorella variabilis* NC64A, transcriptome analyses revealed that components of putative RNAi pathways are upregulated upon infection with the PBCV-1 virus (Rowe *et al.*, 2013). This collective evidence, although limited, does support a role for RNAi in defence responses against viruses and transposons in Archaeplastida microalgae.

In higher eukaryotes, another major function of RNAi is the regulation of gene expression, mediated by miRNAs and/or other endogenous sRNAs. As already described, candidate miRNAs have been identified or predicted in *Chlamydomonas reinhardtii*, *V. carteri*, *P. yezoensis* and *E. denticulatum*, and computational approaches have been used to find putative target RNAs based on sequence complementarity (Molnár *et al.*, 2007; Zhao *et al.*, 2007; Liang *et al.*, 2010; Shu and Hu, 2012; Evers *et al.*, 2015; Voshall *et al.*, 2015; Gao *et al.*, 2016). However, since target gene identification is a challenging problem (Voshall *et al.*, 2015), it is unclear what fraction of these genes might actually correspond to genuine sRNA-regulated transcripts. In *Chlamydomonas*, several miRNAs are differentially expressed under sulfur deprivation and some of these changes were correlated with altered abundance of putative target mRNAs (Shu and Hu, 2012). In *V. carteri*, distinct miRNA populations are enriched in gonidia or somatic cells, suggesting that miRNAs may play a role in regulating cell differentiation (Li *et al.*, 2014). In the glaucophyte *Cyanophora paradoxa*, genome-wide analyses have identified many sRNAs preferentially matching exonic mRNA sequences and expression of these sRNAs seems to be modulated under stress conditions (Gross *et al.*, 2013). These correlative observations are suggestive of a role of miRNAs and other endogenous sRNAs in the regulation of gene expression in microalgae. However, experimental evidence supporting this interpretation, for instance direct validation of miRNA targets, is rather scant.

In land plants, the complementarity of miRNAs and target RNAs is generally near perfect, often resulting in Argonaute-mediated cleavage of target transcripts in the region matching to nucleotides 10 and 11 of an miRNA (Axtell, 2013; Rogers and Chen, 2013; Borges and Martienssen, 2015). In *Chlamydomonas reinhardtii*, expected mRNA cleavage products for a few predicted miRNA targets have been detected by 5′ rapid amplification of cDNA ends (Molnár *et al.*, 2007; Zhao *et al.*, 2007; Voshall *et al.*, 2015), suggesting that miRNAs can indeed trigger endonucleolytic cleavage of complementary transcripts. High throughput 'degradome' sequencing indicated that miRNA-mediated target RNA cleavage might also occur in *V. carteri* (Li *et al.*, 2014). However, not many of the reported *Chlamydomonas* or *Volvox* miRNAs have identifiable targets with near perfect complementarity (Molnár *et al.*, 2007; Zhao *et al.*, 2007; Li *et al.*, 2014; Voshall *et al.*, 2015) and, for a number of predicted targets, RNA products resulting from cleavage within the miRNA pairing region have not been observed (Molnár *et al.*, 2007; Zhao *et al.*, 2007). Moreover, very few predicted cleavage targets showed an expected increase in mRNA abundance in transcriptomic analyses of a *Chlamydomonas* mutant strain virtually devoid of small RNAs (Voshall *et al.*, 2015). Interestingly, recent evidence suggests that miRNA regulation of transcript expression in *C. reinhardtii* may function, at least for some targets, by translation repression, without or with only minimal mRNA destabilization (Ma *et al.*, 2013; Voshall *et al.*, 2015; Yamasaki *et al.*, 2016). Ribosomes associated with sRNA-repressed transcripts showed reduced sensitivity to translation inhibition by some antibiotics, such as cycloheximide, possibly indicating RNAi-mediated alterations to the function/structural conformation of translating ribosomes (Ma *et al.*, 2013).

In summary, RNAi appears to be entirely absent from some microalgae and its biological role(s), in the species that possess core RNAi machinery components, is poorly understood. Limited evidence suggests that an siRNA pathway may operate as a defence mechanism against transposon mobilization and, possibly, in antiviral immunity. An miRNA pathway, when present, may contribute to endogenous gene regulation. However, the identification of genuine miRNA targets remains challenging and, to date, no specific metabolic or physiological process controlled or modulated by miRNAs has been clearly defined in microalgae. Moreover, *Chlamydomonas* miRNA/sRNA-deficient mutants have no obvious abnormal phenotypes when grown under standard laboratory conditions (Voshall *et al.*, 2015; Yamasaki *et al.*, 2016), implying that sRNAs are not essential for growth and survival. As previously proposed (Voshall *et al.*, 2015), miRNAs may have no role in regulating core cellular functions in this microalga but may possibly be involved in modulating more recently evolved responses to specific abiotic and/or biotic stresses. Other possible functions of RNAi in phenomena such as heterochromatin formation, DNA methylation, DNA repair or maintenance of genome stability, to our knowledge, have not been explored in Archaeplastida algae.

## 4.4 Perspective

Algae show enormous diversity and can potentially become a sustainable source of valuable bioproducts. However, for most species, our knowledge of their physiology, metabolism and gene regulation is fairly limited, constraining meaningful biotechnological advances. In eukaryotes, gene silencing mechanisms play important roles in the reversible repression of genes that need to be expressed only under certain developmental or environmental contexts, and in the suppression of genomic parasites such as transposons and viruses. The recent sequencing of various algal genomes is beginning to provide insights into the complexity and uniqueness of some of these mechanisms in algal species. DNA cytosine methyltransferases and core RNAi machinery components, which are well conserved in vertebrates and land plants, are present with patchy distribution in Archaeplastida algae. An intricate evolutionary pattern of lineage-specific losses and duplications appears to have resulted in individual algae often containing subsets of these factors. DNA cytosine methylation and/or RNAi have been implicated in transposon and transgene silencing in at least some algae, although specific molecular pathways may diverge from those in land plants. Furthermore, a possible function of these mechanisms in endogenous gene regulation is poorly understood in most algal species.

Limited evidence also suggests that typical gene silencing mechanisms or epigenetic marks may have acquired unique roles in algae, but many gaps remain in our knowledge. For instance, what is the function of the extreme enrichment of 5mC in the chloroplast DNA of *Chlamydomonas* zygospores? What is the significance of gene body CG methylation in some algal species and of nucleosomal linker methylation in others? What is the biological impact of non-conserved, lineage-specific miRNAs in Archaeplastida algae? More broadly, what is the possible contribution of epigenetic gene silencing to phenotypic plasticity and acclimation of algal species to fluctuating environmental conditions? Investigation of these and other open questions may advance our understanding of the complex biology of these diverse aquatic organisms.

## Acknowledgements

This work was supported in part by a grant from the National Science Foundation (to H.C.). We also acknowledge the support of the Nebraska EPSCoR programme and the Korea Research Institute of Bioscience and Biotechnology.

## References

Axtell, M.J. (2013) Classification and comparison of small RNAs from plants. *Annual Review of Plant Biology* 64, 137–159.

Babinger, P., Völkl, R., Cakstina, I., Maftei, A. and Schmitt, R. (2007) Maintenance DNA methyltransferase (Met1) and silencing of CpG-methylated foreign DNA in *Volvox carteri*. *Plant Molecular Biology* 63, 325–336.

Bannister, A.J. and Kouzarides, T. (2011) Regulation of chromatin by histone modifications. *Cell Research* 21, 381–395.

Bartel, D.P. (2009) MicroRNAs: target recognition and regulatory functions. *Cell* 136, 215–233.

Baulcombe, D. (2004) RNA silencing in plants. *Nature* 431, 356–363.

Bhattacharya, D., Price, D.C., Chan, C.X., Qiu, H., Rose, N., Ball, S., Weber, A.P.M., Arias, M.C., Henrissat, B., Coutinho, P.M., Krishnan, A., Zäuner, S., Morath, S., Hilliou, F., Egizi, A., Perrineau, M.-M. and Yoon, H.S. (2013) Genome of the red alga *Porphyridium purpureum*. *Nature Communications* 4, 1941.

Blanc, G., Duncan, G., Agarkova, I., Borodovsky, M., Gurnon, J., Kuo, A., Lindquist, E., Lucas, S., Pangilinan, J., Polle, J., Salamov, A., Terry, A., Yamada, T., Dunigan, D.D., Grigoriev, I.V., Claverie, J.-M. and Van Etten, J.L. (2010) The *Chlorella variabilis* NC64A genome reveals adaptation to photosymbiosis, coevolution with viruses, and cryptic sex. *The Plant Cell* 22, 2943–2955.

Blanc, G., Agarkova, I., Grimwood, J., Kuo, A., Brueggeman, A., Dunigan, D.D., Gurnon, J., Ladunga, I., Lindquist, E., Lucas, S., Pangilinan, J., Pröschold, T., Salamov, A., Schmutz, J., Weeks, D., Yamada, T., Lomsadze, A., Borodovsky, M., Claverie, J.-M., Grigoriev, I.V. and Van Etten, J.L. (2012) The genome of the polar eukaryotic microalga *Coccomyxa subellipsoidea* reveals traits of cold adaptation. *Genome Biology* 13, R39.

Borges, F. and Martienssen, R.A. (2015) The expanding world of small RNAs in plants. *Nature Reviews Molecular Cell Biology* 16, 727–741.

Breiling, A. and Lyko, F. (2015) Epigenetic regulatory functions of DNA modifications: 5-methylcytosine and beyond. *Epigenetics and Chromatin* 8, 24.

Burroughs, A.M., Ando, Y. and Aravind, L. (2014) New perspectives on the diversification of the RNA interference system: insights from comparative genomics and small RNA sequencing. *Wiley Interdisciplinary Reviews: RNA* 5, 141–181.

Capuano, F., Mülleder, M., Kok, R., Blom, H.J. and Ralser, M. (2014) Cytosine DNA methylation is found in *Drosophila melanogaster* but absent in *Saccharomyces cerevisiae*, *Schizosaccharomyces pombe*, and other yeast species. *Analytical Chemistry* 86, 3697–3702.

Casas-Mollano, J.A., Rohr, J., Kim, E.-J., Balassa, E., van Dijk, K. and Cerutti, H. (2008) Diversification of the core RNA interference machinery in *Chlamydomonas reinhardtii* and the role of DCL1 in transposon silencing. *Genetics* 179, 69–81.

Cerutti, H. and Casas-Mollano, J.A. (2006) On the origin and functions of RNA-mediated silencing: from protists to man. *Current Genetics* 50, 81–99.

Cerutti, H., Johnson, A.M., Gillham, N.W. and Boynton, J.E. (1997) Epigenetic silencing of a foreign gene in nuclear transformants of *Chlamydomonas*. *The Plant Cell* 9, 925–945.

Cerutti, H., Ma, X., Msanne, J. and Repas, T. (2011) RNA-mediated silencing in algae: biological roles and tools for analysis of gene function. *Eukaryotic Cell* 10, 1164–1172.

Du, J., Johnson, L.M., Jacobsen, S.E. and Patel, D.J. (2015) DNA methylation pathways and their crosstalk with histone methylation. *Nature Reviews Molecular Cell Biology* 16, 519–532.

Evers, M., Huttner, M., Dueck, A., Meister, G. and Engelmann, J.C. (2015) miRA: adaptable novel miRNA identification in plants using small RNA sequencing data. *BMC Bioinformatics* 16, 370.

Feng, S., Cokus, S.J., Zhang, X., Chen, P.Y., Bostick, M., Goll, M.G., Hetzel, J., Jain, J., Strauss, S.H., Halpern, M.E., Ukomadu, C., Sadler, K.C., Pradhan, S., Pellegrini, M. and Jacobsen, S.E. (2010) Conservation and divergence of methylation patterning in plants and animals. *Proceedings of the National Academy of Sciences of the United States of America* 107, 8689–8694.

Fire, A., Xu, S., Montgomery, M.K., Kostas, S.A., Driver, S.E. and Mello, C.C. (1998) Potent and specific genetic interference by double-stranded RNA in *Caenorhabditis elegans*. *Nature* 391, 806–811.

Fortier, E. and Belote, J.M. (2000) Temperature-dependent gene silencing by an expressed inverted repeat in *Drosophila*. *Genesis* 26, 240–244.

Fu, Y., Luo, G.-Z., Chen, K., Deng, X., Yu, M., Han, D., Hao, Z., Liu, J., Lu, X., Doré, L.C., Weng, X., Ji, Q., Mets, L. and He, C. (2015) $N^6$-methyldeoxyadenosine marks active transcription start sites in *Chlamydomonas*. *Cell* 161, 879–892.

Gao, F., Nan, F., Feng, J., Lv, J., Liu, Q. and Xie, S. (2016) Identification and characterization of microRNAs in *Eucheuma denticulatum* by high-throughput sequencing and bioinformatics analysis. *RNA Biology* 13, 343–352.

Gardini, A. and Shiekhattar, R. (2015) The many faces of long noncoding RNAs. *The FEBS Journal* 282, 1647–1657.

Gimpel, J.A., Henríquez, V. and Mayfield, S.P. (2015) In metabolic engineering of eukaryotic microalgae: potential and challenges come with great diversity. *Frontiers in Microbiology* 6, 1376.

Goll, M.G. and Bestor, T.H. (2005) Eukaryotic cytosine methyltransferases. *Annual Review of Biochemistry* 74, 481–514.

Gross, J., Wajid, S., Price, D.C., Zelzion, E., Li, J., Chan, C.X. and Bhattacharya, D. (2013) Evidence for widespread exonic small RNAs in the glaucophyte alga *Cyanophora paradoxa*. *PLoS ONE* 8, e67669.

Guarnieri, M.T. and Pienkos, P.T. (2015) Algal omics: unlocking bioproduct diversity in algae cell factories. *Photosynthesis Research* 123, 255–263.

Holoch, D. and Moazed, D. (2015) RNA-mediated epigenetic regulation of gene expression. *Nature Reviews Genetics* 16, 71–84.

Hori, K., Maruyama, F., Fujisawa, T., Togashi, T., Yamamoto, N., Seo, M., Sato, S., Yamada, T., Mori, H., Tajima, N., Moriyama, T., Ikeuchi, M., Watanabe, M., Wada, H., Kobayashi, K., Saito, M., Masuda, T., Sasaki-Sekimoto, Y., Mashiguchi, K., Awai, K., Shimojima, M., Masuda, S., Iwai, M., Nobusawa, T., Narise, T., Kondo, S., Saito, H., Sato, R., Murakawa, M., Ihara, Y., Oshima-Yamada, Y., Ohtaka, K., Satoh, M., Sonobe, K., Ishii, M., Ohtani, R., Kanamori-Sato, M., Honoki, R., Miyazaki, D., Mochizuki, H., Umetsu, J., Higashi, K., Shibata, D., Kamiya, Y., Sato, N., Nakamura, Y., Tabata, S., Ida, S., Kurokawa, K. and Ohta, H. (2014) *Klebsormidium flaccidum* genome reveals primary factors for plant terrestrial adaptation. *Nature Communications* 5, 3978.

Huff, J.T. and Zilberman, D. (2014) Dnmt1-independent CG methylation contributes to nucleosome positioning in diverse eukaryotes. *Cell* 156, 1286–1297.

Iwakawa, H.O. and Tomari, Y. (2015) The functions of microRNAs: mRNA decay and translational repression. *Trends in Cell Biology* 25, 651–665.

Kim, E.-J., Ma, X. and Cerutti, H. (2015) Gene silencing in microalgae: mechanisms and biological roles. *Bioresource Technology* 184, 23–32.

Li, J.R., Wu, Y. and Qi, Y.J. (2014) microRNAs in a multicellular green alga *Volvox carteri*. *Science China Life Sciences* 57, 36–45.

Liang, C., Zhang, X., Zou, J., Xu, D., Su, F. and Ye, N. (2010) Identification of miRNA from *Porphyra yezoensis* by high-throughput sequencing and bioinformatics analysis. *PLoS ONE* 5, e10698.

Lopez, D.A., Hamaji, T., Kropat, J., De Hoff, P., Morselli, M., Rubbi, L., Fitz-Gibbon, S., Gallaher, S.D., Merchant, S.S., Umen, J.G. and Pellegrini, M. (2015) Dynamic changes in the transcriptome and methylome of *Chlamydomonas reinhardtii* throughout its life cycle. *Plant Physiology* 169, 2730–2743.

Ma, X., Kim, E.-J., Kook, I., Ma, F., Voshall, A., Moriyama, E. and Cerutti, H. (2013) Small interfering RNA-mediated translation repression alters ribosome sensitivity to inhibition by cycloheximide in *Chlamydomonas reinhardtii*. *The Plant Cell* 25, 985–998.

Matsuzaki, M., Misumi, O., Shin-I, T., Maruyama, S., Takahara, M., Miyagishima, S., Mori, T., Nishida, K., Yagisawa, F., Nishida, K., Yoshida, Y., Nishimura, Y., Nakao, S., Kobayashi, T., Momoyama, Y., Higashiyama, T., Minoda, A., Sano, M., Nomoto, H., Oishi K., Hayashi, H., Ohta, F., Nishizaka, S., Haga, S., Miura, S., Morishita, T., Kabeya, Y., Terasawa, K., Suzuki, Y., Ishii, Y., Asakawa, S., Takano, H., Ohta, N., Kuroiwa, H., Tanaka, K., Shimizu, N., Sugano, S., Sato, N., Nozaki, H., Ogasawara, N., Kohara, Y. and Kuroiwa, T. (2004) Genome sequence of the ultrasmall unicellular red alga *Cyanidioschyzon merolae* 10D. *Nature* 428, 653–657.

Merchant, S.S., Prochnik, S.E., Vallon, O., Harris, E.H., Karpowicz, S.J., Witman, G.B., Terry, A., Salamov, A., Fritz-Laylin, L.K., Maréchal-Drouard, L., Marshall, W.F., Qu, L.-H., Nelson, D.R., Sanderfoot, A.A., Spalding, M.H., Kapitonov, V.V., Ren, Q., Ferris, P., Lindquist, E., Shapiro, H., Lucas, S.M., Grimwood, J., Schmutz, J., Chlamydomonas Annotation Team, JGI Annotation Team, Grigoriev, I.V., Rokhsar, D.S. and Grossman, A.R. (2007) The *Chlamydomonas* genome reveals the evolution of key animal and plant functions. *Science* 318, 245–251.

Molnár, A., Schwach, F., Studholme, D.J., Thuenemann, E.C. and Baulcombe, D.C. (2007) MiRNAs control gene expression in the single-cell alga *Chlamydomonas reinhardtii*. *Nature* 447, 1126–1129.

Moreau, H., Verhelst, B., Couloux, A., Derelle, E., Rombauts, S., Grimsley, N., Van Bel, M., Poulain, J., Katinka, M., Hohmann-Marriott, M.F., Piganeau, G., Rouzé, P., Silva, C.D., Wincker, P., Van de Peer, Y. and Vandepoele, K. (2012) Gene functionalities and genome structure in *Bathycoccus prasinos* reflect cellular specializations at the base of the green lineage. *Genome Biology* 13, R74.

Ngan, C.Y., Wong, C.-H., Choi, C., Yoshinaga, Y., Louie, K., Jia, J., Chen, C., Bowen, B., Cheng, H., Leonelli, L., Kuo, R., Baran, R., García-Cerdán, J.G., Pratap, A., Wang, M., Lim, J., Tice, H.,

Daum, C., Xu, J., Northen, T., Visel, A., Bristow, J., Niyogi, K.K. and Wei, C.-L. (2015) Lineage-specific chromatin signatures reveal a regulator of lipid metabolism in microalgae. *Nature Plants* 1, 15107.

Nishiyama, R., Wada, Y., Mibu, M., Yamaguchi, Y., Shimogawara, K. and Sano, H. (2004) Role of a nonselective *de novo* DNA methyltransferase in maternal inheritance of chloroplast genes in the green alga, *Chlamydomonas reinhardtii*. *Genetics* 168, 809–816.

Palenik, B., Grimwood, J., Aerts, A., Rouzé, P., Salamov, A., Putnam, N., Dupont, C., Jorgensen, R., Derelle, E., Rombauts, S., Zhou, K., Otillar, R., Merchant, S.S., Podell, S., Gaasterland, T. Napoli, C., Gendler, K., Manuell, A., Tai, V., Vallon, O., Piganeau, G., Jancek, S., Heijde, M., Jabbari, K., Bowler, C., Lohr, M., Robbens, S., Werner, G., Dubchak, I., Pazour, G.J., Ren, Q., Paulsen, I., Delwiche, C., Schmutz, J., Rokhsar, D., Van de Peer, Y., Moreau, H. and Grigoriev, I.V. (2007) The tiny eukaryote *Ostreococcus* provides genomic insights into the paradox of plankton speciation. *Proceedings of the National Academy of Sciences of the United States of America* 104, 7705–7710.

Pikaard, C.S. and Mittelsten Scheid, O. (2014) Epigenetic regulation in plants. *Cold Spring Harbor Perspectives in Biology* 6, a019315.

Ponger, L. and Li, W.-H. (2005) Evolutionary diversification of DNA methyltransferases in eukaryotic genomes. *Molecular Biology and Evolution* 22, 1119–1128.

Price, D.C., Chan, C.X., Yoon, H.S., Yang, E.C., Qiu, H., Weber, A.P.M., Schwacke, R., Gross, J., Blouin, N.A., Lane, C., Reyes-Prieto, A., Durnford, D.G., Neilson, J.A.D., Lang, B.F., Burger, G., Steiner, J.M., Löffelhardt, W., Meuser, J.E., Posewitz, M.C., Ball, S., Arias, M.C., Henrissat, B., Coutinho, P.M., Rensing, S.A., Symeonidi, A., Doddapaneni, H., Green, B.R., Rajah, V.D., Boore, J. and Bhattacharya, D. (2012) *Cyanophora paradoxa* genome elucidates origin of photosynthesis in algae and plants. *Science* 335, 843–847.

Prochnik, S.E., Umen, J., Nedelcu, A.M., Hallmann, A., Miller, S.M., Nishii, I., Ferris, P., Kuo, A., Mitros, T., Fritz-Laylin, L.K., Hellsten, U., Chapman, J., Simakov, O., Rensing, S.A., Terry, A., Pangilinan, J., Kapitonov, V., Jurka, J., Salamov, A., Shapiro, H., Schmutz, J., Grimwood, J., Lindquist, E., Lucas, S., Grigoriev, I.V., Schmitt, R., Kirk, D. and Rokhsar, D.S. (2010) Genomic analysis of organismal complexity in the multicellular green alga *Volvox carteri*. *Science* 329, 223–226.

Rensing, S.A., Lang, D., Zimmer, A.D., Terry, A., Salamov, A., Shapiro, H., Nishiyama, T., Perroud, P.-F., Lindquist, E.A., Kamisugi, Y., Tanahashi, T., Sakakibara, K., Fujita, T., Oishi, K., Shin-I, T., Kuroki, Y., Toyoda, A., Suzuki, Y., Hashimoto, S., Yamaguchi, K., Sugano, S., Kohara, Y., Fujiyama, A., Anterola, A., Aoki, S., Ashton, N., Barbazuk, W.B., Barker, E., Bennetzen, J.L., Blankenship, R., Cho, S.H., Dutcher, S.K., Estelle, M., Fawcett, J.A., Gundlach, H., Hanada, K., Heyl, A., Hicks, K.A., Hughes, J., Lohr, M., Mayer, K., Melkozernov, A., Murata, T., Nelson, D.R., Pils, B., Prigge, M., Reiss, B., Renner, T., Rombauts, S., Rushton, P.J., Sanderfoot, A., Schween, G., Shiu, S.-H., Stueber, K., Theodoulou, F.L., Tu, H., Van de Peer, Y., Verrier, P.J., Waters, E., Wood, A., Yang, L., Cove, D., Cuming, A.C. (2008) The *Physcomitrella* genome reveals evolutionary insights into the conquest of land by plants. *Science* 319, 64–69.

Rogers, K. and Chen, X. (2013) Biogenesis, turnover, and mode of action of plant miRNAs. *The Plant Cell* 25, 2383–2399.

Rowe, J.M., Dunigan, D.D., Blanc, G., Gurnon, J.R., Xia, Y. and Van Etten, J.L. (2013) Evaluation of higher plant virus resistance genes in the green alga, *Chlorella variabilis* NC64A, during the early phase of infection with *Paramecium bursaria chlorella virus*-1. *Virology* 442, 101–113.

Saze, H. and Kakutani, T. (2011) Differentiation of epigenetic modifications between transposons and genes. *Current Opinion in Plant Biology* 14, 81–87.

Shaver, S., Casas-Mollano, J.A., Cerny, R.L. and Cerutti, H. (2010) Origin of the polycomb repressive complex 2 and gene silencing by an E(z) homolog in the unicellular alga *Chlamydomonas*. *Epigenetics* 5, 301–312.

Shu, L. and Hu, Z. (2012) Characterization and differential expression of microRNAs elicited by sulfur deprivation in *Chlamydomonas reinhardtii*. *BMC Genomics* 13, 108.

Swarts, D.C., Makarova, K., Wang, Y., Nakanishi, K., Ketting, R.F., Koonin, E.V., Patel, D.J. and van der Oost, J. (2014) The evolutionary journey of Argonaute proteins. *Nature Structural and Molecular Biology* 21, 743–753.

Szittya, G., Silhavy, D., Molnár, A., Havelda, Z., Lovas, Á., Lakatos, L., Bánfalvi, Z. and Burgyán, J. (2003) Low temperature inhibits RNA silencing-mediated defence by the control of siRNA generation. *The EMBO Journal* 22, 633–640.

Tamura, K., Stecher, G., Peterson, D., Filipski, A. and Kumar, S. (2013) MEGA6: Molecular Evolutionary Genetics Analysis version 6.0. *Molecular Biology* and *Evolution* 30(12), 2725–2729. DOI: 10.1093/molbev/mst197

The Arabidopsis Genome Initiative (2000) Analysis of the genome sequence of the flowering plant *Arabidopsis thaliana*. *Nature* 408, 796–815.

Tirichine, L. and Bowler, C. (2011) Decoding algal genomes: tracing back the history of photosynthetic life on Earth. *The Plant Journal* 66, 45–57.

Venkatesh, S. and Workman, J.L. (2015) Histone exchange, chromatin structure and the regulation of transcription. *Nature Reviews Molecular Cell Biology* 16, 178–189.

Voshall, A., Kim, E.-J., Ma, X., Moriyama, E.N. and Cerutti, H. (2015) Identification of AGO3-associated miRNAs and computational prediction of their targets in the green alga *Chlamydomonas reinhardtii*. *Genetics* 200, 105–121.

Vriet, C., Hennig, L. and Laloi, C. (2015) Stress-induced chromatin changes in plants: of memories, metabolites and crop improvement. *Cellular and Molecular Life Sciences* 72, 1261–1273.

Worden, A.Z. and Allen, A.E. (2010) The voyage of the microbial eukaryote. *Current Opinion in Microbiology* 13, 652–660.

Worden, A.Z., Lee, J.H., Mock, T., Rouzé, P., Simmons, M.P., Aerts, A.L., Allen, A.E., Cuvelier, M.L., Derelle, E., Everett, M.V., Foulon, E., Grimwood, J., Gundlach, H., Henrissat, B., Napoli, C., McDonald, S.M., Parker, M.S., Rombauts, S., Salamov, A., Dassow, P.V., Badger, J.H., Coutinho, P.M., Demir, E., Dubchak, I., Gentemann, C., Eikrem, W., Gready, J.E., John, U., Lanier, W., Lindquist, E.A., Lucas, S., Mayer, K.F.X., Moreau, H., Not, F., Otillar, R., Panaud, O., Pangilinan, J., Paulsen, I., Piegu, B., Poliakov, A., Robbens, S., Schmutz, J., Toulza, E., Wyss, T., Zelensky, A., Zhou, K., Armbrust, E.V., Bhattacharya, D., Goodenough, U.W., Van de Peer, Y. and Grigoriev, I.V. (2009) Green evolution and dynamic adaptations revealed by genomes of the marine picoeukaryotes *Micromonas*. *Science* 324, 268–272.

Yamasaki, T., Kim, E.-J., Cerutti, H. and Ohama, T. (2016) Argonaute3 is a key player in miRNA-mediated target cleavage and translational repression in *Chlamydomonas*. *The Plant Journal* 85, 258–268.

Zemach, A., McDaniel, I.E., Silva, P. and Zilberman, D. (2010) Genome-wide evolutionary analysis of eukaryotic DNA methylation. *Science* 328, 916–919.

Zhao, T., Li, G., Mi, S., Li, S., Hannon, G.J., Wang, X.-J. and Qi, Y. (2007) A complex system of small RNAs in the unicellular green alga *Chlamydomonas reinhardtii*. *Genes and Development* 21, 1190–1203.

# 5 Gene Silencing in Fungi: A Diversity of Pathways and Functions

**Santiago Torres-Martínez and Rosa M. Ruiz-Vázquez***
*University of Murcia, Murcia, Spain*

## 5.1 Introduction

Ribonucleic acids (RNAs) have emerged in the last two decades as key players in the complex networks of regulation of gene expression, with RNA-mediated gene silencing being one of the regulatory mechanisms often found in the eukaryotic world. RNA-mediated gene silencing has been referred to in different ways. Thus, Napoli *et al.* (1990) first reported a phenomenon they called 'co-suppression', in which ectopic transgenes inhibited the expression of exogenous and endogenous copies of homologous genes in plants. The same phenomenon was later described in the fungus *Neurospora crassa* by Romano and Macino (1992), where it was described as 'quelling'. Finally, silencing was described and molecularly characterized in *Caenorhabditis elegans* by the breakthrough research of Fire and Mello (Fire *et al.*, 1998), and termed RNA interference (RNAi). The main signature of this mechanism is the generation of small RNA molecules (sRNAs) of 20–30 nucleotides from double-stranded RNA (dsRNA) precursors by the RNase III Dicer enzyme. These sRNAs are incorporated into an effector RNA-induced silencing complex (RISC), with Argonaute proteins as main protagonists. The action of the RISC complex can result in post-transcriptional gene silencing, via degradation or translational repression of targeted transcripts, or in transcriptional gene silencing via specific chromatin modifications (Ghildiyal and Zamore, 2009). Initially considered a defence mechanism that preserves the integrity of the host genome against exogenous nucleic acids, RNAi is also implicated in the regulation of cellular, developmental and physiological processes in a wide diversity of organisms, including fungi (Chang *et al.*, 2012).

The fungal kingdom is a large and diverse group of eukaryotic organisms with enormous impact on human life at many levels, being valuable models for the study of numerous genetic and biochemical processes. Some fungi also have parasitic life cycles and cause a wide variety of diseases in plants, animals and humans. In fact, fungi are the most important group of plant pathogens, and human diseases caused by some oportunistic infectious species are attracting increased attention due to

* Corresponding author e-mail: rmruiz@um.es

 

the growing number of immuno-deficient patients (Ibrahim *et al.*, 2012). Although RNAi has been described in many fungal species, most of the knowledge on sRNA and RNAi pathways in fungi is centred in two reference organisms: the ascomycete *N. crassa* and the opportunistic human pathogen *Mucor circinelloides*, a basal fungus belonging to the subphylum Mucoromycotina. In this review, we focus on RNAi pathways and their functions in *Neurospora* and *Mucor*, and describe the different classes of sRNAs produced by these and other filamentous fungi and yeasts that participate in transcriptional or post-transcriptional gene silencing.

## 5.2 RNAi as a Genome Defence Mechanism in Fungi

The RNAi mechanism can be triggered by a wide variety of exogenous nucleic acids that represent a threat to genome integrity. In this section we discuss RNAi triggered by integrative and non-integrative transgenes, viruses and transposons, as well as by unpaired DNA, during both vegetative and sexual cycles.

### 5.2.1 Quelling and transposon control in *Neurospora crassa*

As indicated above, the first transgene-induced silencing phenomenon in fungi was discovered in *N. crassa* (Romano and Macino, 1992), which is one of the most important model systems for RNAi studies within the fungal kingdom. The phenomenon, then designated 'quelling', was demonstrated to be triggered by ectopic integration of multiple copies of a transgene into genomic DNA, and to operate at a post-transcriptional level (Romano and Macino, 1992). Isolation and functional analysis of three quelling-defective *(qde)* mutants, *qde-1*, *qde-2* and *qde-3*, was instrumental for the understanding of the RNAi mechanism (Cogoni and Macino, 1997). The first component demonstrated to be involved in the RNAi machinery was the RNA-dependent RNA polymerase (RdRP) enzyme encoded by the *qde-1* gene (Cogoni and Macino, 1999a). This was proposed to convert aberrant RNA (aRNA) derived from repetitive transgenes into dsRNA molecules which triggered RNAi. It was later demonstrated that QDE-1 is a bifunctional enzyme, acting as a DNA-dependent RNA polymerase (DdRP) to produce aRNA from DNA, and as an RdRP for generating dsRNA from aRNA templates (Lee *et al.*, 2010a). The *qde-3* gene encoded a homologue RecQ DNA helicase that, together with QDE-1, is involved in both quelling and biogenesis of qiRNA, a type of small RNA generated from ribosomal DNA (rDNA) after DNA damage (see below) (Cogoni and Macino, 1999b; Lee *et al.*, 2009). We now know that QDE-3 participates in these mechanisms by recruiting QDE-1 to anomalous single-stranded DNA structures formed during recombination of DNA repeats, thus facilitating the production of aRNAs and dsRNAs (Zhang *et al.*, 2013). There are two Dicer proteins in *N. crassa*, which have redundant functions in the production of siRNAs from dsRNAs (Catalanotto *et al.*, 2004). The siRNAs are bound to the Argonaute protein QDE-2, whose discovery was the first genetic evidence of a common RNA silencing mechanism shared by worms and fungi (Catalanotto *et al.*, 2000). Within the RISC complex, the passenger strand of the siRNA is cleaved by the endonucleolytic activity of QDE-2 and removed by the exonucleolytic activity of QIP (a QDE-2-interacting protein), thus promoting activation of RISC (Maiti *et al.*, 2007). QIP protein is also required for meiotic silencing by unpaired DNA (MSUD), an RNAi-related phenomenon identified in *N. crassa* that silences unpaired DNA during the meiotic cycle (Shiu *et al.*, 2001; Xiao *et al.*, 2010) (see below). Following studies in *N. crassa*, the basic RNAi machinery was demonstrated to be involved in RNAi triggered by tandem integration of transgenes in many other fungi, operating both during vegetative growth and sexual reproduction (reviewed in Chang *et al.*, 2012; Nicolás and Ruiz-Vázquez, 2013).

The repetitive nature of transgene integration in silenced strains suggested that RNAi in *N. crassa* could be used as a defence mechanism against invasive repetitive sequences such as transposons. In fact, the first example of the role of RNAi in the maintenance of genome integrity in fungi was described in *N. crassa*, by showing that the RNAi machinery is required to suppress transposon replication (Nolan *et al.*, 2005). sRNAs derived from transposable sequences have been detected in other fungi, such as *M. circinelloides* (Nicolás *et al.*, 2010) and *Magnaporthe oryzae* (Nunes *et al.*, 2011). More recently, transposon-derived siRNAs produced by the MSUD mechanism have been also identified in *N. crassa* (Wang *et al.*, 2015), indicating that canonical RNAi and RNAi-related pathways act as effective defence mechanisms against transposable elements during vegetative and sexual cycles in diverse fungal species.

### 5.2.2 Gene silencing triggered by non-integrative transgenes in *Mucor circinelloides*

*M. circinelloides* has achieved a major place as a fungal model system for studying different molecular processes, including RNAi and its role in endogenous gene regulation (Torres-Martínez and Ruiz-Vázquez, 2015). The existence of a transgene-induced gene silencing mechanism in *M. circinelloides* that is associated with accumulation of two different classes of siRNAs, 21 and 25 nt long, was demonstrated by Nicolás *et al.* (2003). Accumulation of 21 nt siRNAs was related with silencing efficiency and stability, suggesting that this siRNA class could be the guide for mRNA degradation (Nicolás *et al.*, 2009). A distinguishing feature of RNAi in *Mucor* is that transgenes do not integrate into the genome due to the self-replicative condition of the vectors used for transformation. Thus, transgene expression is not affected by position effects or host regulatory sequences at insertion sites, demonstrating that the level of transgene expression correlates with silencing efficiency (de Haro *et al.*, 2009; Nicolás *et al.*, 2009; Calo *et al.*, 2012). *Mucor* also produces secondary siRNAs corresponding to sequences of endogenous target genes downstream of the initial triggering molecules, supporting the existence of an amplification step in this fungus (Nicolás *et al.*, 2003). Amplification of the silencing signal is present in plants and nematodes, where different RdRP enzymes use the processed target transcripts as templates to produce secondary siRNAs (reviewed in Ghildiyal and Zamore, 2009). But, although *rdrp* genes have been identified in many fungal species, including *N. crassa* and *Aspergillus nidulans*, those genes are not required for silencing when dsRNA is used as an inducer, and no secondary siRNAs have been detected, suggesting the absence of an siRNA amplification step mediated by

RdRPs (Chang *et al.*, 2012). Unlike those fungi, *M. circinelloides* accumulates both size classes of siRNAs as secondary siRNA (Nicolás *et al.*, 2003).

The RNAi machinery in *Mucor* includes RdRP, Dicer and Ago proteins. Similar to *N. crassa* QDE-1, RdRP-1 is essential for initiation of silencing by sense transgenes through the production of antisense RNA from transgene transcripts, but it is not necessary for amplification of the silencing signal (Calo *et al.*, 2012). Triggering dsRNA molecules are processed by Dcl-2, which plays the major role in the biogenesis of the two size classes of siRNAs (Nicolás *et al.*, 2007; de Haro *et al.*, 2009). The generated siRNAs are bound to Ago-1, which is the only Argonaute protein that operates during vegetative growth, to recognize target genes and suppress their expression (Cervantes *et al.*, 2013). Finally, a functionally different RdRP protein, RdRP-2, uses the processed target transcripts to generate new dsRNAs that are diced into secondary siRNAs, amplifying the silencing signal (Calo *et al.*, 2012). All these elements participate in the exogenously induced RNAi pathway, suggesting a defensive role of the silencing machinery in *Mucor* against invasive nucleic acids. We should also note that they participate in the regulation of endogenous functions, adding a new dimension to the roles of RNAi in fungi (Nicolás *et al.*, 2010; Cervantes *et al.*, 2013; Trieu *et al.*, 2015) (see below).

### 5.2.3 An antiviral defence mechanism

The defensive role of the fungal RNAi machinery against viruses was first demonstrated in the chestnut blight fungus *Cryphonectria parasitica* (Segers *et al.*, 2007). The RNAi-based viral response of this fungus requires the presence of one Dicer-like protein, Dcl-2, and one Argonaute-like protein, Agl-2 (Zhang and Nuss, 2008; Sun *et al.*, 2009), suggesting that the action of the antiviral RNAi pathway in *C. parasitica* follows the canonical mechanism. Although four *rdrp* genes have been identified in the *C. parasitica* genome, none of the single or multiple *rdrp* knockout mutants show differences from the parental strains upon viral infection, indicating that RdRPs do not play any role in the antiviral defence mechanism (Zhang *et al.*, 2014). The antiviral response has been also found in *A. nidulans*, where infective viruses can be both targets and suppressors of the RNAi pathway (Hammond *et al.*, 2008).

### 5.2.4 Meiotic silencing by unpaired DNA

Meiotic silencing by unpaired DNA (MSUD) is an RNAi-related pathway described in *N. crassa* that occurs during meiosis (Shiu *et al.*, 2001). As a defence mechanism, MSUD protects the host genome by transiently silencing all copies of DNA fragments that remain unpaired during pairing of homologous chromosomes, such as transposons and viruses inserted into the host genome. The mechanism of MSUD is also triggered by dsRNA molecules, which are synthesized from aRNA transcribed from unpaired DNA regions and processed to produce MSUD-associated siRNAs (masiRNAs) (Shiu *et al.*, 2001; Hammond *et al.*, 2013a). MSUD requires the *N. crassa* canonical RNAi components DCL-1 and QIP (Alexander *et al.*, 2008; Xiao *et al.*, 2010). However, the specific RdRP and Argonaute proteins SAD-1 and SMS-2 are required for MSUD, indicating that quelling and MSUD are parallel RNAi pathways acting during vegetative and sexual cycles, respectively (Shiu *et al.*, 2001; Lee *et al.*, 2003). The RdRP SAD-1 converts aRNAs into dsRNAs (Shiu and Metzenberg, 2002) with the help of the helicase SAD-3, also involved in MSUD (Hammond *et al.*, 2011). All these MSUD components form a multiprotein complex at the perinuclear region, where the scaffold protein SAD-2, which is not found in other RNAi pathways, is essential for proper localization of the complex (Shiu *et al.*, 2006; Decker *et al.*, 2015). In the working model, aRNAs produced in the nucleus are exported to the perinuclear region where the large meiotic complex converts aRNAs to dsRNAs and then masiRNAs to trigger gene silencing, suppressing the expression of unpaired transposon and virus sequences during meiosis. The role of MSUD in controlling the expression of transposable elements has recently been demonstrated by the identification of transposon-derived masiRNAs when a DNA transposon was unpaired in a cross between two *N. crassa* wild-type strains (Wang *et al.*, 2015). The sterile phenotype in homozygous crosses involving mutants in MSUD components suggested a role of MSUD in controlling sexual development (Alexander *et al.*, 2008). However, identification of new components of the MSUD machinery required for production of masiRNAs that show a homozygous-fertile phenotype uncouples MSUD from sexual development, and represents a breakthrough in our understanding of MSUD (Hammond *et al.*, 2013b).

## 5.3 Regulatory Endogenous Small RNAs in Fungi

Apart from having a genome defence function, in metazoans RNAi has important roles in the regulation of a wide variety of cellular, developmental and physiological processes through the production of diverse small non-coding RNAs derived from endogenous precursors. For a long time it was thought that regulatory endogenous small RNAs (esRNAs), particularly miRNAs, were absent in fungi; however, the application of deep-sequencing technologies has uncovered an unexpected diversity of esRNAs in fungi that are generated by canonical and non-canonical RNAi pathways.

### 5.3.1 RNAi and heterochromatin formation in yeasts

Demonstration by Volpe *et al.* (2002) that components of the RNAi machinery are required for heterochromatin formation in the centromeric regions of the fission yeast was the first evidence of the role of RNAi in the regulation of endogenous processes in fungi. RdRP, Dicer and Argonaute proteins participate in the production of specific siRNAs derived from centromeric regions. These siRNAs provide a guide to the so-called RNA-induced transcriptional silencing (RITS) complex to target centromeric nascent transcripts. Through interaction of RITS with the RdRP-containing complex, those transcripts are in turn converted into dsRNA and processed into new siRNAs, generating a self-enforcing loop (Verdel *et al.*, 2004; Goto and Nakayama, 2012). Interaction of RITS with the pericentromeric region through homology-dependent association with the nascent centromeric transcripts is reinforced by binding of the chromodomain-containing protein Chp1 (part of the RITS complex) to histone H3 methylated at lysine 9, which is abundant at the pericentromeric region (Sadaie *et al.*, 2004). Recruitment of RITS to these regions allows it to interact with other chromatin-modifying components, forming a major protein complex that spreads heterochromatinization through the region. Although knowledge of the major players and effector complexes for RNAi-mediated heterochromatin formation has advanced considerably over recent years, many questions remain to be answered about the components and functions involved in the maintenance of epigenetic silencing (Goto and Nakayama, 2012).

### 5.3.2 qiRNAs: small RNAs induced by DNA damage in *Neurospora*

DNA damage in *N. crassa* induces the production of an sRNA class known as qiRNAs (QDE-2-interacting sRNAs) (Lee *et al.*, 2009). As regular exogenously derived siRNAs, qiRNAs are 21–23 nt in length, show a strong preference for uracil at the 5′ end and are found interacting with Argonaute proteins. But the most relevant characteristic of qiRNAs is that they mostly derive from repetitive sequences of ribosomal DNA loci as a response to DNA damage (Lee *et al.*, 2009). Biogenesis of qiRNA requires the presence of the RdRP protein QDE-1, RecQ DNA helicase QDE-3 and Dicer proteins, indicating that they are produced through a canonical RNAi pathway upon DNA damage (Lee *et al.*, 2009; Lee *et al.*, 2010a). qiRNAs seem to have a relevant role in the cellular response to DNA damage, since it has been proposed that regulation of rRNA expression by qiRNAs would inhibit protein synthesis, which could help cells arrest the cell cycle and activate DNA repair machinery (Lee *et al.*, 2009). The function of qiRNAs as a DNA damage checkpoint seems to be conserved across species, since DNA damage-induced siRNAs, as well as components of the RNAi machinery, have been demonstrated to play key roles in the DNA damage response both in plants and animals (Francia *et al.*, 2012; Wei *et al.*, 2012).

There are many similarities between quelling and qiRNA pathways. Ribosomal DNA is the only highly repetitive sequence in the *N. crassa* genome and it has been suggested that, similar to tandemly inserted transgenes, aberrant RNA (aRNA) would be produced by QDE-1 from single-stranded DNA structures formed during homologous recombination of repetitive sequences in response to DNA damage. In fact, proteins directly involved in homologous recombination, and chromatin remodelling factors necessary for homologous recombination are critical for quelling and qiRNA biogenesis (Zhang *et al.*, 2013). The proposed model suggests that double-strand breaks induced by DNA damage promotes the formation of aberrant forms of recombination intermediates of tandem repeat sequences, which are recognized by QDE-3 and QDE-1 to produce aRNA and dsRNA (Xue *et al.*, 2012; Yang *et al.*, 2015). In this process, *Neurospora* QDE-1 acts as a DdRP to produce aRNAs and as an RdRP to convert aRNAs to dsRNAs (Lee *et al.*, 2009; Lee *et al.*, 2010a).

### 5.3.3 microRNA-like RNAs in *Neurospora* and other filamentous fungi

Although miRNAs have been found to participate in the regulation of different physiological and developmental processes in a wide variety of eukaryotes, their existence in fungi was questioned for a long time, mainly due to the lack of clear phenotypes in mutants affected in the RNAi machinery. However, Lee *et al.* (2010b) found miRNA-like small RNAs (milRNAs) in *N. crassa* by analysing sRNAs species associated with the Argonaute protein QDE-2. These milRNAs derived from stem-loop RNA precursors were mostly 25 nt long and showed a very strong preference for uracil at their 5′ termini. However, in contrast to canonical miRNAs, biogenesis of the *Neurospora* milRNAs is mediated by at least four different mechanisms (Lee *et al.*, 2010b; reviewed in Chang *et al.*, 2012). Apart from participation of Dicer and/or Argonaute proteins in those mechanisms, the involvement of components different from the RNAi machinery (such as the RNA exosome) highlights the complexity of milRNA biogenesis in fungi (Xue *et al.*, 2012). Several lines of evidence suggest that *Neurospora* milRNAs can target endogenous RNA transcripts with imperfect complementarity, as regular miRNAs do (Lee *et al.*, 2010b). However, it seems that they do not play critical roles in regulation of growth or developmental processes during the vegetative cycle, since mutants affected in genes required for biogenesis of milRNAs do not show defects in those processes (Chang *et al.*, 2012). Nevertheless, the sterile phenotype shown by some of those mutants may suggest a role of milRNAs in the regulation of sexual development (Alexander *et al.*, 2008).

Since the discovery of milRNAs in *N. crassa*, many ascomycete and basidiomycete fungi have been predicted to produce milRNAs, although further functional research is necessary to evaluate their biological roles. Thus, a variable number of loci producing milRNA precursors have been computationally predicted in *Mycosphaerella graminicola* (Goodwin *et al.*, 2011), *C. neoformans* (Jiang *et al.*, 2012), *Sclerotinia sclerotiorum* (Zhou *et al.*, 2012), *Trichoderma reesei* (Kang *et al.*, 2013), *Penicillium marneffei* (Lau *et al.*, 2013), *Fusarium oxysporum* (Chen *et al.*, 2014), *Antrodia cinnamomea* (Lin *et al.*, 2015), *Aspergillus flavus* (Bai *et al.*, 2015) and others. In several cases, the existence of these milRNAs has been experimentally confirmed and, based on the analysis of their expression profiles under different developmental and environmental conditions, it has been suggested that some milRNAs could play a role in the regulation of different processes. However, no target transcripts have been experimentally confirmed for any of these milRNAs which, together with their low abundance and lack of conservation among fungal species, casts doubt on their biological role and the specific conditions under which these milRNAs regulate gene expression.

### 5.3.4 *dicer*-dependent and *dicer*-independent exon-derived esRNAs in *Mucor*

Although milRNAs have not been described in *Mucor circinelloides*, this was the first fungus in which esRNAs regulating the expression of endogenous target genes through the repression of the corresponding mRNAs were identified (Nicolás *et al.*, 2010). The phenotypes shown by *M. circinelloides* mutants affected in the RNAi machinery (see below) were the first indication of a role for this mechanism in the regulation of endogenous genes through the production of esRNAs. In fact, a plethora of regulatory esRNA molecules, produced by both *dicer*-dependent and *dicer*-independent pathways, has been identified in *M. circinelloides* (Nicolás *et al.*, 2010; Cervantes *et al.*, 2013; Trieu *et al.*, 2015). Most of the *dicer*-dependent esRNAs derived from exons of protein coding genes, and were named exonic-esRNAs (ex-siRNAs). These ex-siRNAs were experimentally demonstrated to be functional, since they regulate accumulation of mRNAs of the protein coding genes from which they were derived, by guiding their degradation (Nicolás *et al.*, 2010; Cervantes *et al.*, 2013). The ex-siRNAs are produced by canonical and non-canonical RNAi pathways in which a Dicer enzyme and different combinations of proteins of the silencing machinery participate, allowing their classification in four different classes depending on the proteins required for their biogenesis (reviewed in Torres-Martínez and Ruiz-Vázquez, 2015).

In addition to the *dicer*-dependent ex-siRNA classes, a *dicer*-independent but *rdrp*-dependent esRNA class has also been identified in *M. circinelloides* (Trieu *et al.*, 2015). These esRNAs derive from exons, and their structural characteristics, which include random spread of size distribution, strong strand bias (all of them being exclusively sense to the mRNA) and no preference for uracil at the 5′ end) suggested that *rdrp*-dependent *dicer*-independent esRNAs were small degradation products of specific mRNAs. In fact, a new RNase III-like protein known as R3B2 has been identified that, together with RdRP-1 and RdRP-2, participates in the degradation of specific transcripts by this non-canonical degradation pathway (Trieu *et al.*, 2015). In accordance with their nature, the *rdrp*-dependent *dicer*-independent esRNAs were named *rdrp*-dependent *d*egraded RNA (rdRNA). It was also demonstrated that this non-canonical RNA degradation pathway modulates gene expression, since mRNA levels of target genes increased in mutants affected in the pathway (Trieu *et al.*, 2015). The R3B2 protein not only participates in the production of rdRNAs but is also required for biogenesis of other esRNAs, particularly one of the *dicer*-dependent classes of ex-siRNAs (class 3), which shares structural characteristics with rdRNAs (Nicolás *et al.*, 2010; Trieu *et al.*, 2015). Thus R3B2, which is specific to Mucorales, seems to participate in several RNA degradation pathways that regulate different groups of genes and in which different combinations of silencing proteins take part. In the proposed model for biogenesis of rdRNAs and class 3 ex-siRNAs, the activity of RdRP-1 and/or RdRP-2 on target transcripts would generate discrete dsRNA stretches that could be directly recognized by R3B2, targeting those transcripts for degradation (rdRNAs); or could first be processed by a Dicer enzyme and after the initial cleavage the single-stranded portions of mRNAs would be degraded by R3B2 (class 3 ex-siRNAs) (Torres-Martínez and Ruiz-Vázquez, 2015; Trieu *et al.*, 2015).

The role of *dicer*-dependent RNAi and *dicer*-independent RNA degradation pathways in the regulation of endogenous genes has been confirmed by comparing transcriptomes of *M. circinelloides* wild-type and RNAi mutants (Nicolás *et al.*, 2015). Hundreds of genes showing differential mRNA expression in silencing mutants relative to wild type have been identified. Detailed analysis of the differentially expressed genes allowed identification of putative candidate genes that could be responsible for the phenotypes shown by RNAi mutants, such as defects in vegetative growth, hyphal morphology and sporulation efficiency, sexual development or even differential response to nutritional or oxidative stress (Nicolás *et al.*, 2015; Trieu *et al.*, 2015) (see below).

### 5.3.5 Other fungal regulatory esRNAs

Different classes of esRNAs have been also detected in some other fungal species, such as *Magnaporthe oryzae* (Nunes *et al.*, 2011) and *Trichoderma atroviride* (Carreras-Villaseñor *et al.*, 2013). In the rice blast fungal pathogen *M. oryzae*, vegetative esRNAs mainly derive from LTR retrotransposons, while esRNAs produced from specialized-infection tissues are mainly generated from tRNA loci (Nunes *et al.*, 2011). Although tRNA-derived fragments (tRFs) were initially thought to be degradation products, it is now evident that they play regulatory roles both in prokaryotes and eukaryotes, controlling critical cellular functions (Keam and Hutvagner, 2015). The specific presence of tRNA-derived fragments (tRFs) in infection tissue has been proposed to be part of a mechanism that restricts protein biosynthesis in order to direct cellular metabolism towards infection (Nunes *et al.*, 2011). In addition to these esRNAs, analyses of sRNA profiles following exposure to different physiological stressors have identified coding gene-derived esRNAs that may play a role in regulating expression of a small subset of genes (Raman *et al.*, 2013). Similarly, *dicer*-dependent esRNAs derived from exons have been also identified in *T. atroviride*, an ascomycete known for its biocontrol capabilities against a wide range of phytopathogenic fungi (Carreras-Villaseñor *et al.*, 2013). Altogether, those results underscore the significant roles of esRNAs in the regulation of gene expression and the control of developmental processes in the fungal kingdom, beyond genome protection against invasive nucleic acids (Nicolás and Ruiz-Vázquez, 2013).

## 5.4 Regulation of Physiology and Development by esRNAs in Fungi

### 5.4.1 esRNAs in the response to environmental signals

Most fungal mutants in RNAi genes do not show gross alterations in vegetative growth and morphology, and respond normally to physiological and developmental cues, which initially led to the conclusion that fungal RNAi machinery does not play a role in regulation of biological functions. This was reinforced by the lack of an active RNAi pathway in a small number of fungi, including yeasts (Nicolás *et al.*, 2013). However, careful analysis of RNAi mutants in a wide spectrum of fungal species and environmental conditions has uncovered an essential role of esRNAs in regulating different processes.

*Mucor circinelloides* mutants in RNAi genes display diverse altered phenotypes (reviewed in Ruiz-Vázquez *et al.*, 2015). These phenotypes include reduced growth and altered hyphal morphology, reduced asexual sporulation, an accelerated autolytic response to nutrient starvation, defective sexual interaction and reduced production of zygospores, and differential resistance to oxidative stress. These mutant phenotypes are shown by particular RNAi mutants and are probably a consequence of altered expression of genes directly or indirectly regulated by *dicer*-dependent and *dicer*-independent esRNAs (Fig. 5.1) (Nicolás *et al.*, 2015; Trieu *et al.*, 2015). In fact, by contrasting genes differentially expressed in a particular combination of RNAi mutants with the phenotypes shared by those mutants, specific gene clusters can be associated with the control of specific biological processes. Further, it has been rationalized that biological processes affected in a particular combination of RNAi mutants must be controlled by esRNAs requiring that specific combination of silencing proteins for their biogenesis (Torres-Martínez and Ruiz-Vázquez, 2015). Thus, reduction in sexual interaction and differential response to oxidative stress are shown by mutants affected in the *rdrp*-dependent *dicer*-independent RNA degradation pathway, but not by mutants involved in the canonical RNAi pathway. This suggests that rdRNAs and the non-canonical silencing pathway may control the initial steps of sexual interaction and the response to specific environmental stresses by regulating specific target genes (Fig. 5.1) (Nicolás *et al.*, 2015; Trieu *et al.*, 2015). Similarly, the canonical RNAi machinery in *M. circinelloides*, particularly class 1 ex-siRNAs, directly or indirectly regulates developmental responses to carbon source levels in the medium, since mutants affected in genes required for biogenesis of class 1 ex-siRNAs present defects in asexual sporulation and autolysis induced by nutritional starvation (Fig. 5.1) (Nicolás *et al.*, 2015). The importance of the *M. circinelloides* RNAi machinery in regulating responses to endogenous and

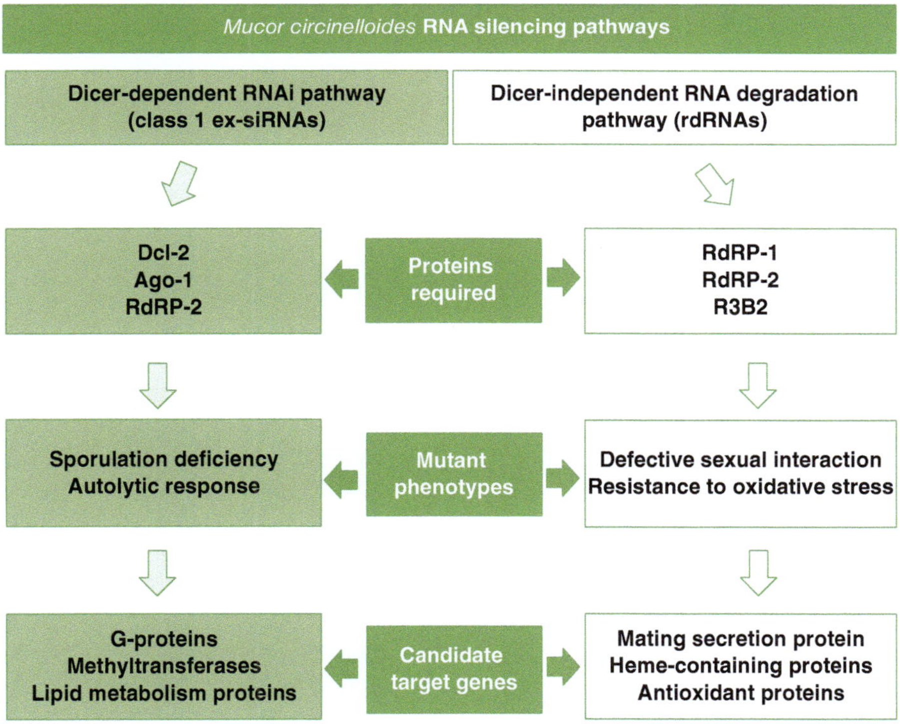


**Fig. 5.1.** Distinct classes of esRNAs regulate different physiological and developmental processes in response to environmental signals in *Mucor circinelloides*. Different combinations of RNAi proteins are involved in *dicer*-dependent RNAi and *dicer*-independent RNA degradation pathways by producing distinct esRNAs that regulate different groups of genes. Differential expression of specific target genes in RNAi mutants is probably responsible for the phenotypes shared by mutants affected in each pathway (reviewed in Ruiz-Vázquez *et al.*, 2015).

environmental signals is highlighted by the ability of this organism to adapt to a changing environment using an RNAi-mediated mechanism (Calo *et al.*, 2014). Through this novel epigenetic mechanism, isolates resistant to the antifungal drug FK506 are obtained by spontaneously triggering RNAi to silence FK506 target genes, giving rise to drug-resistant epimutants. This resistance is unstable, since those epimutants reverted to the drug-sensitive wild-type phenotype when grown in the absence of drugs (Calo *et al.*, 2014), emphasizing the relevance of the RNAi mechanism in the control of phenotypic plasticity.

The role of the canonical RNAi machinery in the control of vegetative growth and development can be extended to other fungi since, in the ascomycetes *T. atroviride* and *Magnaporthe oryzae*, mutants in *dicer* genes and other components of the silencing machinery also show defects in vegetative growth, hyphal morphology and/or reduction in asexual sporulation (Kadotani *et al.*, 2004; Carreras-Villaseñor *et al.*, 2013). All this suggests that the RNAi mechanism in fungi regulates responses to environmental signals, modulating the expression of genes involved in these responses.

### 5.4.2 esRNAs in fungal pathogenesis

Regulatory sRNAs and the RNAi machinery of eukaryotic pathogens have been demonstrated to be involved not only in the control of developmental responses but also in pathogenicity. The oomycete plant pathogen *Phytophthora sojae* produces *cis*-acting esRNAs that silence an avirulent gene, which encodes an effector protein that triggers immunity in plants carrying appropriated resistance genes (Qutob *et al.*, 2013). Silencing of the effector gene is trans-generationally inherited and results in gains in virulence, indicating a role of fungal esRNAs in pathogenesis control. *P. sojae* also encodes RNA silencing suppressors which inhibit plant sRNA biogenesis, acting as effector proteins that subvert host immunity (Qiao *et al.*, 2013). Cross-kingdom RNAi in host–pathogen interaction has received particular attention in recent years, as has the identification of the sRNAs involved (reviewed in Weiberg and Jin, 2015). The existence of plant-derived esRNAs that are used by plant hosts to regulate innate immunity and to target fungal pathogens is well known. But cross-kingdom RNAi can also take place from the fungal parasite to the plant host. In a seminal work, Weiberg *et al.* (2013) demonstrated that the RNAi mechanism can be used by plant fungal pathogens to facilitate host infection by producing specific sRNAs. In fact, sRNAs produced by *Botrytis cinerea*, the causative agent of grey mould disease, are able to hijack *Arabidopsis* and tomato RNAi machineries by binding to host Argonaute protein and selectively silencing host immunity genes. This cross-kingdom hijacking of RNAi seems to act as a virulence mechanism, since *B. cinerea* mutants affected in the production of esRNA have reduced virulence relative to wild type; and, conversely, *Arabidopsis* and tomato *ago* mutants are more resistant to fungal infection. Cross-kingdom RNAi can be also extended to other hosts and parasites, indicating a relevant role of RNAi as a virulence factor (Weiberg and Jin, 2015).

Apart from the examples given above, expression profiles of fungal milRNAs during infection suggest a role for some of these milRNAs in pathogenicity, although the functional role of the esRNA molecules has to be confirmed experimentally. Even though the number of examples of esRNAs involved in virulence is continuously increasing, the real implications of the RNA silencing mechanism in fungal pathogenesis remain unclear.

## Acknowledgements

Work on RNAi pathways in *Mucor* in our laboratory is supported by the Ministerio de Ciencia e Innovación, Spain (BFU2009-07220, co-financed by FEDER) and the Ministerio de Economía y Competitividad, Spain (BFU2012-32246, co-financed by FEDER), the Fundación Séneca de la Comunidad Antónoma de la Región de Murcia, Spain (19339/PI/14) and the DOE Joint Genome Institute, USA (DE-ACO2-05CH11231).

## References

Alexander, W.G., Raju, N.B., Xiao, H., Hammond, T.M., Perdue, T.D., Metzenberg, R.L., Pukkila, P.J. and Shiu, P.K. (2008) DCL-1 colocalizes with other components of the MSUD machinery and is required for silencing. *Fungal Genetics and Biology* 45, 719–727.

Bai, Y., Lan, F., Yang, W., Zhang, F., Yang, K., Li, Z., Gao, P. and Wang, S. (2015) sRNA profiling in *Aspergillus flavus* reveals differentially expressed miRNA-like RNAs response to water activity and temperature. *Fungal Genetics and Biology* 81, 113–119.

Calo, S., Nicolás, F.E., Vila, A., Torres-Martínez, S. and Ruiz-Vázquez, R.M. (2012) Two distinct RNA-dependent RNA polymerases are required for initiation and amplification of RNA silencing in the basal fungus *Mucor circinelloides*. *Molecular Microbiology* 83, 379–394.

Calo, S., Shertz-Wall, C., Lee, S.C., Bastidas, R.J., Nicolás, F.E., Granek, J.A., Mieczkowski, P., Torres-Martínez, S., Ruiz-Vázquez, R.M., Cardenas, M.E. and Heitman, J. (2014) Antifungal drug resistance evoked via RNAi-dependent epimutations. *Nature* 513, 555–558.

Carreras-Villaseñor, N., Esquivel-Naranjo, E.U., Villalobos-Escobedo, J.M., Abreu-Goodger, C. and Herrera-Estrella, A. (2013) The RNAi machinery regulates growth and development in the filamentous fungus *Trichoderma atroviride*. *Molecular Microbiology* 89, 96–112.

Catalanotto, C., Azzalin, G., Macino, G. and Cogoni, C. (2000) Gene silencing in worms and fungi. *Nature* 404, 245.

Catalanotto, C., Pallotta, M., ReFalo, P., Sachs, M.S., Vayssie, L., Macino, G. and Cogoni, C. (2004) Redundancy of the two *dicer* genes in transgene-induced posttranscriptional gene silencing in *Neurospora crassa*. *Molecular and Cellular Biology* 24, 2536–2545.

Cervantes, M., Vila, A., Nicolás, F.E., Moxon, S., de Haro, J.P., Dalmay, T., Torres-Martínez, S. and Ruiz-Vázquez, R.M. (2013) A single *argonaute* gene participates in exogenous and endogenous RNAi and controls cellular functions in the basal fungus *Mucor circinelloides*. *PLOS ONE* 8, e69283.

Chang, S.S., Zhang, Z. and Liu, Y. (2012) RNA interference pathways in fungi: mechanisms and functions. *Annual Reviews of Microbiology* 66, 305–323.

Chen, R., Jiang, N., Jiang, Q., Sun, X., Wang, Y., Zhang, H. and Hu, Z. (2014) Exploring microRNA like small RNAs in the filamentous fungus *Fusarium oxysporum*. *PLOS ONE* 9(8), e104956.

Cogoni, C. and Macino, G. (1997) Isolation of quelling-defective (*qde*) mutants impaired in posttranscriptional transgene-induced gene silencing in *Neurospora crassa*. *Proceedings of the National Academy of Sciences USA* 94, 10233–10238.

Cogoni, C. and Macino, G. (1999a) Gene silencing in *Neurospora crassa* requires a protein homologous to RNA-dependent RNA polymerase. *Nature* 399, 166–169.

Cogoni, C. and Macino, G. (1999b) Posttranscriptional gene silencing in *Neurospora* by a RecQ DNA helicase. *Science* 286, 2342–2344.

de Haro, J.P., Calo, S., Cervantes, M., Nicolás, F.E., Torres-Martínez, S. and Ruiz-Vázquez, R.M. (2009) A single *dicer* gene is required for efficient gene silencing associated with two classes of small antisense RNAs in *Mucor circinelloides*. *Eukaryotic Cell* 8, 1486–1497.

Decker, L.M., Boone, E.C., Xiao, H., Shanker, B.S., Boone, S.F., Kingston, S.L., Lee, S.A., Hammond, T.M. and Shiu, P.K. (2015) Complex formation of RNA silencing proteins in the perinuclear region of *Neurospora crassa*. *Genetics* 199, 1017–1021.

Fire, A., Xu, S., Montgomery, M.K., Kostas, S.A., Driver, S.E. and Mello, C.C. (1998) Potent and specific genetic interference by double-stranded RNA in *Caenorhabditis elegans*. *Nature* 391, 806–811.

Francia, S., Michelini, F., Saxena, A., Tang, D., de Hoon, M., Anelli, V., Mione, M., Carninci, P. and d'Adda di Fagagna, F. (2012) Site-specific DICER and DROSHA RNA products control the DNA damage response. *Nature* 488, 231–235.

Ghildiyal, M. and Zamore, P.D. (2009) Small silencing RNAs: an expanding universe. *Nature Reviews Genetics* 10, 94–108.

Goodwin, S.B., Ben M'Barek, S., Dhillon, B., Wittenberg, A.H.J., Crane, C.F., Hane, J.K., Foster, A.J., Van der Lee, T.A., Grimwood, J., Aerts, A., Antoniw, J., Bailey, A., Bluhm, B., Bowler, J., Bristow, J., van der Burgt, A., Canto-Canché, B., Churchill, A.C., Conde-Ferràez, L., Cools, H.J., Coutinho, P.M., Csukai, M., Dehal, P., De Wit, P., Donzelli, B., van de Geest, H.C., van Ham, R.C., Hammond-Kosack, K.E., Henrissat, B., Kilian, A., Kobayashi, A.K., Koopmann, E., Kourmpetis, Y., Kuzniar, A., Lindquist, E., Lombard, V., Maliepaard, C., Martins, N., Mehrabi, R., Nap, J.P., Ponomarenko, A., Rudd, J.J., Salamov, A., Schmutz, J., Schouten, H.J., Shapiro, H., Stergiopoulos, I., Torriani, S.F., Tu, H., de Vries, R.P., Waalwijk, C., Ware, S.B., Wiebenga, A., Zwiers, L.H., Oliver, R.P., Grigoriev, I.V. and Kema, G.H. (2011) Finished genome of the fungal wheat pathogen *Mycosphaerella graminicola* reveals dispensome structure, chromosome plasticity, and stealth pathogenesis. *PLOS Genetics* 7, e1002070.

Goto, D.B. and Nakayama, J.-I. (2012) RNA and epigenetic silencing: insight from fission yeast. *Development, Growth and Differentiation* 54, 129–141.

Hammond, T.M., Andrewski, M.D., Roossinck, M.J. and Keller, N.P. (2008) *Aspergillus* mycoviruses are targets and suppressors of RNA silencing. *Eukaryotic Cell* 7, 350–357.

Hammond, T.M., Xiao, H., Boone, E.C., Perdue, T.D., Pukkila, P.J. and Shiu, P.K. (2011) SAD-3, a putative helicase required for meiotic silencing by unpaired DNA, interacts with other components of the silencing machinery. *G3 (Bethesda)* 1, 369–376.

Hammond, T.M., Spollen, W.G., Decker, L.M., Blake, S.M., Springer, G.K. and Shiu, P.K. (2013a) Identification of small RNAs associated with meiotic silencing by unpaired DNA. *Genetics* 194, 279–284.

Hammond, T.M., Xiao, H., Boone, E.C., Decker, L.M., Lee, S.A., Perdue, T.D., Pukkila, P.J. and Shiu, P.K. (2013b) Novel proteins required for meiotic silencing by unpaired DNA and siRNA generation in *Neurospora crassa*. *Genetics* 194, 91–100.

Ibrahim, A.S., Spellberg, B., Walsh, T.J. and Kontoyiannis, D.P. (2012) Pathogenesis of mucormycosis. *Clinical Infectious Diseases* 54, S16–S22.

Jiang, N., Yang, Y., Janbon, G., Pan, J. and Zhu, X. (2012) Identification and functional demonstration of miRNAs in the fungus *Cryptococcus neoformans*. *PLOS ONE* 7, e52734.

Kadotani, N., Nakayashiki, H., Tosa, Y. and Mayama, S. (2004) One of the two Dicer-like proteins in the filamentous fungi *Magnaporthe oryzae* genome is responsible for hairpin RNA-triggered RNA silencing and related small interfering RNA accumulation. *Journal of Biological Chemistry* 279, 44467–44474.

Kang, K., Zhong, J., Jiang, L., Liu, G., Gou, C.Y., Wu, Q., Wang, Y., Luo, J. and Gou, D. (2013) Identification of microRNA-like RNAs in the filamentous fungus *Trichoderma reesei* by solexa sequencing. *PLOS ONE* 8, e76288.

Keam, S.P. and Hutvagner, G. (2015) tRNA-derived fragments (tRFs): emerging new roles for an ancient RNA in the regulation of gene expression. *Life (Basel)* 5, 1638–1651.

Lau, S.K., Chow, W.N., Wong, A.Y., Yeung, J.M., Bao, J., Zhang, N., Lok, S., Woo, P.C. and Yuen, K.Y. (2013) Identification of microRNA-like RNAs in mycelial and yeast phases of the thermal dimorphic fungus *Penicillium marneffei*. *PLOS Neglected Tropical Diseases* 7, e2398.

Lee, D.W., Pratt, R.J., McLaughlin, M. and Aramayo, R. (2003) An Argonaute-like protein is required for meiotic silencing. *Genetics* 164, 821–828.

Lee, H.C., Chang, S.S., Choudhary, S., Aalto, A.P., Maiti, M., Bamford, D.H. and Liu, Y. (2009) qiRNA is a new type of small interfering RNA induced by DNA damage. *Nature* 459, 274–277.

Lee, H.C., Aalto, A.P., Yang, Q., Chang, S.S., Huang, G., Fisher, D., Cha, J., Poranen, M.M., Bamford, D.H. and Liu, Y. (2010a) The DNA/RNA-dependent RNA polymerase QDE-1 generates aberrant RNA and dsRNA for RNAi in a process requiring replication protein A and a DNA helicase. *PLOS Biology* 8, e1000496.

Lee, H.C., Li, L., Gu, W., Xue, Z., Crosthwaite, S.K., Pertsemlidis, A., Lewis, Z.A., Freitag, M., Selker, E.U., Mello, C.C. and Liu, Y. (2010b) Diverse pathways generate microRNA-like RNAs and Dicer-independent small interfering RNAs in fungi. *Molecular Cell* 6, 803–814.

Lin, Y.L., Ma, L.T., Lee, Y.R., Lin, S.S., Wang, S.Y., Chang, T.T., Shaw, J.F., Li, W.H. and Chu, F.H. (2015) MicroRNA-like small RNAs prediction in the development of *Antrodia cinnamomea*. *PLOS ONE* 10, e0123245.

Maiti, M., Lee, H.C. and Liu, Y. (2007) QIP, a putative exonuclease, interacts with the *Neurospora* Argonaute protein and facilitates conversion of duplex siRNA into single strands. *Genes and Development* 21, 590–600.

Napoli, C., Lemieux, C. and Jorgensen, R. (1990) Introduction of a chimeric chalcone synthase gene into petunia results in reversible co-suppression of homologous genes in trans. *Plant Cell* 2, 279–289.

Nicolás, F.E. and Ruiz-Vázquez, R.M. (2013) Functional diversity of RNAi-associated sRNAs in fungi. *International Journal of Molecular Sciences* 14, 15348–15360.

Nicolás, F.E., Torres-Martínez, S. and Ruiz-Vázquez, R.M. (2003) Two classes of small antisense RNAs in fungal RNA silencing triggered by non-integrative transgenes. *The EMBO Journal* 22, 3983–3991.

Nicolás, F.E., de Haro, J.P., Torres-Martínez, S. and Ruiz-Vázquez, R.M. (2007) Mutants defective in a *Mucor circinelloides dicer*-like gene are not compromised in siRNA silencing but display developmental defects. *Fungal Genetics and Biology* 44, 504–516.

Nicolás, F.E., Torres-Martínez, S. and Ruiz-Vázquez, R.M. (2009) Transcriptional activation increases RNA silencing efficiency and stability in the fungus *Mucor circinelloides*. *Journal of Biotechnology* 142, 123–126.

Nicolás, F.E., Moxon, S., de Haro, J.P., Calo, S., Grigoriev, I.V., Torres-Martínez, S., Moulton, V., Ruiz-Vázquez, R.M. and Dalmay, T. (2010) Endogenous short RNAs generated by Dicer 2 and RNA-dependent RNA polymerase 1 regulate mRNAs in the basal fungus *Mucor circinelloides*. *Nucleic Acids Research* 38, 5535–5541.

Nicolás, F.E., Torres-Martínez, S. and Ruiz-Vázquez, R.M. (2013) Loss and retention of RNA interference in fungi and parasites. *PLOS Pathogens* 9, e1003089. DOI: 10.1371/journal.ppat.1003089

Nicolás, F.E., Vila, A., Moxon, S., Cascales, M.D., Torres-Martínez, S., Moulton, V., Ruiz-Vázquez, R.M. and Dalmay, T. (2015) The RNAi machinery controls distinct responses to environmental signals in the basal fungus *Mucor circinelloides*. *BMC Genomics* 16, 237.

Nolan, T., Braccini, L., Azzalin, G., De Toni, A., Macino, G. and Cogoni, C. (2005) The post-transcriptional gene silencing machinery functions independently of DNA methylation to repress a LINE1-like retrotransposon in *Neurospora crassa*. *Nucleic Acids Research* 33, 1564–1573.

Nunes, C.C., Gowda, M., Sailsbery, J., Xue, M., Chen, F., Brown, D.E., Oh, Y., Mitchell, T.K. and Dean, R.A. (2011) Diverse and tissue-enriched small RNAs in the plant pathogenic fungus *Magnaporthe oryzae*. *BMC Genomics* 12, 288.

Qiao, Y., Liu, L., Xiong, Q., Flores, C., Wong, J., Shi, J., Wang, X., Liu, X., Xiang, Q., Jiang, S., Zhang, F., Wang, Y., Judelson, H.S., Chen, X. and Ma, W. (2013) Oomycete pathogens encode RNA silencing suppressors. *Nature Genetics* 45, 330–333.

Qutob, D., Chapman, B.P. and Gijzen, M. (2013) Transgenerational gene silencing causes gain of virulence in a plant pathogen. *Nature Communications* 4, 1349.

Raman, V., Simon, S.A., Romag, A., Demirci, F., Mathioni, S.M., Zhai, J., Meyers, B.C. and Donofrio, N.M. (2013) Physiological stressors and invasive plant infections alter the small RNA transcriptome of the rice blast fungus, *Magnaporthe oryzae*. *BMC Genomics* 14, 326.

Romano, N. and Macino, G. (1992) Quelling: transient inactivation of gene expression in *Neurospora crassa* by transformation with homologous sequences. *Molecular Microbiology* 6, 3343–3353.

Ruiz-Vázquez, R.M., Nicolás, F.E., Torres-Martínez, S. and Garre, V. (2015) Distinct RNAi pathways in the regulation of physiology and development in the fungus *Mucor circinelloides*. *Advances in Genetics* 91, 55–102.

Sadaie, M., Iida, T., Urano, T. and Nakayama, J.-I. (2004) A chromodomain protein, Chp1, is required for the establishment of heterochromatin in fission yeast. *The EMBO Journal* 23, 3825–3835.

Segers, G.C., Zhang, X., Deng, F., Sun, Q. and Nuss, D.L. (2007) Evidence that RNA silencing functions as an antiviral defense mechanism in fungi. *Proceedings of the National Academy of Sciences USA* 104, 12902–12906.

Shiu, P.K. and Metzenberg, R.L. (2002) Meiotic silencing by unpaired DNA: properties, regulation and suppression. *Genetics* 161, 1483–1495.

Shiu, P.K., Raju, N.B., Zickler, D. and Metzenberg, R.L. (2001) Meiotic silencing by unpaired DNA. *Cell* 107, 905–916.

Shiu, P.K., Zickler, D., Raju, N.B., Ruprich-Robert, G. and Metzenberg, R.L. (2006) SAD-2 is required for meiotic silencing by unpaired DNA and perinuclear localization of SAD-1 RNA-directed RNA polymerase. *Proceedings of the National Academy of Sciences USA* 103, 2243–2248.

Sun, Q., Choi, G.H. and Nuss, D.L. (2009) A single Argonaute gene is required for induction of RNA silencing antiviral defense and promotes viral RNA recombination. *Proceedings of the National Academy of Sciences USA* 106, 17927–17932.

Torres-Martínez, S. and Ruiz-Vázquez, R.M. (2015) RNAi pathways in *Mucor*: a tale of proteins, small RNAs and functional diversity. *Fungal Genetics and Biology* 90, 44–52.

Trieu, T.A., Calo, S., Nicolás, F.E., Vila, A., Moxon, S., Dalmay, T., Torres-Martínez, S., Garre, V. and Ruiz-Vázquez, R.M. (2015) A non-canonical RNA silencing pathway promotes mRNA degradation in basal fungi. *PLOS Genetics* 11, e1005168.

Verdel, A., Jia, S., Gerber, S., Sugiyama, T., Gygi, S., Grewal, S.I. and Moazed, D. (2004) RNAi-mediated targeting of heterochromatin by the RITS complex. *Science* 303, 672–676.

Volpe, T.A., Kidner, C., Hall, I.M., Teng, G., Grewal, S.I.S. and Martienssen, R.A. (2002) Regulation of heterochromatic silencing and histone H3 lysine-9 methylation by RNAi. *Science* 297, 1833–1837.

Wang, Y., Smith, K.M., Taylor, J.W., Freitag, M. and Stajich, J.E. (2015) Endogenous small RNA mediates meiotic silencing of a novel DNA transposon. *G3 (Bethesda)* 5, 1949–1960.

Wei, W., Ba, Z., Gao, M., Wu, Y., Ma, Y., Amiard, S., White, C.I., Rendtlew Danielsen, J.M., Yang, Y.G. and Qi, Y. (2012) A role for small RNAs in DNA double-strand break repair. *Cell* 149, 101–112.

Weiberg, A. and Jin, H. (2015) Small RNAs – the secret agents in the plant-pathogen interactions. *Current Opinions in Plant Biology* 26, 87–94.

Weiberg, A., Wang, M., Lin, F.M., Zhao, H., Zhang, Z., Kaloshian, I., Huang, H.D. and Jin, H. (2013) Fungal small RNAs suppress plant immunity by hijacking host RNA interference pathways. *Science* 342, 118–123.

Xiao, H., Alexander, W.G., Hammond, T.M., Boone, E.C., Perdue, T.D., Pukkila, P.J. and Shiu, P.K. (2010) QIP, a protein that converts duplex siRNA into single strands, is required for meiotic silencing by unpaired DNA. *Genetics* 186, 119–126.

Xue, Z., Yuan, H., Guo, J. and Liu, Y. (2012) Reconstitution of an Argonaute-dependent small RNA biogenesis pathway reveals a handover mechanism involving the RNA exosome and the exonuclease QIP. *Molecular Cell* 46, 299–310.

Yang, Q., Ye, Q.A. and Liu, Y. (2015) Mechanism of siRNA production from repetitive DNA. *Genes and Development* 29, 526–537.

Zhang, D.X., Spiering, M.J. and Nuss, D.L. (2014) Characterizing the roles of *Cryphonectria parasitica* RNA-dependent RNA polymerase-like genes in antiviral defense, viral recombination and transposon transcript accumulation. PLOS ONE 9, e108653. DOI: 10.1371/journal.pone.0108653

Zhang, X. and Nuss, D.L. (2008) A host Dicer is required for defective viral RNA production and recombinant virus vector RNA instability for a positive sense RNA virus. *Proceedings of the National Academy of Sciences USA* 105, 16749–16754.

Zhang, Z., Chang, S.-S., Zhang, Z., Xue, Z., Zhang, H., Li, S. and Liu, Y. (2013) Homologous recombination as a mechanism to recognize repetitive DNA sequences in an RNAi pathway. *Genes and Development* 27, 145–150.

Zhou, J., Fu, Y., Xie, J., Li, B., Jiang, D., Li, G. and Cheng, J. (2012) Identification of microRNA-like RNAs in a plant pathogenic fungus *Sclerotinia sclerotiorum* by high-throughput sequencing. *Molecular Genetics and Genomics* 287, 275–282.

# 6 Artificial Small RNA-based Strategies for Effective and Specific Gene Silencing in Plants

**Alberto Carbonell***

*Instituto de Biología Molecular y Celular de Plantas (CSIC-UPV), Valencia, Spain*

## 6.1 Introduction

In plants, small RNAs (sRNAs) function in diverse RNA silencing pathways to regulate development, control genome integrity and protect against viruses (Borges and Martienssen, 2015). MicroRNAs (miRNAs) and small interfering RNAs (siRNAs) are the two main classes of plant sRNAs acting in post-transcriptional gene silencing pathways. Both associate with an ARGONAUTE (AGO) protein to target and silence highly sequence-complementary transcripts through direct AGO-mediated endonucleolytic cleavage or through their translational repression (Axtell, 2013). MiRNAs and siRNAs differ in their biogenesis pathways as well as in the spectrum of their target transcripts. MiRNAs arise from endogenous miRNA transcripts with imperfect self-complementary foldback structures processed by DICER-LIKE1 (DCL1), and target other cellular transcripts. SiRNAs originate from transposons, centromeres or exogenous nucleic acids such as transgene inverted repeats or viral RNAs, are processed by DCL2, DCL3 or DCL4 and usually silence the transcript from which they derive. *Trans*-acting siRNAs (tasiRNAs) are a particular subclass of plant siRNAs that are produced in a sophisticated way. In *Arabidopsis thaliana*, the cleavage of a transcript from the *TAS* family by a miRNA/AGO complex triggers the RNA-DEPENDENT RNA POLYMERASE6 (RDR6)-dependent synthesis of double-stranded RNA (dsRNA) from one of the cleavage products. DCL4 sequentially processes the dsRNA into 21 nucleotide (nt) tasiRNA duplexes in register with the miRNA-guided cleavage site. In all cases, one strand of the miRNA or tasiRNA duplex (named the 'guide strand') is selectively incorporated into an AGO protein to direct specific silencing of cognate transcripts, while the other strand (named the 'star strand', or *) is generally degraded (Bologna and Voinnet, 2014).

Endogenous sRNA-directed silencing pathways have been exploited in plants to selectively silence genes of interest in gene function studies, and to generate antiviral resistance or other type of crop improvement. Classic RNA interference (RNAi) approaches such as virus induced gene silencing (VIGS) or hairpin-based silencing are based on the expression of dsRNA or dsRNA-like precursors containing

* Corresponding author e-mail: acarbonell@ibmcp.upv.es

  

sequences corresponding to the target transcript. The expression *inplanta* of such type of precursors triggers the DCL-dependent production of siRNAs which will silence complementary target sequences (Ossowski *et al.*, 2008; Baykal and Zhang, 2010). In VIGS, a sequence corresponding to the target transcript is inserted into a viral genome. Replication of the virus in a susceptible host triggers the synthesis by endogenous RDRs of dsRNAs of viral sequence, including those derived from the target RNA sequence that was inserted in the VIGS construct. In hairpin-based silencing, a transgene containing an inverted repeat sequence of the target transcript is introduced into plants and expresses a hairpin RNA (hpRNA) with a characteristic stem-loop secondary structure. Unfortunately, and despite their massive use in recent years, both approaches have a relatively high risk of inducing undesired off-target effects, as the populations of siRNA species produced from dsRNAs or hpRNAs may lead to the accidental targeting of other cellular transcripts sharing high sequence complementarity with that of certain siRNAs.

## 6.2 Plant Artificial Small RNAs

The limited specificity of classic plant RNAi approaches was overcome by the recent development of 'second-generation RNAi' strategies based on artificial sRNAs. These strategies include artificial microRNAs (amiRNAs) and synthetic tasiRNAs (syn-tasiRNAs). In both cases the artificial sRNA is designed to specifically silence the desired target(s) with no off-target effects, and is produced *in planta* by expressing a functional miRNA or tasiRNA precursor with modified miRNA/miRNA* or tasiRNA sequences, respectively. Details of the biogenesis, action and application of both classes of plant artificial sRNAs are provided below.

### 6.2.1 Artificial microRNAs

AmiRNAs have been the most extensively used plant artificial sRNAs. AmiRNAs are generated *in planta* by expressing amiRNA transgenes that include the sequence of a plant miRNA precursor in which the endogenous miRNA sequence is substituted by the sequence of the designed amiRNA (Fig. 6.1). Other sequences of the miRNA precursor (including the miRNA*) are also modified to preserve the original secondary structure of the miRNA precursor and allow its accurate processing by DCL1 to produce the 21 nt amiRNA duplex. The amiRNA guide strand is usually designed with a 5′U to associate with AGO1 and direct the endonucleolytic cleavage or the translational repression of target genes (Li *et al.*, 2013; Tiwari *et al.*, 2014; Yu and Pilot, 2014).

Since their initial applications (Alvarez *et al.*, 2006; Schwab *et al.*, 2006) amiRNAs produced from different miRNA precursors have been used in multiple plant species – including eudicots, monocots, mosses or algae – (Table 6.1) to silence endogenous genes and non-coding RNAs, reporter transgenes and viruses (reviewed in Tiwari *et al.*, 2014). AmiRNAs are typically designed to target a single transcript, but in some cases it is possible to design amiRNAs to target multiple transcripts simultaneously in cases where these share enough sequence similarity (e.g. members of the same gene family). For example, a single amiRNA was used

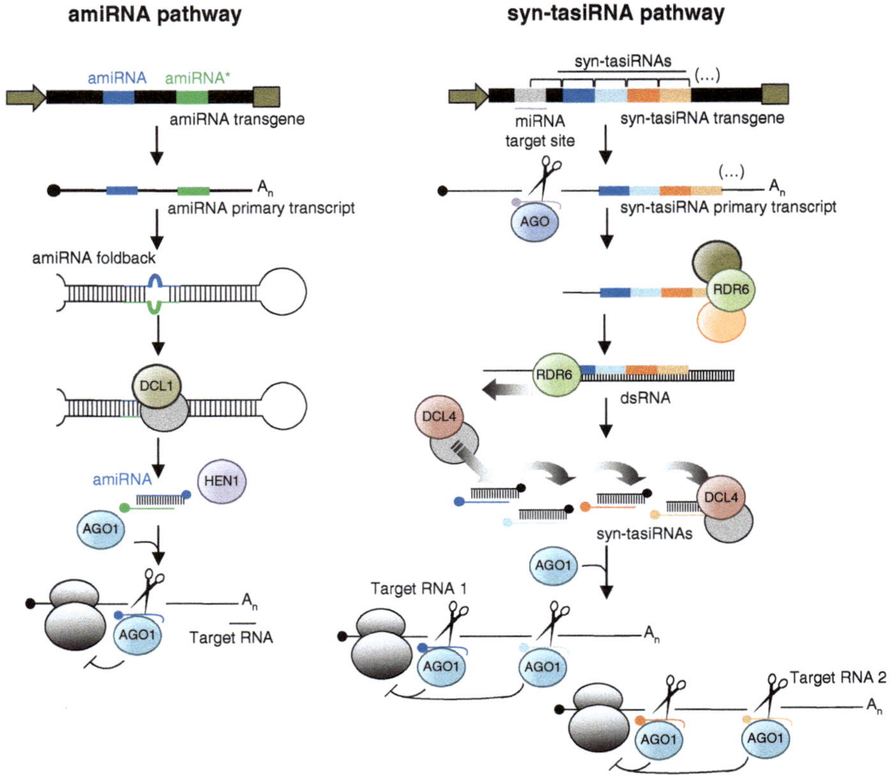


**Fig. 6.1.** Artificial sRNA pathways in plants. Left, the amiRNA pathway. An amiRNA transgene, containing a plant miRNA precursor in which the original miRNA/miRNA* sequences have been substituted by the amiRNA/amiRNA* sequences, is introduced into plants to express an amiRNA primary transcript processed into an amiRNA foldback. A rational amiRNA design requires that the amiRNA foldback preserves the original secondary structure of the endogenous precursor. DCL1 further processes the amiRNA foldback to produce the amiRNA duplex methylated by HEN1. The 5′U amiRNA strand is usually incorporated into AGO1 to silence highly complementary transcripts by direct slicing or by repressing their translation. Right, syn-tasiRNA pathway. A syn-tasiRNA transgene, containing a plant *TAS* precursor in which a subset of the original tasiRNA sequences has been substituted by several syn-tasiRNA sequences in tandem, is introduced into plants to express a syn-tasiRNA primary transcript. An endogenous miRNA cleaves this primary transcript, a process that triggers the recruitment of RDR6 complexes to synthesize a dsRNA from one of the cleavage products. DCL4 processes the dsRNA into phased tasiRNA duplexes in 21 nt register with the miRNA cleavage site. Syn-tasiRNA guide strands with a 5′U are incorporated into AGO1 to direct specific silencing of sequence-unrelated target transcripts at one or multiple sites.

to downregulate three *MYB* transcripts (*TRIPTYCHON* (*TRY*), *CAPRICE* (*CPC*) and *ENHANCER OF TRIPTYCHON AND CAPRICE2* (*ETC2*)) and induce high numbers of clustered trichomes in *A. thaliana* rosette leaves (Schwab *et al.*, 2006; Liang *et al.*, 2012; Carbonell *et al.*, 2014). Multiple amiRNAs can be expressed from a

**Table 6.1.** Endogenous precursors used to produce artificial sRNAs in plants.

| Artificial sRNA | Precursor | Species tested | Original reference |
|---|---|---|---|
| amiRNA | *Ath-miR159a* | *Arabidopsis thaliana* | Niu *et al*., 2006 |
| | | *Nicotiana benthamiana* | Mitter *et al*., 2016 |
| | | *Nicotiana tabacum* | Mitter *et al*., 2016 |
| | | *Solanum lycopersicum* | Zhang *et al*., 2011 |
| | *Ath-miR159b* | *Arabidopsis thaliana* | Eamens *et al*., 2011 |
| | *Ath-miR164a* | *Arabidopsis thaliana* | Alvarez *et al*., 2006 |
| | | *Solanum lycopersicum* | Alvarez *et al*., 2006 |
| | *Ath-miR164b* | *Nicotiana tabacum* | Alvarez *et al*., 2006 |
| | *Ath-miR168a* | *Solanum tuberosum* | Bhagwat *et al*., 2013 |
| | *Ath-miR169d* | *Arabidopsis thaliana* | Liu *et al*., 2010 |
| | *Ath-miR171a* | *Arabidopsis thaliana* | Qu *et al*., 2007 |
| | *Ath-miR172a* | *Arabidopsis thaliana* | Schwab *et al*., 2006 |
| | *Ath-miR319a* | *Arabidopsis thaliana* | Schwab *et al*., 2006 |
| | | *Catharanthus roseus* | Li *et al*., 2013 |
| | | *Corchorus olitorius* | Shafrin *et al*., 2015 |
| | | *Glycine max* | Melito *et al*., 2010 |
| | | *Helianthus annuus* | Li *et al*., 2013 |
| | | *Medicago sativa* | Verdonk and Sulllivan, 2013 |
| | | *Nicotiana benthamiana* | Li *et al*., 2013 |
| | | *Nicotiana tabacum* | Vu *et al*., 2013 |
| | | *Petunia hybrida* | Guo *et al*., 2014 |
| | | *Phaeodactylum tricornutum* | Kaur and Spillane, 2015 |
| | | *Physcomitrella patens* | Khraiwesh *et al*., 2008 |
| | | *Solanum lycopersicum* | Fernandez *et al*., 2009 |
| | | *Solanum melongena* | Toppino *et al*., 2011 |
| | | *Solanum tuberosum* | Wyrzykowska *et al*., 2016 |
| | | *Vitis vinifera* | Jelly *et al*., 2012 |
| | | *Zea mays* | Li *et al*., 2013 |
| | *Ath-miR390a* | *Arabidopsis thaliana* | Montgomery *et al*., 2008a |
| | | *Nicotiana benthamiana* | Montgomery *et al*., 2008a |
| | *Ath-miR395a* | *Arabidopsis thaliana* | Liang *et al*., 2012 |
| | *Cre-miR1157* | *Chlamydomonas reinhardtii* | Molnar *et al*., 2009 |
| | *Cre-miR1162* | *Chlamydomonas reinhardtii* | Zhao *et al*., 2009 |
| | *Ghb-miR169a* | *Nicotiana benthamiana* | Ali *et al*., 2013 |
| | *Gma-miR159a* | *Glycine max* | Yamada *et al*., 2014 |
| | *Hvu-miR171* | *Hordeum vulgare* | Kis *et al*., 2015 |
| | | *Nicotiana benthamiana* | Kis *et al*., 2015 |
| | *Lgi-miR166a* | *Lemna minor* | Canto-Pastor *et al*., 2015 |
| | *Mpo-miR160* | *Marchantia polymorpha* | Flores-Sandoval *et al*., 2015 |
| | *Mtr-miR159b* | *Medicago truncatula* | Devers *et al*., 2013 |
| | *Osa-miR390* | *Brachypodium distachyon* | Carbonell *et al*., 2015 |
| | *Osa-miR395* | *Triticum aestivum* | Fahim *et al*., 2012 |
| | *Osa-miR528* | *Oryza sativa* | Warthmann *et al*., 2008 |

*Continued*

**Table 6.1.** Continued.

| Artificial sRNA | Precursor | Species tested | Original reference |
|---|---|---|---|
| | *Ptc-miR408* | *Populus trichocarpa* | Shi *et al.*, 2010 |
| | *Skr-miR166* | *Marchantia polymorpha* | Flores-Sandoval *et al.*, 2015 |
| | *Sly-miR159* | *Nicotiana tabacum* | Vu *et al.*, 2013 |
| | | *Solanum lycopersicum* | Vu *et al.*, 2013 |
| | *Sly-miR168a* | *Nicotiana tabacum* | Vu *et al.*, 2013 |
| | | *Solanum lycopersicum* | Vu *et al.*, 2013 |
| | *Vvi-miR166f* | *Nicotiana benthamiana* | Roumi *et al.*, 2012 |
| syn-tasiRNA | *Ath-TAS1a* | *Arabidopsis thaliana* | Felippes and Weigel, 2009 |
| | *Ath-TAS1c* | *Arabidopsis thaliana* | de la Luz Gutierrez-Nava *et al.*, 2008 |
| | | *Nicotiana benthamiana* | Montgomery *et al.*, 2008b |
| | *Ath-TAS3a* | *Arabidopsis thaliana* | Montgomery *et al.*, 2008a |
| | | *Nicotiana benthamiana* | Montgomery *et al.*, 2008a |

single construct when they are produced from an endogenous polycistronic precursor. For instance, rice *Osa-miR395* precursor containing five stem-loop structures, each of which produces a functional miRNA, was used to produce five different amiRNAs and induce resistance to *Wheat streak virus* in wheat (Fahim *et al.*, 2012). Alternatively, multiple amiRNAs can be produced from a single transcript containing the same precursor sequence in tandem (Liang *et al.*, 2012). In all cases, the expression of multiple amiRNAs in a single plant favours multi-targeting and can be used to enhance the silencing of a particular target (if multiple amiRNAs target different sites in the same target transcript) and/or to silence different sequence-unrelated target transcripts (if each amiRNA targets a different sequence-unrelated target transcript).

### 6.2.2 Synthetic trans-acting siRNAs

Syn-tasiRNA transgenes include the sequence of a *TAS* precursor in which a region corresponding to various endogenous tasiRNAs is substituted by a fragment containing multiple syn-tasiRNA sequences (Fig. 6.1). When transcribed, the syn-tasiRNA primary transcript is cleaved by a miRNA/AGO complex, and one of the cleaved products is used by RDR6 as a template for dsRNA synthesis. DsRNA is processed by DCL4 in several syn-tasiRNA duplexes in register with the miRNA cleavage site. Twenty-one nt syn-tasiRNA guide strands designed to have an AGO1-preferred 5′U direct silencing of cognate transcript(s) (Fig. 6.1). Initially, syn-tasiRNAs were used in *A. thaliana* when expressed from *TAS1a* (Felippes and Weigel, 2009), *TAS1c* (de la Luz Gutierrez-Nava et al., 2008; Montgomery et al., 2008b) and *TAS3a* precursors (Montgomery *et al.*, 2008a) (Table 6.1) to silence single genes (for a recent review see Zhang, 2014). However, the main advantage of the syn-tasiRNA approach is the possibility of multiplexing multiple syn-tasiRNAs in the same construct, which allows multi-targeting, as shown recently. For example, the simultaneous targeting of *TRY/CPC/ETC2* and sequence-unrelated *FLOWERING LOCUS T (FT)* endogenous transcripts in transgenic *A. thaliana* expressing two different *TAS1c*-based syn-tasiRNAs produced the expected combined phenotype of increased clustering of trichomes in rosette leaves and delay in flowering (Carbonell *et al.*, 2014). In another recent study, five different syn-tasiRNAs expressed transgenically from *TAS3a* precursors in *A. thaliana* were used against *Turnip mosaic virus* and *Cucumber mosaic virus*, two sequence-unrelated RNA viruses, to confer multiple antiviral resistance (Chen *et al.*, 2016). Importantly, *TAS1/TAS2*- and *TAS3a*-based syn-tasiRNA biogenesis depends on the presence of miR173 and miR390a, respectively. Thus, because miR173 is unique to *A. thaliana* (and close relatives), it must be co-expressed with *TAS1/TAS2*-based syn-tasiRNA transgenes to trigger syn-tasiRNA biogenesis in species different from *A. thaliana*.

A strategy similar to syn-tasiRNAs was described and named MIGS (for MiRNA Induced Gene Silencing) (Felippes *et al.*, 2012). In this case, a transgene including a fragment of the target gene fused to an upstream miR173 target site was expressed in *A. thaliana* and induced the accumulation of tasiRNAs derived from the target gene sequences. These tasiRNAs have been shown to direct effective silencing of the desired target (Felippes *et al.*, 2012). It is important to clarify that the MIGS technology does not produce authentic syn-tasiRNAs, but rather generates a series of non-designed tasiRNAs which can induce undesired off-target effects, as observed in a recent study (Han *et al.*, 2015).

## 6.3 Design of Plant Artificial Small RNAs

The main difference between plant artificial sRNAs and other sRNAs produced in certain RNAi approaches is that the former are designed to be specific for the intended target(s). The rational design of artificial sRNAs must optimize both the effectiveness and the specificity of the designed sRNA. First, an effective artificial sRNA is required to have high sequence complementarity with the target RNA. Although the exact base-pairing requirements for productive sRNA/target RNA interactions are not fully known, it is generally accepted that mismatches within the sRNA seed region (nts 2–14) drastically reduce the sRNA activity, while mismatches in positions 1 or 14–21 have a much more moderate effect. Second, the artificial sRNA must be highly specific and silence the intended target(s) exclusively. The specificity of an artificial sRNA is assessed through the analysis of all possible base-pairing interactions between the candidate artificial sRNA and the complete set of cellular transcripts. These off-target analyses are only possible for species with annotated transcriptomes or expressed sequence tag (EST) collections, and thus must be computational. The two main tools used for plant artificial sRNA design are WMD3 (from Web MicroRNA Designer 3) (Ossowski *et al.*, 2008; Schwab *et al.*, 2010) and P-SAMS (from Plant Small RNA Maker Suite) (Fahlgren *et al.*, 2016), and are described below.

WMD3 (http://wmd3.weigelworld.org/cgi-bin/webapp.cgi) is an amiRNA designer tool initially implemented for *A. thaliana* (Schwab *et al.*, 2006) but which has now been extended to more than 200 species. It has been extensively used as a general amiRNA designer tool, and also to obtain the sequences of the six oligonucleotides necessary for amiRNA cloning in *Ath-miR319a*, *Osa-miR528* or *Cre-miR1157* precursors (Schwab *et al.*, 2006; Warthmann *et al.*, 2008; Molnar *et al.*, 2009). The user inputs the ID or sequence of the target transcript, selects the plant species where the amiRNA is expressed, submits the job and several hours later receives an e-mail with a link to a web page where a list of candidate amiRNAs is displayed. Candidate amiRNAs are ranked based on a cumulative score which depends on different empirical rules regarding the hybridization energy of the amiRNA/target RNA base-pairing, the base-pair composition of the amiRNA and the specificity criteria. WMD3 recommends selecting an amiRNA (i) with no mismatches at positions 2–12 when paired to the desired target; (ii) one or two mismatches at positions 18–21 to avoid transitivity (due to priming and extension by RDRs) and preserve specificity; and (iii) targeting the coding region, as these are usually better annotated than the untranslated regions (UTRs). In addition, it is recommended that the hybridization energy of the binding between the amiRNA and the target ranges between −35 and −40 kcal/mole.

P-SAMS (http://p-sams.carringtonlab.org) was developed more recently, and contains two applications, P-SAMS amiRNA Designer and P-SAMS syn-tasiRNA Designer, for the simple and automated design of amiRNAs and syn-tasiRNAs, respectively (Fahlgren *et al.*, 2016). Key features of P-SAMS are its simplicity and speed. It has a user-friendly, modern interface and wizard-assisted navigation which guides the user through the whole design process. Simple questions are answered by the user to advance to the next step of the design process, with help boxes appearing when requested. Median job time for single-targeting amiRNA design is around 3 min, which represents a considerable improvement in speed compared to other tools. The on-screen results page includes the sequence of the designed sRNA, the sequence of the two oligonucleotides required for cloning into compatible *AtMIR390a*- or *AtTAS1c*-based B/c vectors (see next section) and a summary of the off-targeting analysis. P-SAMS outputs designed sRNAs as 'Optimal Results' or as 'Suboptimal Results' if they have or do not have off-targets, respectively. Up to three Optimal and/or Suboptimal results are displayed with no specific ranking criteria. Regarding the computational design of artificial sRNAs, P-SAMS first catalogues all target sites that do not contain a 15-nt sequence from positions 6–20 perfectly matching a transcript not contained in the input set. An sRNA is then designed to target each target site from the input transcript with the additional criteria that the sRNA has: (i) an AGO1-preferred 5′U nucleotide; (ii) a C in position 19 to produce a star strand with an AGO1 non-preferred 5′G, thus limiting the competition for AGO1 association with the guide strand; and (iii) an intentional mismatch with the target transcript at position 21 to reduce transitivity.

## 6.4 Engineering Artificial Small RNA Constructs

The selection of the endogenous precursor to be used to express the artificial sRNAs is a critical step when engineering plant vectors for artificial sRNA cloning. The ideal precursor should accumulate to high levels and be processed accurately in multiple plant species for broad, efficient and specific gene silencing.

In addition, the sequence and/or structural features of a given precursor may benefit a particular cloning strategy, while the particular processing pathway of a given precursor could determine the accuracy of its processing. For example, *Ath-miR390a* was selected as the preferred precursor for amiRNA cloning and expression in eudicots (Carbonell *et al.*, 2014) because: (i) the *MIR390* family is deeply conserved in different plant species, hence *Ath-miR390a* is likely to be processed accurately in multiple species; (ii) miR390 precursors are processed base to loop, which should favour a more accurate processing, resulting in reduced off-target effects compared to precursors that follow a multi-step loop to base processing such as *Ath-miR319a*; and (iii) it contains a short stem-loop region (Fig. 6.2A) compared to other conserved miRNA precursors, facilitating the synthesis of the amiRNA insert (see below).

### 6.4.1 AmiRNA cloning

The methodology most used for amiRNA cloning was initially described for cloning amiRNAs in *Ath-miR319a* precursors (Schwab *et al.*, 2006), and later adapted for generating *Osa-miR528-* (Warthmann *et al.*, 2008) and *Cre-miR1157*-based (Molnar *et al.*, 2009) amiRNA constructs (Table 6.1). In this strategy, PCR-based mutagenesis is used to amplify the sequence of the miRNA precursor while substituting the original miRNA/miRNA* sequence with the corresponding amiRNA/amiRNA* sequence. Six amiRNA-specific oligonucleotides are needed in three different PCRs to amplify the complete amiRNA precursor sequence in three pieces when using a plasmid containing the wild-type miRNA precursor sequence. After gel purifying the three PCR fragments, an additional recombinant PCR with two generic oligonucleotides binding at the ends of the precursor and including the three PCR purified fragments produces the final amplicon containing the whole amiRNA precursor sequence. This DNA fragment is digested with specific restriction enzymes and cloned into an intermediate vector linearized with the same restriction enzymes. The DNA fragment containing the amiRNA precursor is finally transferred by restriction enzyme digestion and DNA ligase ligation into the expression vector of interest.

More recent methodologies for amiRNA cloning have aimed to reduce the number of reactions and treatments for the generation of amiRNA constructs, and are listed in Table 6.2. In all cases, a sequence of the precursor lacking the miRNA/loop/miRNA* region was already introduced in the cloning plasmid. Of particular interest is the simple, fast and cost-effective cloning strategy recently described for high-throughput generation of amiRNA constructs for efficient gene silencing in eudicots (Carbonell *et al.*, 2014) and monocots (Carbonell *et al.*, 2015). This strategy is based on the Golden Gate cloning method (Engler *et al.*, 2008), uses positive insert selection for zero-background cloning and eliminates PCR steps, gel-based DNA purifications, restriction digestion and sub-cloning of amiRNA inserts (Table 6.2). The amiRNA insert is synthesized by annealing two overlapping and partially complementary oligonucleotides obtained with the P-SAMS tool and containing the amiRNA/stem-loop/amiRNA* sequence, and has 4-nt specific 5′ overhangs for direct cloning in the corresponding amiRNA vector (Fig. 6.2B).

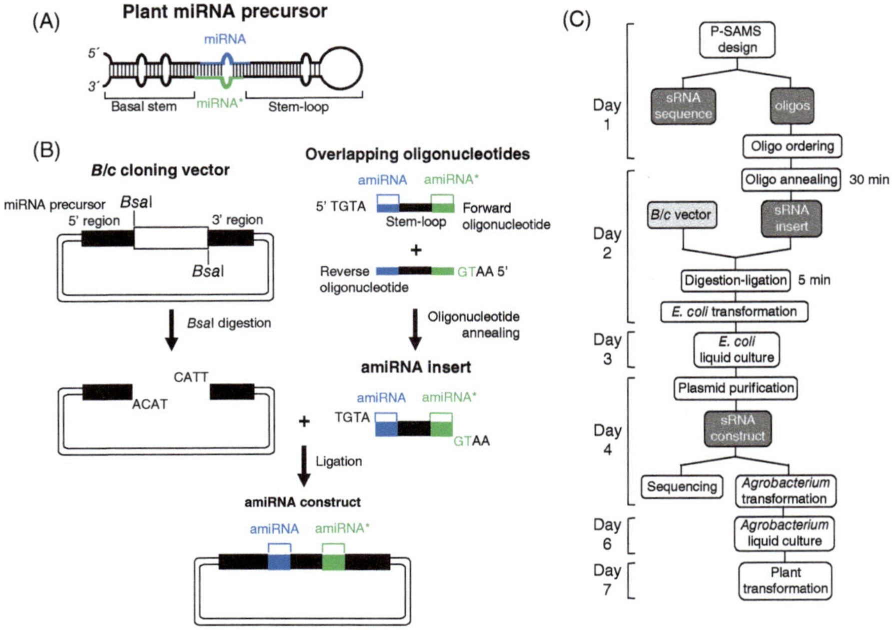


**Fig. 6.2.** High-throughput cloning of plant artificial sRNAs in B/c vectors (*BsaI*/*ccd*B or 'B/c' vectors). A, Diagram of a canonical plant miRNA precursor. B, Diagram of the main steps for amiRNA cloning in B/c vectors containing a modified version of *Ath-miR390a* precursor that includes a *ccd*B cassette flanked by two inverted *Bsa*I sites. The amiRNA insert obtained after annealing the two partially complementary and overlapping oligonucleotides has 5′ overhangs compatible with those resulting from the *Bsa*I digestion of the B/c cloning vector where it is inserted directionally. The DNA fragments corresponding to the miRNA precursor, the amiRNA and the amiRNA* are in black, blue and green, respectively. C, Flowchart of the main steps from sRNA design to plant transformation using P-SAMS and B/c cloning vectors. (Adapted from Carbonell *et al*., 2014.)

The amiRNA cloning vector or 'B/c' (from BsaI/ccdB) vector contains a truncated miRNA precursor sequence whose miRNA/stem-loop/amiRNA* region was replaced by a DNA cassette containing the *ccd*B gene (Bernard and Couturier, 1992) flanked by two *Bsa*I sites (one inverted with respect to the other) to have both *Bsa*I recognition sites outside of the miRNA precursor sequence (Fig. 6.2B). AmiRNA inserts are ligated directionally into B/c cloning vectors in a short 5 min digestion-ligation reaction including the amiRNA cloning vector, the amiRNA insert, *Bsa*I restriction enzyme and T4 DNA ligase. An aliquot of the digestion-ligation reaction is used to transform a *ccd*B-sensitive *Escherichia coli* strain (e.g. DH5α). Typically, all colonies growing on plates containing the appropriate bacterial resistance antibiotic for selection are positive. One or two colonies are usually sent for sequencing to confirm that the sequence of the insert does not contain any mutation that could have accumulated during the oligonucleotide synthesis.

Table 6.2. Published methods to generate plant artificial sRNA constructs.

| Artificial sRNA | Plant use | Precursor | Cloning in expression vector | Positive bacterial selection/ directional cloning[a] | Number of reactions/ treatments[b] | Types of reactions/ treatments[c] | Reference | Software for oligonucleotide design[d] |
|---|---|---|---|---|---|---|---|---|
| amiRNA | *Chlamydomonas* | *Cre-miR1157* | + | –/– | 7 | OA, PR, P, DP, L | Molnar *et al.*, 2009 | WMD3 |
| | | | + | –/+ | 5 | OA, PR, D, L | Hu *et al.*, 2014 | – |
| | Eudicots | *Ath-miR159a* | – | –/+ | 6 | PCR, PR, D, L, R | Niu *et al.*, 2006 | – |
| | | | – | GFP/+ | 6 | PCR, PR, R, M | Yan *et al.*, 2011a | – |
| | | *Ath-miR159b* | + | lacZ/– | 5 | PCR, PR, D, L | Eamens *et al.*, 2011 | amiRNA primer designer |
| | | *Ath-miR164a* | – | GFP/– | 6 | PCR, PR, R, M | Yan *et al.*, 2011a | – |
| | | *Ath-miR319a* | – | ? /– | 11 | PCR, PR, D, L | Schwab *et al.*, 2006 | WMD3 |
| | | | – | lacO/– | 5 | PCR, PR, D, L, M | Yan *et al.*, 2011b | – |
| | | | – | GFP/– | 6 | PCR, PR, R, M | Yan *et al.*, 2011a | – |
| | | | – | –/– | 6 | PCR, PR, D, L | Liang *et al.*, 2012 | – |
| | | | + | –/– | 3 | PCR, PR, D, L | Wang *et al.*, 2012 | – |
| | | | + | *ccd*B/– | 6 | D, K, PCR, PR, L | Zhou *et al.*, 2013 | – |
| | | *Ath-miR390a* | + | *ccd*B/+ | 2 | OA, D-L | Carbonell *et al.*, 2014 | P-SAMS |
| | | *Ath-miR395* | – | –/– | 6 | PCR, PR, D, L | Liang *et al.*, 2012 | – |
| | Monocots | *Osa-miR390a* | + | *ccd*B/+ | 2 | OA, D-L | Carbonell *et al.*, 2015 | P-SAMS |

*Continued*

**Table 6.2.** Continued.

| Artificial sRNA | Plant use | Precursor | Cloning in expression vector | Positive bacterial selection/ directional cloning[a] | Number of reactions/ treatments[b] | Types of reactions/ treatments[c] | Reference | Software for oligonucleotide design[d] |
|---|---|---|---|---|---|---|---|---|
| | | *Osa-miR528* | – | ?/– | 11 | PCR, PR, D, L | Warthmann *et al*., 2008 | WMD3 |
| | | | + | *ccd*B/– | 3 | PCR, D, L | Chen *et al*., 2009 | – |
| | | | + | –/– | 4 | OA, PCR, D, L | Wang *et al*., 2010 | – |
| | | | + | –/– | 6 | PCR, PR, D, L | Yan *et al*., 2012 | – |
| | | | + | *ccd*B/– | 6 | D, K, PCR, PR, L | Zhou *et al*., 2013 | – |
| | | | + | *ccd*B/+ | 2 | OA, D-L | Li *et al*., 2014 | – |
| syn-tasiRNA | *A. thaliana* | *Ath-TAS1c* | + | *ccd*B/+ | 2 | OA, D-L | Carbonell *et al*., 2014 | P-SAMS |

[a]Refers to the first vector used to clone the insert containing the amiRNA/amiRNA* sequence. ?, depends on the intermediate cloning vector of choice; *ccd*B, toxin from the *ccd* system of *Escherichia coli*; GFP, green fluorescent protein; lacO, operator site involved in the transcriptional regulation of the lac operon of *E. coli*; lacZ, structural gene of the lac operon coding β-galactosidase

[b]In each case the number was estimated by counting the basic reactions or treatments preceding bacterial transformation and described in the standard protocol detailed in the given reference

[c]D, digestion; DP, dephosphorylation; K, Klenow fill-in; L, ligation; M, mating; OA, oligo annealing; P, phosphorylation; PCR, polymerase chain reaction; PR, purification; R, recombination

[d]amiRNA primer designer is available at http://sydney.edu.au/science/molecular_bioscience/waterhouse/amiR_new/amiRcat2.html, P-SAMS at http://p-sams.carringtonlab.org, and WMD3 at http://wmd3.weigelworld.org

Importantly, non-linearized B/c cloning vectors with no amiRNA insert will express the *ccd*B toxin and will not propagate in *ccd*B-sensitive *E. coli* strains.

B/c amiRNA vectors contain the *Ath-miR390a* (Carbonell *et al.*, 2014) or the *Osa-miR390* (Carbonell *et al.*, 2015) precursor for use in eudicots or monocots, respectively (Table 6.3). Most amiRNA B/c vectors are expression vectors, with a unique combination of bacterial and plant antibiotic resistance genes. Others are intermediate Gateway-compatible entry vectors used for cloning the amiRNA insert, and subsequently recombine the complete amiRNA precursor sequence to the preferred Gateway expression vector containing a promoter, terminator or other features of choice (Table 6.3). In summary, this methodology based on a new generation of B/c amiRNA vectors is simple, fast and cost-effective compared to other described strategies for cloning amiRNAs. Indeed, the whole process from amiRNA design to plant transformation can be completed in just 1 week (Fig. 6.2C).

### 6.4.2 Syn-tasiRNA cloning

Initial syn-tasiRNA cloning methods (de la Luz Gutierrez-Nava *et al.*, 2008; Montgomery *et al.*, 2008a; Felippes and Weigel, 2009) were not adapted for efficient cloning. Similar to amiRNA B/c vectors, a new generation of syn-tasiRNA vectors also named 'B/'c was recently developed for high-throughput cloning of syn-tasiRNAs (Carbonell *et al.*, 2014) (Table 6.3). B/c syn-tasiRNA vectors include a modified *TAS1c* gene from *A. thaliana* whose region including endogenous tasiRNAs D3[+] and D4[+] was substituted with a *ccd*B cassette flanked with two *Bsa*I sites, in the same configuration described for B/c amiRNA vectors. B/c syn-tasiRNA vectors allow the multiplexing of several syn-tasiRNA sequences in a single construct, which represents an attractive option when multiple and unrelated sequences need to be targeted. The cloning methodology is very similar to that described for generating B/c-based amiRNA constructs. Briefly, two overlapping and partially complementary oligonucleotides containing the syn-tasiRNA sequences are designed with the P-SAMS tool. The syn-tasiRNA insert results from the annealing of both oligonucleotides, contains two 4-nt specific 5′ overhangs and is cloned directionally in a B/c syn-tasiRNA vector in a 5 min digestion-ligation reaction (Table 6.2). Several *Ath-TAS1c*-based B/c syn-tasiRNA expression vectors were developed, as well as a Gateway-compatible entry vector (Table 6.3).

## 6.5 Validation of Artificial Small RNA Constructs

The accuracy of the processing of artificial sRNA precursors has typically been evaluated by Northern blot analysis only, but should also be analysed by other complementary and more sensitive technologies such as sRNA deep sequencing. An accurate processing of the precursor results in the accumulation of the artificial sRNA as a single RNA species revealed by Northern blot hybridization, and in the sequencing of a high proportion of reads corresponding to the desired artificial sRNA sequence and not to other sequences derived from the precursor. For example, the accuracy of processing *Ath-miR390a*/*Osa-miR390* amiRNA and

**Table 6.3.** New generation of plant 'B/c' vectors for high-throughput cloning of amiRNAs and syn-tasiRNAs.

| Artificial sRNA | Plant use | Vector | Bacterial antibiotic resistance | Plant antibiotic resistance | Gateway use | Promoter | Terminator |
|---|---|---|---|---|---|---|---|
| amiRNA | Eudicots | *pENTR-AtMIR390a-B/c* | Kanamycin | – | Donor | – | – |
| | | *pFK210B-AtMIR390a-B/c* | Spectinomycin | BASTA | – | *CaMV 35S* | *rbcS* |
| | | *pMDC123SB-AtMIR390a-B/c* | Kanamycin | BASTA | – | *CaMV* 2x35S | *nos* |
| | | *pMDC32B-AtMIR390a-B/c* | Kanamycin | Hygromycin | – | *CaMV* 2x35S | *nos* |
| | Monocots | *pENTR-OsMIR390-B/c* | Kanamycin | – | Donor | – | – |
| | | *pH7WG2-OsMIR390-B/c* | Spectinomycin | Hygromycin | – | *Os Ubiquitin* | *CaMV* |
| | | *pMDC123SB-OsMIR390-B/c* | Kanamycin | BASTA | – | *CaMV* 2x35S | *nos* |
| | | *pMDC32B-OsMIR390-B/c* | Kanamycin | Hygromycin | – | *CaMV* 2x35S | *nos* |
| syn-tasiRNA | *Arabidopsis*[a] | *pENTR-AtTAS1c-B/c* | Kanamycin | – | Donor | – | – |
| | | *pMDC123SB-AtTAS1c-B/c* | Kanamycin | BASTA | – | *CaMV* 2x35S | *nos* |
| | | *pMDC32B-AtTAS1c-B/c* | Kanamycin | Hygromycin | – | *CaMV* 2x35S | *nos* |

[a]As miR173 is a non-conserved miRNA, a construct expressing miR173 has to be co-expressed with the syn-tasiRNA construct to trigger syn-tasiRNA biogenesis in species other than *Arabidopsis thaliana* (or close relatives)

*CaMV*, *Cauliflower mosaic virus*; nos, nopaline synthase; *Os*, *Oryza sativa*; *rbcS*, Rubisco small subunit

*Ath-TAS1c/Ath-TAS3a* syn-tasiRNA precursors was confirmed by both methodologies (Montgomery *et al.*, 2008a; Montgomery *et al.*, 2008b; Carbonell *et al.*, 2014; Carbonell *et al.*, 2015). In particular, sRNA library analysis was also used to confirm the correct phasing of *Ath-TAS1c-* and *Ath-TAS3a*-derived syn-tasiRNAs (Montgomery *et al.*, 2008a; Montgomery *et al.*, 2008b; Carbonell *et al.*, 2014). In contrast, sRNA deep sequencing analyses in amiRNA-expressing *Petunia* revealed that multiple sRNA species of different size are generated from different regions of *Ath-miR319a* precursors (Guo *et al.*, 2014), and that many of these sRNAs meet the required criteria for amiRNA design (Schwab *et al.*, 2006) and therefore are very likely to induce off-target effects. It is suggested that the reason for the accumulation of these additional sRNAs is the multi-step loop to base processing of *Ath-miR319a* precursors by DCL1 (Guo *et al.*, 2014).

The efficacy of individual sRNA candidates can easily be screened in *N. benthamiana* transient assays by co-agroinfiltrating each artificial sRNA with the target RNA and quantifying target silencing (Yu and Pilot, 2014). The most effective amiRNAs are then selected for stable expression in transgenic plants. Indeed, a positive correlation was observed between the activity of amiRNAs tested transiently in *N. benthamiana* and their activity in amiRNA-expressing *A. thaliana* transgenic lines, which supports the use of this type of preliminary assay (Yu and Pilot, 2014). Another alternative to test artificial sRNA efficacy is the epitope-tagged protein-based amiRNA (ETPamiR) screens. In these assays, target mRNAs encoding epitope-tagged proteins are constitutively or inducibly co-expressed in protoplasts with amiRNA candidates targeting single or multiple genes (Li *et al.*, 2013; Li *et al.*, 2014). Finally, when using syn-tasiRNAs, testing multiple amiRNAs in a rapid assay first (e.g. agroinfiltration in *N. benthamiana*) is suggested, then selection of the most effective and expression of them all in the same construct as syn-tasiRNAs (Carbonell *et al.*, 2016; Carbonell and Daròs, 2017).

The quantification of target gene silencing is usually done by quantitative RT-PCR analysis of target RNAs preferentially using oligonucleotides spanning the amiRNA-sRNA-guided cleavage site, as it is still considered that amiRNA effects are mostly evident at the transcript level. Indeed, amiRNA-guided cleavage sites can be mapped by 5′RLM-RACE analysis. The specificity of plant artificial sRNAs has been evaluated in just a few cases. In particular, amiRNA specificity was confirmed for *Ath-miR319a*-based amiRNAs in *A. thaliana* by genome-wide expression profiles (Schwab *et al.*, 2006) and, more recently, for *Osa-miR390*-based amiRNAs in *B. distachyon* by genome-wide transcriptome profiling combined with 5′RLM-RACE analysis (Carbonell *et al.*, 2015).

## 6.6 Conclusions and Future Challenges

Despite the recent emergence of the potent CRISPR (from 'Clustered Regularly Interspaced Short Palindromic Repeats') knock-out technology for genome editing, plant artificial sRNAs should still be used extensively in the next few years because of their unique features:

**1.** Plant artificial sRNAs are highly specific.
**2.** sRNA-based silencing is a knock-down approach that allows the study of genes whose complete knock-out causes lethality.

**3.** A fine-tuned regulation of artificial sRNA activity either by adjusting the artificial sRNA spatio-temporal activity (e.g. using tissue-specific and/or inducible promoters), efficacy (e.g. by varying the degree of base-pairing of the artificial sRNA with the target RNA) or accumulation (e.g. varying promoter strength) allows for the generation of an allelic series for a knock-down gene. For instance, the generation of an allelic series of individuals differing in time of flowering could be economically relevant for certain crops.
**4.** A single artificial sRNA construct should allow for the targeting of duplicated genes (and gene families), antisense transcripts or individual isoforms.
**5.** Plant artificial sRNAs can be used to engineer disease resistance against pathogens with RNA genomes such as viruses.

In conclusion, the recent development of high-throughput cloning strategies and automated design tools for the generation of plant artificial sRNA constructs should accelerate gene function studies and crop improvement (for a recent review see Kamthan *et al.*, 2015). However, a better knowledge of the rules governing the biogenesis, efficacy and mode of action (cleavage or translational repression) of endogenous sRNAs will undoubtedly help to further improve and refine plant artificial sRNA-based approaches.

## References


Ali, I., Amin, I., Briddon, R.W. and Mansoor, S. (2013) Artificial microRNA-mediated resistance against the monopartite begomovirus *cotton leaf curl Burewala virus*. *Virology Journal* 10, 231.

Alvarez, J.P., Pekker, I., Goldshmidt, A., Blum, E., Amsellem, Z. and Eshed, Y. (2006) Endogenous and synthetic microRNAs stimulate simultaneous, efficient, and localized regulation of multiple targets in diverse species. *The Plant Cell* 18, 1134–1151.

Axtell, M.J. (2013) Classification and comparison of small RNAs from plants. *Annual Review of Plant Biology* 64, 137–159.

Baykal, U. and Zhang, Z. (2010) Small RNA-mediated gene silencing for plant biotechnology. In: Catalano, A.J. (ed.) *Gene Silencing: Theory, Techniques and Applications.* Nova Science Publishers, Hauppauge, New York, pp. 255–269.

Bernard, P. and Couturier, M. (1992) Cell killing by the F plasmid CcdB protein involves poisoning of DNA-topoisomerase II complexes. *Journal of Molecular Biology* 226, 735–745.

Bhagwat, B., Chi, M., Su, L., Tang, H., Tang, G. and Xiang, Y. (2013) An in vivo transient expression system can be applied for rapid and effective selection of artificial microRNA constructs for plant stable genetic transformation. *Journal of Genetics and Genomics* 40, 261–270.

Bologna, N.G. and Voinnet, O. (2014) The diversity, biogenesis, and activities of endogenous silencing small RNAs in *Arabidopsis*. *Annual Review of Plant Biology* 65, 473–503.

Borges, F. and Martienssen, R.A. (2015) The expanding world of small RNAs in plants. *Nature Reviews Molecular Cell Biology* 16, 727–741.

Canto-Pastor, A., Molla-Morales, A., Ernst, E., Dahl, W., Zhai, J., Yan, Y., Meyers, B.C., Shanklin, J. and Martienssen, R. (2015) Efficient transformation and artificial miRNA gene silencing in *Lemna minor*. *Plant Biology* 17, 59–65.

Carbonell, A. and Daròs, J.A. (2017) Artificial microRNAs and synthetic *trans*-acting small interfering RNAs interfere with viroid infection. *Molecular Plant Pathology* (in press). DOI: 10.1111/mpp.12529

Carbonell, A., Takeda, A., Fahlgren, N., Johnson, S.C., Cuperus, J.T. and Carrington, J.C. (2014) New generation of artificial MicroRNA and synthetic trans-acting small interfering RNA vectors for efficient gene silencing in *Arabidopsis*. *Plant Physiology* 165, 15–29.

Carbonell, A., Fahlgren, N., Mitchell, S., Cox, K.L. Jr, Reilly, K.C., Mockler, T.C. and Carrington, J.C. (2015) Highly specific gene silencing in a monocot species by artificial microRNAs derived from chimeric miRNA precursors. *The Plant Journal* 82, 1061–1075.

Carbonell, A., Carrington, J.C. and Daròs, J.-A. (2016) Fast-forward generation of effective artificial small RNAs for enhanced antiviral defense in plants. *RNA and Disease* 3, e1130.

Chen, L., Cheng, X., Cai, J., Zhan, L., Wu, X., Liu, Q. and Wu, X. (2016) Multiple virus resistance using artificial trans-acting siRNAs. *Journal of Virological Methods* 228, 16–20.

Chen, S., Songkumarn, P., Liu, J. and Wang, G.L. (2009) A versatile zero background T-vector system for gene cloning and functional genomics. *Plant Physiology* 150, 1111–1121.

de la Luz Gutierrez-Nava, M., Aukerman, M.J., Sakai, H., Tingey, S.V. and Williams, R.W. (2008) Artificial trans-acting siRNAs confer consistent and effective gene silencing. *Plant Physiology* 147, 543–551.

Devers, E.A., Teply, J., Reinert, A., Gaude, N. and Krajinski, F. (2013) An endogenous artificial microRNA system for unraveling the function of root endosymbioses related genes in *Medicago truncatula*. *BMC Plant Biology* 13, 82.

Eamens, A.L., Agius, C., Smith, N.A., Waterhouse, P.M. and Wang, M.B. (2011) Efficient silencing of endogenous microRNAs using artificial microRNAs in *Arabidopsis thaliana*. *Molecular Plant* 4, 157–170.

Engler, C., Kandzia, R. and Marillonnet, S. (2008) A one pot, one step, precision cloning method with high throughput capability. *PLoS ONE* 3, e3647.

Fahim, M., Millar, A.A., Wood, C.C. and Larkin, P.J. (2012) Resistance to *Wheat streak mosaic virus* generated by expression of an artificial polycistronic microRNA in wheat. *Plant Biotechnology Journal* 10, 150–163.

Fahlgren, N., Hill, S.T., Carrington, J.C. and Carbonell, A. (2016) P-SAMS: a web site for plant artificial microRNA and synthetic trans-acting small interfering RNA design. *Bioinformatics* 32, 157–158.

Felippes, F.F. and Weigel, D. (2009) Triggering the formation of tasiRNAs in *Arabidopsis thaliana*: the role of microRNA miR173. *EMBO Reports* 10, 264–270.

Felippes, F.F., Wang, J.W. and Weigel, D. (2012) MIGS: miRNA-induced gene silencing. *The Plant Journal* 70, 541–547.

Fernandez, A.I., Viron, N., Alhagdow, M., Karimi, M., Jones, M., Amsellem, Z., Sicard, A., Czerednik, A., Angenent, G., Grierson, D., May, S., Seymour, G., Eshed, Y., Lemaire-Chamley, M., Rothan, C. and Hilson, P. (2009) Flexible tools for gene expression and silencing in tomato. *Plant Physiology* 151, 1729–1740.

Flores-Sandoval, E., Dierschke, T., Fisher, T.J. and Bowman, J.L. (2015) Efficient and inducible use of artificial MicroRNAs in *Marchantia polymorpha*. *Plant and Cell Physiology* 57, 281–290.

Guo, Y., Han, Y., Ma, J., Wang, H., Sang, X. and Li, M. (2014) Undesired small RNAs originate from an artificial microRNA precursor in transgenic petunia (*Petunia hybrida*). *PLoS ONE* 9, e98783.

Han, Y., Zhang, B., Qin, X., Li, M. and Guo, Y. (2015) Investigation of a miRNA-induced gene silencing technique in petunia reveals alterations in miR173 precursor processing and the accumulation of secondary siRNAs from endogenous genes. *PLoS ONE* 10, e0144909.

Hu, J., Deng, X., Shao, N., Wang, G. and Huang, K. (2014) Rapid construction and screening of artificial microRNA systems in *Chlamydomonas reinhardtii*. *The Plant Journal* 79, 1052–1064.

Jelly, N.S., Schellenbaum, P., Walter, B. and Maillot, P. (2012) Transient expression of artificial microRNAs targeting grapevine fanleaf virus and evidence for RNA silencing in grapevine somatic embryos. *Transgenic Research* 21, 1319–1327.

Kamthan, A., Chaudhuri, A., Kamthan, M. and Datta, A. (2015) Small RNAs in plants: recent development and application for crop improvement. *Frontiers in Plant Science* 6, 208.

Kaur, S. and Spillane, C. (2015) Reduction in carotenoid levels in the marine diatom *Phaeodactylum tricornutum* by artificial microRNAs targeted against the endogenous phytoene synthase gene. *Marine Biotechnology* 17, 1–7.

Khraiwesh, B., Ossowski, S., Weigel, D., Reski, R. and Frank, W. (2008) Specific gene silencing by artificial MicroRNAs in *Physcomitrella patens*: an alternative to targeted gene knockouts. *Plant Physiology* 148, 684–693.

Kis, A., Tholt, G., Ivanics, M., Varallyay, E., Jenes, B. and Havelda, Z. (2015) Polycistronic artificial miRNA-mediated resistance to *wheat dwarf virus* in barley is highly efficient at low temperature. *Molecular Plant Pathology* 17, 427–437.

Li, J.F., Chung, H.S., Niu, Y., Bush, J., McCormack, M. and Sheen, J. (2013) Comprehensive protein-based artificial microRNA screens for effective gene silencing in plants. *The Plant Cell* 25, 1507–1522.

Li, J.F., Zhang, D. and Sheen, J. (2014) Epitope-tagged protein-based artificial miRNA screens for optimized gene silencing in plants. *Nature Protocols* 9, 939–949.

Liang, G., He, H., Li, Y. and Yu, D. (2012) A new strategy for construction of artificial miRNA vectors in *Arabidopsis*. *Planta* 235, 1421–1429.

Liu, C., Zhang, L., Sun, J., Luo, Y., Wang, M.B., Fan, Y.L. and Wang, L. (2010) A simple artificial microRNA vector based on ath-miR169d precursor from *Arabidopsis*. *Molecular Biology Reports* 37, 903–909.

Melito, S., Heuberger, A.L., Cook, D., Diers, B.W., MacGuidwin, A.E. and Bent, A.F. (2010) A nematode demographics assay in transgenic roots reveals no significant impacts of the Rhg1 locus LRR-Kinase on soybean cyst nematode resistance. *BMC Plant Biology* 10, 104.

Mitter, N., Zhai, Y., Bai, A.X., Chua, K., Eid, S., Constantin, M., Mitchell, R. and Pappu, H.R. (2016) Evaluation and identification of candidate genes for artificial microRNA-mediated resistance to *tomato spotted wilt virus*. *Virus Research* 211, 151–158.

Molnar, A., Bassett, A., Thuenemann, E., Schwach, F., Karkare, S., Ossowski, S., Weigel, D. and Baulcombe, D. (2009) Highly specific gene silencing by artificial microRNAs in the unicellular alga *Chlamydomonas reinhardtii*. *The Plant Journal* 58, 165–174.

Montgomery, T.A., Howell, M.D., Cuperus, J.T., Li, D., Hansen, J.E., Alexander, A.L., Chapman, E.J., Fahlgren, N., Allen, E. and Carrington, J.C. (2008a) Specificity of ARGONAUTE7-miR390 interaction and dual functionality in TAS3 trans-acting siRNA formation. *Cell* 133, 128–141.

Montgomery, T.A., Yoo, S.J., Fahlgren, N., Gilbert, S.D., Howell, M.D., Sullivan, C.M., Alexander, A., Nguyen, G., Allen, E., Ahn, J.H. and Carrington, J.C. (2008b) AGO1-miR173 complex initiates phased siRNA formation in plants. *Proceedings of the National Academy of Sciences of the USA* 105, 20055–20062.

Niu, Q.W., Lin, S.S., Reyes, J.L., Chen, K.C., Wu, H.W., Yeh, S.D. and Chua, N.H. (2006) Expression of artificial microRNAs in transgenic *Arabidopsis thaliana* confers virus resistance. *Nature Biotechnology* 24, 1420–1428.

Ossowski, S., Schwab, R. and Weigel, D. (2008) Gene silencing in plants using artificial microRNAs and other small RNAs. *The Plant Journal* 53, 674–690.

Qu, J., Ye, J. and Fang, R. (2007) Artificial microRNA-mediated virus resistance in plants. *Journal of Virology* 81, 6690–6699.

Roumi, V., Afsharifar, A., Saldarelli, P., Niazi, A., Martelli, G.P. and Izadpanah, K. (2012) Transient expression of artificial microRNAs confers resistance to *grapevine virus A* in *Nicotiana benthamiana*. *Journal of Plant Pathology* 94, 643–649.

Schwab, R., Ossowski, S., Riester, M., Warthmann, N. and Weigel, D. (2006) Highly specific gene silencing by artificial microRNAs in *Arabidopsis*. *The Plant Cell* 18, 1121–1133.

Schwab, R., Ossowski, S., Warthmann, N. and Weigel, D. (2010) Directed gene silencing with artificial microRNAs. *Methods in Molecular Biology* 592, 71–88.

Shafrin, F., Das, S.S., Sanan-Mishra, N. and Khan, H. (2015) Artificial miRNA-mediated down-regulation of two monolignoid biosynthetic genes (C3H and F5H) cause reduction in lignin content in jute. *Plant Molecular Biology* 89, 511–527.

Shi, R., Yang, C., Lu, S., Sederoff, R. and Chiang, V.L. (2010) Specific down-regulation of PAL genes by artificial microRNAs in *Populus trichocarpa*. *Planta* 232, 1281–1288.

Tiwari, M., Sharma, D. and Trivedi, P.K. (2014) Artificial microRNA mediated gene silencing in plants: progress and perspectives. *Plant Molecular Biology* 86, 1–18.

Toppino, L., Kooiker, M., Lindner, M., Dreni, L., Rotino, G.L. and Kater, M.M. (2011) Reversible male sterility in eggplant (*Solanum melongena* L.) by artificial microRNA-mediated silencing of general transcription factor genes. *Plant Biotechnology Journal* 9, 684–692.

Verdonk, J. and Sulllivan, M. (2013) Artificial microRNA (amiRNA) induced gene silencing in alfalfa (*Medicago sativa*). *Botany* 91, 117–122.

Vu, T.V., Choudhury, N.R. and Mukherjee, S.K. (2013) Transgenic tomato plants expressing artificial microRNAs for silencing the pre-coat and coat proteins of a begomovirus, *Tomato leaf curl New Delhi virus*, show tolerance to virus infection. *Virus Research* 172, 35–45.

Wang, X., Yang, Y., Yu, C., Zhou, J., Cheng, Y., Yan, C. and Chen, J. (2010) A highly efficient method for construction of rice artificial MicroRNA vectors. *Molecular Biotechnology* 46, 211–218.

Wang, X., Yang, Y., Zhou, J., Yu, C., Cheng, Y., Yan, C. and Chen, J. (2012) Two-step method for constructing *Arabidopsis* artificial microRNA vectors. *Biotechnology Letters* 34, 1343–1349.

Warthmann, N., Chen, H., Ossowski, S., Weigel, D. and Herve, P. (2008) Highly specific gene silencing by artificial miRNAs in rice. *PLoS ONE* 3, e1829.

Wyrzykowska, A., Pieczynski, M. and Szweykowska-Kulinska, Z. (2016) Construction of artificial miRNAs to prevent drought stress in *Solanum tuberosum*. *Methods in Molecular Biology* 1398, 271–290.

Yamada, T., Mori, Y., Yasue, K., Maruyama, N., Kitamura, K. and Abe, J. (2014) Knockdown of the 7S globulin subunits shifts distribution of nitrogen sources to the residual protein fraction in transgenic soybean seeds. *Plant Cell Reports* 33, 1963–1976.

Yan, H., Deng, X., Cao, Y., Huang, J., Ma, L. and Zhao, B. (2011a) A novel approach for the construction of plant amiRNA expression vectors. *Journal of Biotechnology* 151, 9–14.

Yan, H., Zhong, X., Jiang, S., Zhai, C. and Ma, L. (2011b) Improved method for constructing plant amiRNA vectors with blue-white screening and MAGIC. *Biotechnology Letters* 33, 1683–1688.

Yan, F., Lu, Y., Wu, G., Peng, J., Zheng, H., Lin, L. and Chen, J. (2012) A simplified method for constructing artificial microRNAs based on the osa-MIR528 precursor. *Journal of Biotechnology* 160, 146–150.

Yu, S. and Pilot, G. (2014) Testing the efficiency of plant artificial microRNAs by transient expression in *Nicotiana benthamiana* reveals additional action at the translational level. *Frontiers in Plant Science* 5, 622.

Zhang, X., Li, H., Zhang, J., Zhang, C., Gong, P., Ziaf, K., Xiao, F. and Ye, Z. (2011) Expression of artificial microRNAs in tomato confers efficient and stable virus resistance in a cell-autonomous manner. *Transgenic Research* 20, 569–581.

Zhang, Z.J. (2014) Artificial trans-acting small interfering RNA: a tool for plant biology study and crop improvements. *Planta* 239, 1139–1146.

Zhao, T., Wang, W., Bai, X. and Qi, Y. (2009) Gene silencing by artificial microRNAs in *Chlamydomonas*. *The Plant Journal* 58, 157–164.

Zhou, J., Yu, F., Chen, B., Wang, X., Yang, Y., Cheng, Y., Yan, C. and Chen, J. (2013) Universal vectors for constructing artificial microRNAs in plants. *Biotechnology Letters* 35, 1127–1133.

# 7 Application of RNA Silencing in Improving Plant Traits for Industrial Use

**Sumit Ghosh[1], Mohammad Irfan[2] and Asis Datta[2,*]**

*[1]CSIR-Central Institute of Medicinal and Aromatic Plants, Lucknow, India; [2]National Institute of Plant Genome Research, New Delhi, India*

## 7.1 Introduction

Besides being a major food source to the world's ever-growing population, plants also serve as sustainable resources for petrochemicals, therapeutics, pharmaceuticals and fibres (Fesenko and Edwards, 2014). Optimization and exploitation of the huge renewable plant biomass for the efficient and cost-effective production of industrial materials are present and future challenges. Biotechnology and synthetic biology approaches are used in engineering biochemical pathways to produce as much photosynthate as possible for fuel and chemicals (Mussgnug *et al*., 2007; Kalluri *et al*., 2014).The aim of the engineering pathway is either to over-express the gene(s) whose expression has a positive effect on product accumulation or to suppress expression of the gene(s) that negatively affect product accumulation. It is possible that both approaches may be combined to obtain superior product yield.

The discovery of RNA silencing mechanisms has transformed research into gene function and has also allowed novel traits to be developed in plants, including some for industrial applications (Fig. 7.1; Baulcombe, 2004; Eamens *et al*., 2008; Kamthan *et al*., 2015). To silence the expression of an undesirable gene, RNA silencing may be preferable over direct gene mutation as it leads to a dominant trait. This helps in straightforward screening of the desirable phenotypes. Moreover, RNA silencing is useful in tissue, organ and time-specific silencing of the genes that may play essential roles in plant life and, therefore, direct gene mutation may not be effective. This chapter summarizes the implication of RNA silencing in improving plant traits for industrial applications.

## 7.2 Biomass Recalcitrance in Industrial Processing

Plant biomass is used for commercial production of textile, paper, chemicals and energy. For industrial purposes, a range of plant species including fibre crops

* Corresponding author e-mail: asisdatta@hotmail.com

 

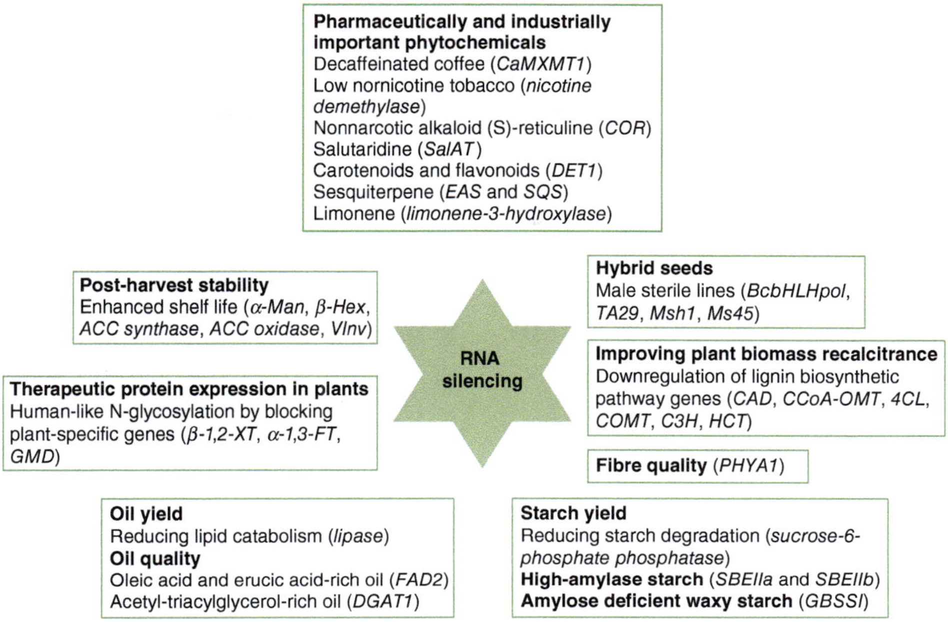


**Fig. 7.1.** Examples of improved industrial traits through RNA silencing of plant genes.

(cotton, jute and flax), woody plants (poplar, eucalyptus and conifers), food crops (maize, sorghum, sugarbeet, sugarcane, barley, cassava and potato) and grasses are exploited for their biomass to generate fuels and fibres. From the perspective of sustainable and environmentally friendly use, plant-based products, which are biodegradable and renewable, are advantageous over synthetic products made of plastic and glass. Moreover the scarcity of fossil fuels, and concerns over global climate change, have led us to search for alternative fuels from sustainable and renewable resources (Rubin, 2008). Renewable sources of energy from plant biomass are alternatives to fossil fuels and can reverse environmental degradation through carbon sequestration (Trumbo *et al.*, 2015). Ethanol produced from fermentation of plant starch and sugars can be used as fuel in its pure form or as an additive to petroleum. At present most bioethanol is produced from sugar and starch-rich food crops through fermentation. However, industrial use of storage carbohydrates from edible parts of plants raises concerns about food security (Karp and Shields, 2008). Production of bioethanol from non-food plants or waste plant biomass is, therefore, an important strategy. The plant cell wall contains a significant proportion of lignocellulosic matter which can be bioconverted to ethanol, and its fibres can be extracted for use in the paper and textile industries. The major bottleneck for these processes is the recalcitrance of lignocellulosic matter to microbial and enzymatic digestion during their industrial processing (Himmel *et al.*, 2007). This affects the saccharification process (and eventual fuel yields) and the pulping process for the extraction of plant fibres. Therefore, non-environmentally friendly thermochemical pretreatments are required to make the lignocellulosic biomass accessible to microbial and enzymatic digestion. Considerable efforts have been made to improve the digestibility of plant biomass by modifying the structure and composition of plant cell walls and by manipulating the genes involved in the biosynthesis and organization of cell wall components.

Plant cell walls are composed of a heterogeneous polysaccharide network of cellulose, hemicellulose, pectin and lignin. The primary cell walls are mainly composed of cellulose, hemicelluloses and pectin, while secondary cell walls are largely made of cellulose, hemicelluloses and lignin. Lignin deposition in secondary cell walls provides strength to the cell walls and maintains the integrity of the plant. The content, composition and structure of lignin influence the physical and chemical properties of plant cell walls, and so have been targeted for genetic manipulation to improve biomass recalcitrance to industrial processing (Chen and Dixon, 2007). Lignin is made of three monomers, *para*-hydroxyphenyl (H), guaiacyl (G) and syringyl (S) that are synthesized through the phenylpropanoid pathway (Fig. 7.2). The content of these three in lignin polymers differs among species and also among cell types of a plant species. The biosynthesis of lignin has been studied extensively, and most of the enzymes directly involved in lignin biosynthesis have been identified and characterized (Weng and Chapple, 2010).

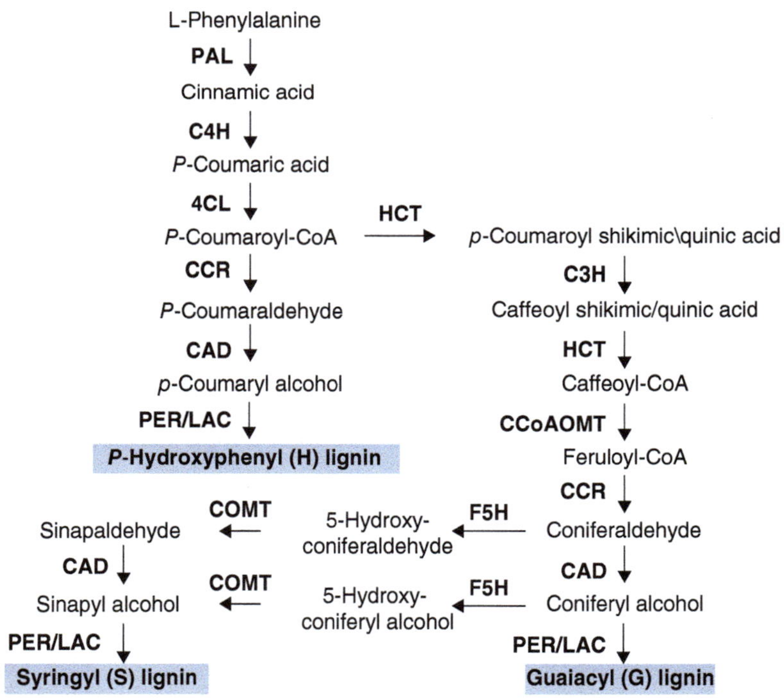


**Fig. 7.2.** Lignin biosynthesis pathway. 4CL, 4-coumarate:CoA ligase; C3H, p-coumarate 3-hydroxylase; C4H, Cinnamate 4-hydroxylase; CAD, Cinnamyl alcohol dehydrogenase; CCoAOMT, Caffeoyl-CoA O-methyltransferase; CCR, Cinnamoyl-CoA reductase; COMT, Caffeic acid O-methyltransferase; F5H(CAld5H), Ferulate 5-hydroxylase (coniferaldehyde 5-hydroxylase); HCT, p-Hydroxycinnamoyl-CoA:quinate/shikimate p-hydroxycinnamoyltransferase; PAL, Phenylalanine ammonia-lyase; PER/LAC, Peroxidase/Laccase.

Because lignin limits the use of plant biomass for use in the production of energy, chemicals and fibres, efforts have been made to modify lignin content and composition in plants. Recent work on RNA silencing of maize cinnamyl alcohol dehydrogenase (CAD), a key enzyme for monolignol biosynthesis, resulted in modified lignin composition in the plant stem. Although total lignin content in the maize stem was not altered, a reduction in the S/G ratio was observed, and cell walls accumulated higher levels of cellulose and arabinoxylans. These RNAi plants produced higher amounts of biomass that were more digestible and resulted in higher levels of ethanol production compared to wild-type plants (Fornale *et al.*, 2012). Improved saccharification efficiency was also reported in switchgrass by RNA silencing of CAD (Fu *et al.*, 2011a; Saathoff *et al.*, 2011). Antisense-mediated downregulation of CAD expression in poplar has been shown to improve efficiency of industrial kraft pulping by allowing easier delignification and using smaller amounts of chemicals during pretreatment, while yielding more high-quality pulp from the wood (Pilate *et al.*, 2002). Recently, RNA silencing of CAD in flax has been shown to reduce lignin content and increase the cellulose and pectin content of the cell wall. This genetic manipulation had positive effect on fibre processing and improved tensile strength was also recorded (Preisner *et al.*, 2014). Similarly, RNA silencing of maize caffeoyl-CoA O-methyltransferase (CCoA-OMT), another enzyme of the lignin pathway, led to reduced lignin and higher cellulose contents, and a higher S/G ratio than in wild-type plants (Fig. 7.3; Li *et al.*, 2013). In switchgrass, RNA silencing of 4-coumarate:coenzyme A ligase (4CL), one of the key enzymes involved in the monolignol biosynthetic pathway, has been shown to improve fermentable sugar yields for biofuel production by reducing lignin content and decreasing guaiacyl unit composition (Xu *et al.*, 2011). In sugarcane, recalcitrance of the lignocellulosic biomass was improved by silencing the expression of caffeic acid O-methyltransferase (COMT), reducing the lignin content and S/G ratio, and increasing saccharification efficiency (Jung *et al.*, 2013). Similarly, downregulation of COMT in switchgrass resulted in reduced biomass recalcitrance and improved ethanol yield (Fu *et al.*, 2011b). In an another RNA silencing effort, downregulation of a poplar glycosyltransferase (GAUT12) resulted in reduced cell wall xylan and pectin content during wood formation, reduced recalcitrance to enzymatic saccharification, more easily extractable cell walls and increased plant growth (Biswal *et al.*, 2015). Similarly, silencing of *p*-coumaroyl-CoA 3′-hydroxylase (C3H), which catalyses 3′-hydroxylation of *p*-coumaroyl shikimate and *p*-coumaroyl quinate, resulted in reduced lignin levels and altered the composition of monolignols (Coleman *et al.*, 2008). Downregulation of C3H has been shown to enhance the digestibility and pulping performance of transgenic alfalfa (Ralph *et al.*, 2006). In conifer *Pinus radiata*, RNA silencing of *p*-hydroxycinnamoyl-CoA:shikimate *p*-hydroxycinnamoyltransferase (HCT) affected lignin content, monolignol composition and interunit linkage distribution. In the monolignol biosynthesis pathway, HCT catalyses the transfer of coumarate from *p*-coumaroyl-CoA to shikimate (Fig. 7.2). Silencing of HCT in *P. radiata* resulted in a reduction of up to 42% in lignin content and an increase in *p*-hydroxyphenyl units (Wagner *et al.*, 2007). Moreover, expression of microRNAs such as miR156, which are components of the RNA silencing pathway, also led to enhanced plant biomass yield and reduced biomass recalcitrance (Trumbo *et al.*, 2015). All this work provides evidence for the application of RNA silencing in improving plant biomass recalcitrance in industrial processing.

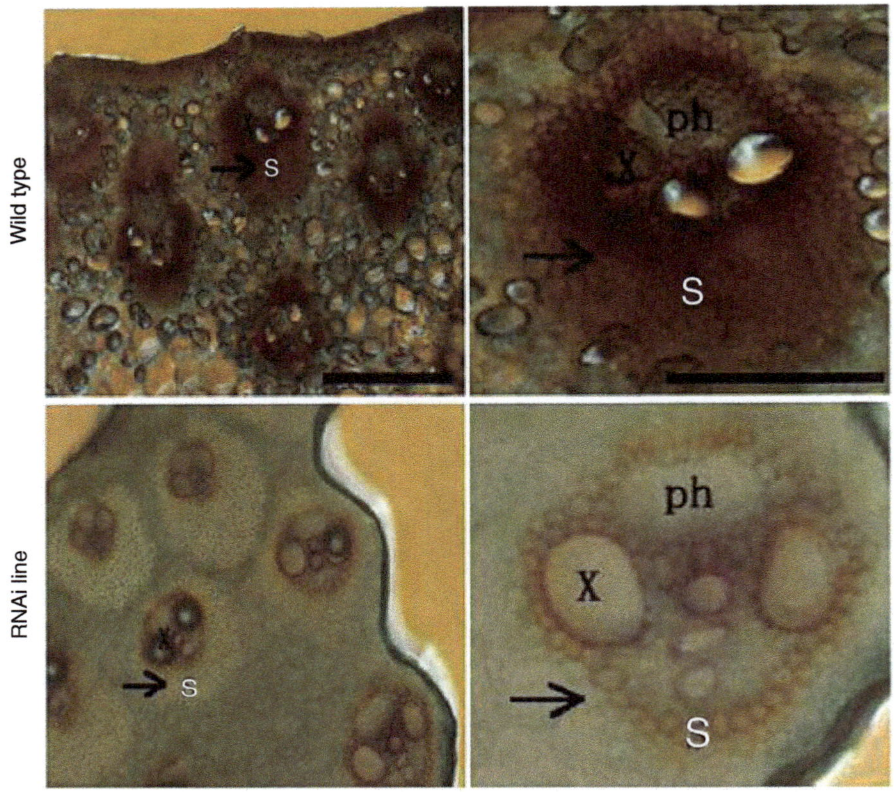


Fig. 7.3. RNA silencing of CCoAOMT reduced lignin production in maize straw. Lignin deposition was visualized following staining with Wiesner reagent. Arrows indicate lignin staining. X, xylem; S, sclerenchyma; ph, phloem. (Reproduced, with permission, from Li *et al.*, 2013.)

## 7.3 Oil Yield and Quality

Plant fats and oils are used in industry for the production of biodiesel, lubricants and oil-based solvents, and for the formulation of paints, inks, resins, varnishes and plasticizers. Considerable research efforts have been made to identify and manipulate the genes of fatty acid biosynthetic pathways to improve the yield and quality of plant oils. The oil-rich jatropha seed is an important feedstock for the production of biodiesel, which is a mono-alkyl ester of fatty acid. Edible oils of rapeseed, soybean, coconut, sunflowers and palm are also exploited for biodiesel. Photosynthetic algae such as *Chlamydomonas reinhardtii* are being studied because the ease with which they can be cultivated and their oil extracted makes them of interest for oil production. In these organisms research focus is shifted towards biosynthesis and degradation pathways of fatty acids. Some strategies to engineer fatty acid biosynthesis towards more suitable lipid profiles are overexpression of the biosynthetic pathway enzymes; downregulation of the catabolic pathway; and modification of saturation profiles and chain length of fatty acids, as a higher proportion of short-chain fatty acids is more desirable in biodiesel production. Downregulation of the genes involved in lipid catabolism, such as long-chain acyl-CoA synthetase and 3-ketoacyl-CoA thiolase, may be effective in increasing oil yield in plants. However, spatio-temporal silencing of gene expression is essential because lipid catabolism is essential for proper seedling development (Germain *et al.*, 2001; Fulda *et al.*, 2004). In contrast, downregulation of lipid catabolism by targeted knockdown of a multifunctional lipase/phospholipase/acyltransferase increased lipid yields without affecting growth in the diatom *Thalassiosira pseudonana* (Trentacoste *et al.*, 2013). In jatropha and rapeseed, RNA silencing of sugar-dependent 1 triacylglycerol lipases (SDP1) was effective in increasing seed oil yield up to 8–30% higher than in the normal cultivar (Kelly *et al.*, 2013; Kim *et al.*, 2014).

Optimum use of light energy for photosynthesis is central to increase biomass yield, and hence production of biofuels and chemicals from photosynthetic organisms. Under high light levels, light-harvesting complex (LHC) proteins facilitate the dissipation of a large proportion of light energy (~80–95% photons) as heat or fluorescence. To hinder this energy loss process, RNAi technology has been effective in downregulating the expression of all 20 genes encoding for LHCI, LHCII, CP26 and CP29 of *C. reinhardtii* (Mussgnug *et al.*, 2007). This strategy resulted in reduced fluorescence losses, increased photosynthetic quantum yield, increased resistance to photoinhibition and faster algal growth rate at elevated light levels. Moreover, by RNA silencing of a major lipid droplet protein (MLDP), the size of lipid droplets (the storage body of triacylglycerols) was increased in *C. reinhardtii* (Moellering and Benning, 2010).

Oils rich in oleic acid are beneficial to health and are also suitable for producing industrial lubricants because of their high oxidative stability. Oleic acid is also used for production of azelaic acid and ricinoleic acid, which have significant industrial applications (Chen *et al.*, 2015). RNA silencing of FAD2 genes, which encode oleoyl-phosphatidylcholine ω6-desaturase enzymes, has been successful in increasing oleic acid content (up to 80%) in rapeseed, Indian mustard (*B. juncea*), soybean, cottonseed and flax (Liu *et al.*, 2000; Stoutjesdijk *et al.*, 2000; Liu *et al.*, 2002; Mroczka *et al.*, 2010; Wagner *et al.*, 2011; Chen *et al.*, 2015). FAD2 catalyses the metabolism of oleic acid (18:1(Δ9)) to linoleic acid (18:2(Δ9,Δ12)) by inserting a double bond at the Δ12 position. Erucic acid (22:1(Δ9)) is an important feedstock for manufacturing plastics, nylon13-13 and high-temperature lubricants. Following overexpression of lysophosphatidic acid acyltransferase (LdLPAAT) and fatty acid elongase (BnFAE1), and silencing of FAD2, ultra-high erucic acid oil has been developed in the industrial oil crop *Crambe abyssinica* (Li *et al.*, 2012). Acetyl-triacylglycerols (acetyl-TAGs), which have reduced viscosity and low melting temperatures compared to conventional triacylglycerols, confer benefits as food emulsifiers, lubricants, plasticizers and 'drop-in' fuels for diesel engines. RNAi suppression of type 1 diacylglycerol acyltransferase (DGAT1) in combination with overexpression of diacylglycerol acetyltransferase (EaDAcT) increased acetyl-TAG levels to up to 85 mol% in field-grown transgenic *Camelina sativa*, without affecting seed size and viability (Liu *et al.*, 2015). Seed-specific RNA silencing of β-ketoacyl-acyl carrier protein synthase II (KASII) was effective in *Arabidopsis* to produce more palm-like tropical oil with many industrial applications (Pidkowich *et al.*, 2007). These examples demonstrate the utility of RNA silencing in improving oil yield and quality for industrial use.

## 7.4 Therapeutic Proteins

Therapeutic proteins including monoclonal antibodies, enzymes, antigens and human growth factors, have become very important in the diagnosis and treatment of life-threatening non-communicable diseases and infectious diseases, and are also used as vaccines. Several expression systems such as mammalian cells, yeasts, *Escherichia coli* and plants are exploited for the industrial-scale production of therapeutic proteins. However, plants have emerged as a convenient, safe and economically attractive expression system because of low production costs and flexible scalability (Lai *et al.*, 2010; Nagels *et al.*, 2012). Remarkable progress has been made in producing-plant made pharmaceuticals over the last two decades. The production of vaccines for hepatitis B, rabies, cholera, diarrhoea and influenza; antibodies; human enzymes; and therapeutic proteins for the treatment of diabetes, cystic fibrosis, Fabry's disease, thrombosis, Gaucher's disease, etc., have already been documented in plants (Ma *et al.*, 1995; Vezina *et al.*, 2009; Nagels *et al.*, 2012). Many plant species, such as tobacco, tomato, potato, *Arabidopsis*, lettuce, spinach, maize, duckweed, carrot, rice, sunflower, barley and the moss *Physcomitrella patens* have been tested for these purposes. However, correct post-translation modification of the expressed proteins is absolutely necessary for proper folding, maintaining stability, activity and also therapeutic efficiency. Significant success has already been achieved in this respect by manipulating the plant N-glycan processing pathway for incorporation of human-like N-glycan structures in plant-expressed therapeutic proteins. Because identical enzymes are shared in the early steps of N-glycan processing, oligomannose-type N-glycan structures are identical in humans and plants. However, in the later stages, N-glycan processing differs in these eukaryotes. This leads to different complex-type N-glycan structures in plants and humans (Fig. 7.4). Plant-specific complex N-glycans have α(1,3)-fucose and β(1,2)-xylose residues, which are absent in humans (Strasser *et al.*, 2012). These specific modifications in plant N-glycoproteins have been shown to elicit allergic and immune responses in humans and so are not wanted in proteins for therapeutic use (Bardor *et al.*, 2003). Moreover, unlike in humans, plant complex-type N-glycans lack a terminally sialylated Lewis X epitope (i.e. Neu5Acα(2-3)Galβ(1-4)[Fucα(1-3)]GlcNAc) (Fig. 7.4). Instead, plant complex-type N-glycans have a Lewis A epitope (i.e. Galβ(1-3)[Fucα(1-4)]GlcNAc) (Nagels *et al.*, 2012). These differences between plant and human N-glycan processing represent a major drawback to the therapeutic use of the plant-expressed proteins and, therefore, the humanization of the plant N-glycosylation pathway remains a prerequisite.

To inhibit the incorporation of plant-specific β-1,2-xylose and core α-1,3-fucose residues in the plant-expressed therapeutic proteins, the expression of *β*1, 2-xylosyltransferase (β-1,2-XT) and α-1,3-fucosyltransferase (α-1,3-FT) that are responsible for these N-glycan modifications have been downregulated in plants, following an RNA silencing approach. In alfalfa, a reduced level of plant-specific glycoepitopes was achieved following RNA silencing of β-1,2-XT and α-1,3-FT, along with overexpression of human β-1,4-galactosyltransferase, which completely inhibited accumulation of the plant-specific Lewis A glycoepitope on complex *N*-glycans (Sourrouille *et al.*, 2008). RNA silencing of β-1,2-XT and α-1,3-FT along with overexpression of

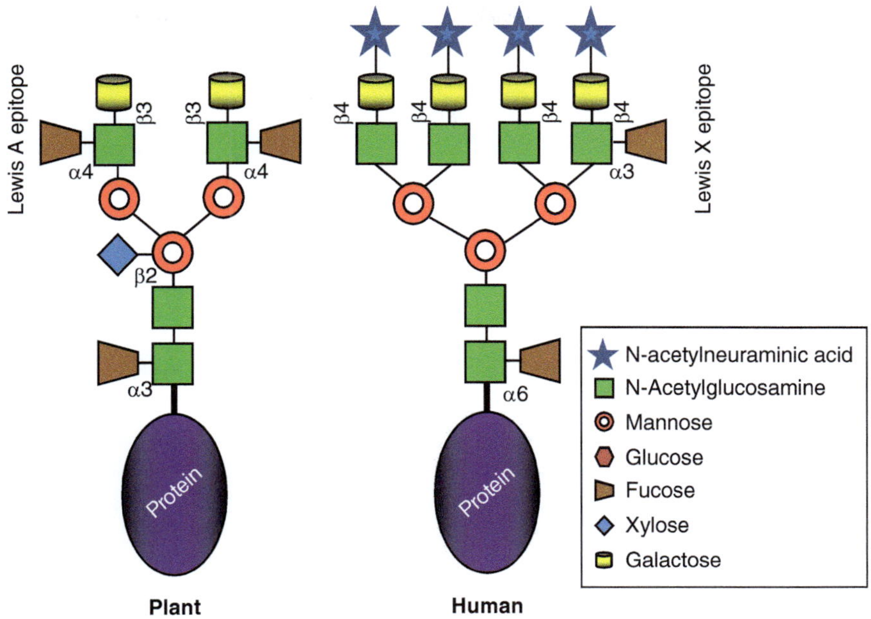


**Fig. 7.4.** Complex-type N-glycans found in plants and humans.

mammalian glycosyltransferases has also been effective in tobacco (Strasser *et al.*, 2008; Castilho *et al.*, 2010; Nagels *et al.*, 2011), duckweed (Cox *et al.*, 2006) and rice (Shin *et al.*, 2011) in producing human-like *N*-glycan structures. By following a different strategy in tobacco, plant-specific core α-1,3-fucose and α-1,4-fucose residues in the Lewis A epitopes were eliminated by repressing the expression of the guanosine5′-diphosphate (GDP)-D-mannose 4,6-dehydratase (GMD) gene. That gene is associated with the biosynthesis of GDP-L-fucose, which serves as a substrate for α-1,3-FT and α-1,4-FT enzymes (Matsuo and Matsumura, 2011). RNA silencing might be effective in indirectly increasing the yield of therapeutic proteins in seeds by repression of storage protein accumulation (Shigemitsu *et al.*, 2012).

## 7.5 Phytochemicals of Pharmaceutical and Industrial Importance

Plants produce thousands of low-molecular-weight organic compounds in ways that are specific to developmental age, tissue, organ and species. Most of these phytochemicals do not play a direct role in plant growth and development, but they facilitate plant adaptation to the challenging conditions encountered under biotic and abiotic stresses and are, therefore, referred to as secondary or specialized metabolites (Gershenzon and Dudareva, 2007). Many of these secondary metabolites have been exploited to treat several diseases in traditional systems of medicine, and several phytochemicals are also used as flavours, fragrances and pesticides. Several phytochemicals are used directly as drugs, and many are leading models for the development of semi-synthetic and synthetic drugs (De Luca *et al.*, 2012). About 25% of modern medicines originated from a natural source. Most importantly, developing countries – which comprise 80% of the world's population – primarily depend on herbal medicines for their basic health care (De Luca *et al.*, 2012). During the year 2000, the total global market for herbal drugs was estimated to be about US$62 billion, and might grow US$5 trillion by the year 2050 (Leone *et al.*, 2007). A few examples of commercially important pharmaceuticals include the anti-carcinogenic taxol from *Taxus baccata*; vinca alkaloids from *Catharanthus roseus*; the antimalarial artemisinin from *Artemisia annua*; and the analgesics morphine and codeine from *Papaver somniferum*. RNA silencing technology has been useful, not only in characterizing phytochemical biosynthetic pathways, but also in manipulating them to develop desired chemotypes in medicinal and aromatic plants (Ogita *et al.*, 2003; Hileman *et al.*, 2005; Lewis *et al.*, 2008).

The use of decaffeinated coffee is recommended for caffeine-sensitive individuals because caffeine can cause palpitations, increased blood pressure and insomnia. Although coffee may be decaffeinated industrially, the process is costly and also affects the flavour of the coffee. RNA silencing of *theobromine synthase* (*CaMXMT1*), a biosynthetic pathway gene, was found to be effective in significant reduction (up to 70%) of the caffeine content in transgenic coffee plants (Ogita *et al.*, 2003). Similarly, RNA silencing of the biosynthetic pathway gene (*nicotine demethylase*) was useful in reducing the level of a secondary tobacco alkaloid, nornicotine, which serves as the precursor for the synthesis of a carcinogen (*N*-nitrosonornicotine) in tobacco during curing and processing (Lewis *et al.*, 2008). Allen and co-workers have shown the use of RNA silencing in modulating the alkaloid biosynthetic pathway in poppy (Allen *et al.*, 2004). By silencing the members of the codeinone reductase (COR), they have demonstrated the replacement of morphine with the non-narcotic alkaloid (S)-reticuline, a potentially valuable pathway intermediate. Metabolic engineering of morphinan alkaloids in poppy was also reported by silencing salutaridinol 7-*O*-acetyltransferase (SalAT), which resulted in accumulation of the alkaloid salutaridine at levels of up to 23% of the total alkaloid. However, this alkaloid was below the detectable limit in the parental genotype (Allen *et al.*, 2008). Liu and co-workers reported a reduction of up to 65% in detrimental glucosinolate progoitrin and increased accumulation of the beneficial glucosinolate glucoraphanin (42.6 μmol $g^{-1}$ seed) in transgenic *Brassica napus* seeds following silencing of the GSL-ALK gene family (Liu *et al.*, 2012). In tomato fruits, the levels of health-promoting phytochemicals such as carotenoids and flavonoids were increased by RNAi-mediated fruit-specific silencing of a photomorphogenesis regulatory gene, *DE-ETIOLATED1* (Davuluri *et al.*, 2005).

RNA silencing of genes of the endogenous competing pathways has been useful in accumulating higher levels of precursors for conversion to desired phytochemicals. In *N. benthamiana*, silencing of the endogenous 5-*epi*-aristolochene synthase (*EAS*) and squalene synthase (*SQS*) has been effective in increasing the level of farnesyl pyrophosphate, the precursor for sesquiterpenes (Cankar *et al.*, 2015). Further, combined silencing of *EAS* and *SQS*, and overexpression of (+)-valencene synthase (*CnVS*) was found to be effective in producing increased levels of (+)-valencene, an aromatic sesquiterpene (Cankar *et al.*, 2015). Another example of secondary metabolic pathway engineering is the accumulation of limonene in peppermint (up to 80% of the essential oil compared to about 2% of the oil in wild-type plants) following cosuppression of limonene-3-hydroxylase (Mahmoud *et al.*, 2004).

## 7.6 Starch for Industrial Use

Starch obtained from potato, maize, wheat and cassava is one of the most preferred raw materials for use in several industries including paper and pulp, textiles, cosmetics, pharmaceuticals, biodegradable plastic film, construction and mining (Leone *et al.*, 2007). Starches are made of two types of carbohydrates, amylose and amylopectin. Amylose is linear, a polysaccharide of α-1,4-linked glucose residues that is occasionally (about one branch per 1000 glucose residues) branched with α-1,6-linkages. However, amylopectin is highly branched (about one branch per 25 α-1,4-linked glucose residues) with α-1,6-linkages. Amylose-deficient, or waxy starch containing high amylopectin, is preferable for textiles, corrugated packaging and adhesive industries. However, starch with elevated amylose concentrations (≥40%) is used for gums and sweets, and in the paper and pulp industries (Leone *et al.*, 2007). Moreover, high-amylose starch is considered to be resistant starch (RS) that is not digested in the small intestine, but is used by bacteria in the colon. RS works as dietary fibre, reducing calorie uptake, and so has a low glycaemic index and is beneficial for colon health.

The RNA silencing approach has been effective in improving starch quality and quantity in plants. The RNAi of sucrose-6-phosphate phosphatase, which catalyses the last step in the sucrose biosynthetic pathway by converting sucrose-6-phosphate to sucrose, results in 3–5 times more starch in tobacco (Chen *et al.*, 2005). Similarly, silencing of a β-amylase has been shown to affect the breakdown of starch into maltose, leading to starch accumulation in *Arabidopsis* (Kaplan and Guy, 2005). High-amylase starches were also developed by RNA silencing of the starch branching enzyme II isoforms (SBEIIa and SBEIIb) in common wheat (Regina *et al.*, 2006), durum wheat (Sestili *et al.*, 2010) and barley (Regina *et al.*, 2010). This approach was found to be effective in improving large bowel indices in rats, and has potential in the design of healthy starches for human consumptions (Regina *et al.*, 2006). Transgenic sweet potato and cassava with low-amylase (waxy) starch have been developed following RNAi-mediated downregulation of granule-bound starch synthase I (GBSSI), an enzyme involved in amylose synthesis in the amyloplasts (Otani *et al.*, 2007; Zhao *et al.*, 2011). Because waxy starch has improved paste clarity and stability, it is preferable for thickening fruit pies, improving smoothness and creaminess of canned food and dairy products and the freeze–thaw stability of frozen foods. In the adhesive industry, waxy starch is useful in making printer paper, concrete adhesives and textiles (Zhao *et al.*, 2011).

## 7.7 Post-harvest Stability

Food crops, besides being important components in the human diet, are also used as raw materials for the production of processed foods, animal feeds and for several other industrial purposes. The post-harvest decay process of food crops that perish quickly, such as fleshy fruits and vegetables, affects their shelf life and limits transportation and storage. This accounts for huge post-harvest losses which may reach 50% of the total produce (Meli *et al.*, 2010). Therefore, enhancement of fruit shelf life by slowing down the post-harvest process of decay is a target of genetic crop improvement efforts. Two N-glycan-processing enzymes (α-mannosidase (α-Man) and β-D-N-acetylhexosaminidase (β-Hex)), which express at high levels during the fruit softening process, were targeted for the genetic manipulation of fruits and vegetables through the RNAi approach. Downregulating *α-Man* and *β-Hex* expression resulted in enhanced fruit shelf life, due to the reduced rate of fruit softening, in both climacteric tomato (requires ethylene to complete ripening process) and non-climacteric capsicum (does not require ethylene to complete ripening process) (Fig. 7.5; Meli *et al.*, 2010; Ghosh *et al.*, 2011). The RNAi approach has also been effective in enhancing fruit shelf life by silencing the genes involved in the biosynthesis of plant hormones such as ethylene (1-aminocyclopropane-1-carboxylic acid (ACC) synthase and ACC oxidase) and abscisic acid (9-cis-epoxycarotenoid dioxygenase) which initiate and accelerate fruit ripening, and genes (polygalacturonase, expasin) involved in fruit softening by degrading cell wall components (Lopez-Gomez *et al.*, 2009; Sun *et al.*, 2012; Gupta *et al.*, 2013; Molina-Hidalgo *et al.*, 2013).

Cold storage of fruits and vegetables delays post-harvest decay; however, it also results in accumulation of undesired metabolites. For example potato tubers – one of the most important food sources – are stored in the cold to avoid sprouting, prevent losses from post-harvest decay and to maintain a steady supply to consumers and the processing industries. However, cold storage leads to the accumulation of reducing sugars, and subsequent high-temperature processing results in dark-coloured, bitter-tasting products which consumers are unwilling to accept. During processing, cold-stored potato products also accumulate elevated amounts of acrylamide, a neurotoxin and potential carcinogen. It has been demonstrated that RNA silencing of the vacuolar acid invertase gene (*VInv*) can prevent reducing sugar accumulation during cold storage and so prevents cold-induced sweetening, improves processing quality and lowers acrylamide formation (Bhaskar *et al.*, 2010; Ye *et al.*, 2010).

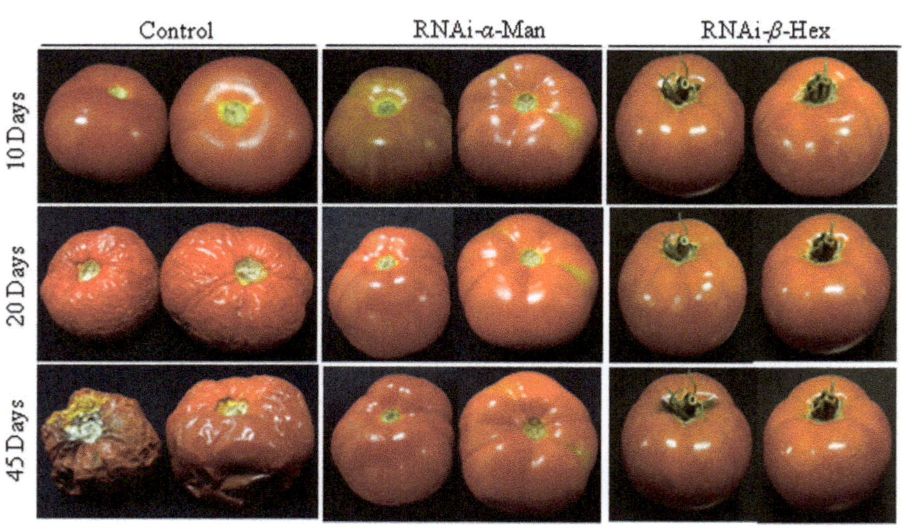


Fig. 7.5. Enhancement of tomato shelf life by silencing of α-Man or β-Hex. (Reproduced, with permission, from Meli *et al.*, 2010.)

## 7.8 Inducing Male Sterility

Heterozygous F1 hybrids of inbred lines often show superior agronomic performance and produce higher yields as compared to the parental cultivars. A few examples of improved yield include hybrid rice, which yields 15–20% more than most inbred cultivars (Cheng *et al.*, 2007), hybrid mustard and maize (DuPont, 2015a, b). Globally, hybrid maize is grown in about 65% of the area under maize cultivation. However, in the USA about 95% of the maize acreage is planted to hybrids (Hochholdinger and Hoecker, 2007). The phenomenon of heterosis (hybrid vigour) is being exploited by the seed industry for the production of superior hybrids (Nawaz-ul-Rehman *et al.*, 2007; Birchler *et al.*, 2010). Hybrid seed production demands that the male sterile line is used as a female parent. Male sterility prevents seed set following selfing and only outcrossing leads to seed sets, i.e. hybrid seed production. The application of RNA silencing technology in inducing male sterility has been demonstrated in several plants. Male sterility was reported in cabbage by silencing a basic helix–loop–helix transcription factor, *BcbHLHpol* (Liu *et al.*, 2014); in tobacco by silencing a male-specific gene (*TA29*) expressed exclusively in anthers at the time of microspore development (Nawaz-ul-Rehman *et al.*, 2007); in petunia by silencing a Tapetum-Specific Zinc Finger Gene *TAZ1* (Kapoor *et al.*, 2002); and in tomato and tobacco by silencing *Msh1* which is involved in the suppression of illegitimate recombination in plant mitochondria (Sandhu *et al.*, 2007). The transcriptional gene silencing approach was also exploited in maize, for inducing male sterility, by targeting the promoter of the genes (such as *Ms45*) that determine male fertility (Cigan *et al.*, 2014).

## 7.9 Other Industrial Traits

RNA silencing has also been useful in several other industries. These include improved fibre quality in jute (Abdurakhmonov *et al.*, 2014); reduced levels of tear-inducing lachrymatory factor in onion (Eady *et al.*, 2008); increased levels of health-promoting gamma-aminobutyric acid (GABA) in rice kernels (Shimajiri *et al.*, 2013) and low-gliadin in common wheat (Gil-Humanes *et al.*, 2014). Abdurakhmonov and co-workers have demonstrated that silencing of a photoreceptor (phytochrome *PHYA1*) improves fibre quality and agromonic performance of cotton (Fig. 7.6). To reduce the level of tear-inducing lachrymatory factor in onion, silencing of the lachrymatory factor synthase gene using RNAi was carried out. This genetic manipulation also resulted in increased levels of health-promoting secondary sulfur metabolites in onion (Eady *et al.*, 2008). To increase GABA levels in rice kernels, Shimajiri and co-workers silenced GABA-transaminase and overexpressed glutamate decarboxylase in a seed-specific manner (Shimajiri *et al.*, 2013). Ingestion of wheat gliadin and related proteins from other food crops can cause coeliac disease, a food-sensitive enteropathy. To combat this problem, low-gliadin common wheat has been developed by RNA silencing of gliadin genes (Gil-Humanes *et al.*, 2014). Gliadin-silenced wheat flours have shown increased stability and better tolerance to over-mixing (Gil-Humanes *et al.*, 2014).

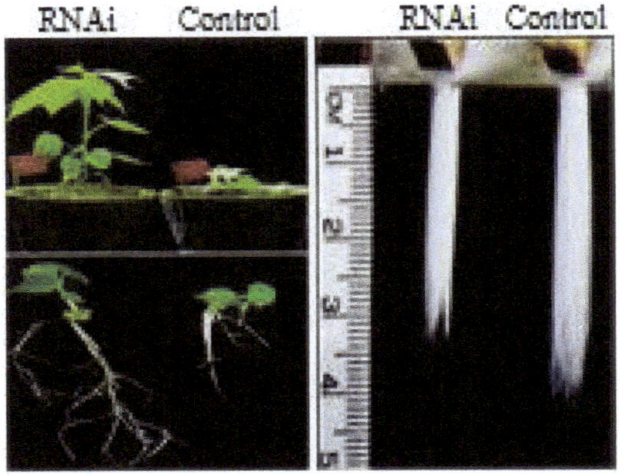


Fig. 7.6. RNA silencing of PHYA1 in cotton improved fibre quality and agronomic traits. (Reproduced, with permission, from Abdurakhmonov *et al.*, 2014.)

## 7.10 Conclusion

A detailed understanding of RNA silencing mechanisms has revolutionized our studies of functional genomics and facilitated the introduction of desired traits in plants, including industrial traits. Plants developed after RNA silencing of genes that act on cellular organization or metabolism are emerging as convenient and renewable raw materials for industries producing fuels, oils, pharmaceuticals, fibres, papers, seeds and processed foods. Several genetically modified plants, which may not have use in human consumption, have nevertheless been developed for specific industrial applications and await commercial-scale cultivation. Many more industrial crops are under different stages of development by industry and research organizations.

## References

Abdurakhmonov, I.Y., Buriev, Z.T., Saha, S., Jenkins, J.N., Abdukarimov, A. and Pepper, A.E. (2014) Phytochrome RNAi enhances major fibre quality and agronomic traits of the cotton *Gossypium hirsutum* L. *Nature Communications* 5, 3062.

Allen, R.S., Millgate, A.G., Chitty, J.A., Thisleton, J., Miller, J.A., Fist, A.J., Gerlach, W.L. and Larkin, P.J. (2004) RNAi-mediated replacement of morphine with the nonnarcotic alkaloid reticuline in opium poppy. *Nature Biotechnology* 22(12), 1559–1566.

Allen, R.S., Miller, J.A., Chitty, J.A., Fist, A.J., Gerlach, W.L. and Larkin, P.J. (2008) Metabolic engineering of morphinan alkaloids by over-expression and RNAi suppression of salutaridinol 7-O-acetyltransferase in Opium poppy. *Plant Biotechnology Journal* 6, 22–30.

Bardor, M., Faveeuw, C., Fitchette, A.C., Gilbert, D., Galas, L., Trottein, F., Faye, L. and Lerouge, P. (2003) Immunoreactivity in mammals of two typical plant glyco-epitopes, core alpha(1,3)-fucose and core xylose. *Glycobiology* 13, 427–434.

Baulcombe D. (2004) RNA silencing in plants. *Nature* 431, 356–363.

Bhaskar, P.B., Wu, L., Busse, J.S., Whitty, B.R., Hamernik, A.J., Jansky, S.H., Buell, C.R., Bethke, P.C. and Jiang, J. (2010) Suppression of the vacuolar invertase gene prevents cold-induced sweetening in potato. *Plant Physiology* 154, 939–948.

Birchler, J.A., Yao, H., Chudalayandi, S., Vaiman, D. and Veitia, R.A. (2010) Heterosis. *Plant Cell* 22, 2105–2112.

Biswal, A.K., Hao, Z., Pattathil, S., Yang, X., Winkeler, K., Collins, C., Mohanty, S.S., Richardson, E.A., Gelineo-Albersheim, I., Hunt, K., Ryno, D., Sykes, R.W., Turner, G.B., Ziebell, A., Gjersing, E., Lukowitz, W., Davis, M.F., Decker, S.R., Hahn, M.G. and Mohnen, D. (2015) Downregulation of GAUT12 in *Populus deltoides* by RNA silencing results in reduced recalcitrance, increased growth and reduced xylan and pectin in a woody biofuel feedstock. *Biotechnology for Biofuels* 8(41). DOI: 10.1186/s13068-015-0218-y

Cankar, K., Jongedijk, E., Klompmaker, M., Majdic, T., Mumm, R., Bouwmeester, H., Bosch, D. and Beekwilder, J. (2015) (+)-Valencene production in *Nicotiana benthamiana* is increased by down-regulation of competing pathways. *Biotechnology Journal* 10, 180–189.

Castilho, A., Strasser, R., Stadlmann, J., Grass, J., Jez, J., Gattinger, P., Kunert, R., Quendler, H., Pabst, M., Leonard, R., Altmann, F. and Steinkellner, H. (2010) *In planta* protein sialylation through overexpression of the respective mammalian pathway. *The Journal of Biological Chemistry* 285, 15923–15930.

Chen, F. and Dixon, R.A. (2007) Lignin modification improves fermentable sugar yields for biofuel production. *Nature Biotechnology* 25, 759–761.

Chen, S., Hajirezaei, M., Peisker, M., Tschiersch, H., Sonnewald, U. and Börnke, F. (2005) Decreased sucrose-6-phosphate phosphatase level in transgenic tobacco inhibits photosynthesis, alters carbohydrate partitioning, and reduces growth. *Planta* 221, 479–492.

Chen, Y., Zhou, X.R., Zhang, Z.J., Dribnenki, P., Singh, S. and Green, A. (2015) Development of high oleic oil crop platform in flax through RNAi-mediated multiple FAD2 gene silencing. *Plant Cell Reports* 34, 643–653.

Cheng, S.H., Zhuang, J.Y., Fan, Y.Y., Du, J.H. and Cao, L.Y. (2007) Progress in research and development on hybrid rice: a super-domesticate in China. *Annals of Botany* 100, 959–966.

Cigan, A.M., Haug-Collet, K. and Clapp, J. (2014) Transcriptional silencing of heterologous anther promoters in maize: a genetic method to replace detasseling for seed production. *Plant Reproduction* 27, 109–120.

Coleman, H.D., Park, J.Y., Nair, R., Chapple, C. and Mansfield, S.D. (2008) RNAi-mediated suppression of p-coumaroyl-CoA 3′-hydroxylase in hybrid poplar impacts lignin deposition and soluble secondary metabolism. *Proceedings of the National Academy of Sciences of the United States of America* 105, 4501–4506.

Cox, K.M., Sterling, J.D., Regan, J.T., Gasdaska, J.R., Frantz, K.K., Peele, C.G., Black, A., Passmore, D., Moldovan-Loomis, C., Srinivasan, M., Cuison, S., Cardarelli, P.M. and Dickey, L.F. (2006) Glycan optimization of a human monoclonal antibody in the aquatic plant *Lemna minor*. *Nature Biotechnology* 24, 1591–1597.

Davuluri, G.R., van Tuinen, A., Fraser, P.D., Manfredonia, A., Newman, R., Burgess, D., Brummell, D.A., King S.R., Palys, J., Uhlig, J., Bramley, P.M., Pennings, H.M. and Bowler, C. (2005) Fruit-specific RNAi-mediated suppression of DET1 enhances carotenoid and flavonoid content in tomatoes. *Nature Biotechnology* 23, 890–895.

De Luca, V., Salim, V., Atsumi, S.M. and Yu, F. (2012) Mining the biodiversity of plants: a revolution in the making. *Science* 336, 1658–1661.

DuPont (2015a) DuPont™ Pioneer® hybrid mustard seed for high yield advantage. Available at: http://www.dupont.co.in/corporate-functions/our-approach/global-challenges/shunya/articles/pioneer-hybrid-mustard-seeds.html (accessed 21 December 2016).

DuPont (2015b) Pioneer ® Brand Corn Hybrids. Available at: https://www.pioneer.com/web/site/india/Corn/ (accessed 21 December 2016).

Eady, C.C., Kamoi, T., Kato, M., Porter, N.G., Davis, S., Shaw, M., Kamoi, A. and Imai, S. (2008) Silencing onion lachrymatory factor synthase causes a significant change in the sulfur secondary metabolite profile. *Plant Physiology* 147, 2096–2106.

Eamens, A., Wang, M.B., Smith, N.A. and Waterhouse, P.M. (2008) RNA silencing in plants: yesterday, today, and tomorrow. *Plant Physiology* 147, 456–468.

Fesenko, E. and Edwards, R. (2014) Plant synthetic biology: a new platform for industrial biotechnology. *Journal of Experimental Botany* 65, 1927–1937.

Fornale, S., Capellades, M., Encina, A., Wang, K., Irar, S., Lapierre, C., Ruel, K., Joseleau, J.P., Berenguer, J., Puigdomènech, P., Rigau, J. and Caparros-Ruiz, D. (2012) Altered lignin biosynthesis improves cellulosic bioethanol production in transgenic maize plants down-regulated for cinnamyl alcohol dehydrogenase. *Molecular Plant* 5, 817–830.

Fu, C.X., Xiao, X.R., Xi, Y.J., Ge, Y.X., Chen, F., Bouton, J., Dixon, R.A. and Wang, Z.Y. (2011a) Downregulation of cinnamyl alcohol dehydrogenase (CAD) leads to improved saccharification efficiency in switchgrass. *Bioenergy Research* 4, 153–164.

Fu, C., Mielenz, J.R., Xiao, X., Ge, Y., Hamilton, C.Y., Rodriguez, M. Jr, Chen, F., Foston, M., Ragauskas, A., Bouton, J., Dixon, R.A. and Wang, Z.Y. (2011b) Genetic manipulation of lignin reduces recalcitrance and improves ethanol production from switchgrass. *Proceedings of the National Academy of Sciences of the United States of America* 108, 3803–3808.

Fulda, M., Schnurr, J., Abbadi, A. and Heinz, E. (2004) Peroxisomal acyl-CoA synthetase activity is essential for seedling development in *Arabidopsis thaliana*. *Plant Cell* 16, 394–405.

Germain, V., Rylott, E.L., Larson, T.R., Sherson, S.M., Bechtold, N., Carde, J.P., Bryce, J.H., Graham, I.A. and Smith, S.M. (2001) Requirement for 3-ketoacyl-CoA thiolase-2 in peroxisome development, fatty acid beta oxidation and breakdown of triacylglycerol in lipid bodies of *Arabidopsis* seedlings. *The Plant Journal* 28, 1–12.

Gershenzon, J. and Dudareva, N. (2007) The function of terpene natural products in the natural world. *Nature Chemical Biology* 3, 408–414.

Ghosh, S., Meli, V.S., Kumar, A., Thakur, A., Chakraborty, N., Chakraborty, S. and Datta, A. (2011) The N-glycan processing enzymes alpha-mannosidase and beta-D-N-acetylhexosaminidase are involved in ripening-associated softening in the non-climacteric fruits of capsicum. *Journal of Experimental Botany* 62, 571–582.

Gil-Humanes, J., Piston, F., Barro, F. and Rosell, C.M. (2014) The shutdown of celiac disease-related gliadin epitopes in bread wheat by RNAi provides flours with increased stability and better tolerance to over-mixing. *PLoS ONE* 9, e91931.

Gupta, A., Pal, R.K. and Rajam, M.V. (2013) Delayed ripening and improved fruit processing quality in tomato by RNAi-mediated silencing of three homologsof 1-aminopropane-1-carboxylate synthase gene. *Journal of Plant Physiology* 170, 987–995.

Hileman, L.C., Drea, S., Martino, G., Litt, A. and Irish, V.F. (2005) Virus-induced gene silencing is an effective tool for assaying gene function in the basal eudicot species *Papaver somniferum* (opium poppy). *The Plant Journal* 44, 334–341.

Himmel, M.E., Ding, S.Y., Johnson, D.K., Adney, W.S., Nimlos, M.R., Brady, J.W. and Foust, T.D. (2007) Biomass recalcitrance: engineering plants and enzymes for biofuels production. *Science* 315, 804–807.

Hochholdinger, F. and Hoecker, N. (2007) Towards the molecular basis of heterosis. *Trends in Plant Science* 12, 427–432.

Jung, J.H., Vermerris, W., Gallo, M., Fedenko, J.R., Erickson, J.E. and Altpeter, F. (2013) RNA interference suppression of lignin biosynthesis increases fermentable sugar yields for biofuel production from field-grown sugarcane. *Plant Biotechnology Journal* 11, 709–716.

Kalluri, U.C., Yin, H., Yang, X. and Davison, B.H. (2014) Systems and synthetic biology approaches to alter plant cell walls and reduce biomass recalcitrance. *Plant Biotechnology Journal* 12, 1207–1216.

Kamthan, A., Chaudhuri, A., Kamthan, M. and Datta, A. (2015) Small RNAs in plants: recent development and application for crop improvement. *Frontiers in Plant Science* 6, 208.

Kaplan, F. and Guy, C.L. (2005) RNA interference of *Arabidopsis* beta-amylase8 prevents maltose accumulation upon cold shock and increases sensitivity of PSII photochemical efficiency to freezing stress. *The Plant Journal* 44, 730–743.

Kapoor, S., Kobayashi, A. and Takatsuji, H. (2002) Silencing of the tapetum-specific zinc finger gene TAZ1 causes premature degeneration of tapetum and pollen abortion in petunia. *Plant Cell* 14, 2353–2367.

Karp, A. and Shields, I. (2008) Bioenergy from plants and the sustainable yield challenge. *New Phytologist* 178, 473–485.

Kelly, A.A., Shaw, E., Powers, S.J., Kurup, S. and Eastmond, P.J. (2013) Suppression of the SUGAR-DEPENDENT1 triacylglycerol lipase family during seed development enhances oil yield in oilseed rape (*Brassica napus* L.). *Plant Biotechnology Journal* 11(3), 355–361.

Kim, M.J., Yang, S.W., Mao, H.Z., Veena, S.P., Yin, J.L. and Chua, N.H. (2014) Gene silencing of sugar-dependent 1 (JcSDP1), encoding a patatin-domain triacylglycerol lipase, enhances seedoil accumulation in *Jatropha curcas*. *Biotechnology for Biofuels* 7, 36.

Lai, H., Engle, M., Fuchs, A., Keller, T., Johnson, S., Gorlatov, S., Diamond, M.S. and Chen, Q. (2010) Monoclonal antibody produced in plants efficiently treats *West Nile virus* infection in mice. *Proceedings of the National Academy of Sciences of the United States of America* 107, 2419–2424.

Leone, A., Grillo, S., Monti, L. and Cardi, T. (2007) Molecular tailoring and boosting of bioactive secondary metabolites in medicinal plants. In: Ranalli, P. (ed.) *Improvement of Crop Plants for Industrial End Uses*. Springer, Dordrecht, The Netherlands, pp. 471–507.

Lewis, R.S., Jack, A.M., Morris, J.W., Robert, V.J., Gavilano, L.B., Siminszky, B., Bush, L.P., Hayes, A.J. and Dewey, R.E. (2008) RNA interference (RNAi)-induced suppression of nicotine demethylase activity reduces levels of a key carcinogen in cured tobacco leaves. *Plant Biotechnology Journal* 6, 346–354.

Li, X., van Loo, E.N., Gruber, J., Fan, J., Guan, R., Frentzen, M., Stymne, S. and Zhu, L.H. (2012) Development of ultra-high erucic acid oil in the industrial oil crop *Crambe abyssinica*. *Plant Biotechnology Journal* 10, 862–870.

Li, X., Chen, W., Zhao, Y., Xiang, Y., Jiang, H., Zhu, S. and Cheng, B. (2013) Downregulation of caffeoyl-CoA O-methyltransferase (CCoAOMT) by RNA interference leads to reduced lignin production in maize straw. *Genetics and Molecular Biology* 36, 540–546.

Liu, Z.I., Hirani, A.H., McVetty P.B., Daayf, F., Quiros, C.F. and Li, G. (2012) Reducing progoitrin and enriching glucoraphanin in *Brassica napus* seeds through silencing of the GSL-ALK gene family. *Plant Molecular Biology* 79(1–2),179–189.

Liu, J., Rice, A., McGlew, K., Shaw, V., Park, H., Clemente, T., Pollard, M., Ohlrogge, J. and Durrett, T.P. (2015) Metabolic engineering of oilseed crops to produce high levels of novel acetyl glyceride oils with reduced viscosity, freezing point and calorific value. *Plant Biotechnology Journal* 13, 858–865.

Liu, Q., Singh, S. and Green, A. (2000) Genetic modification of cotton seed oil using inverted-repeat gene-silencing techniques. *Biochemical Society Transactions* 28, 927–929.

Liu, Q., Singh, S.P. and Green, A.G. (2002) High-stearic and high oleic cottonseed oils produced by hairpin RNA-mediated post-transcriptional gene silencing. *Plant Physiology* 129, 1732–1743.

Liu, T., Li, Y., Zhang, C., Duan, W., Huang, F. and Hou, X. (2014) Basic helix-loop-helix transcription factor BcbHLHpol functions as a positive regulator of pollen development in non-heading Chinese cabbage. *Functional and Integrative Genomics* 14, 731–739.

Lopez-Gomez, R., Cabrera-Ponce, J.L., Saucedo-Arias, L.J., Carreto-Montoya, L., Villanueva-Arce, R., Díaz-Perez, J.C., Gómez-Lim, M.A. and Herrera-Estrella, L. (2009) Ripening in papaya fruit is altered by ACC oxidase cosuppression. *Transgenic Research* 18, 89–97.

Ma, J.K., Hiatt, A., Hein, M., Vine, N.D., Wang, F., Stabila, P., van Dolleweerd, C., Mostov, K. and Lehner, T. (1995) Generation and assembly of secretory antibodies in plants. *Science* 268, 716–719.

Mahmoud, S.S., Williams, M. and Croteau, R. (2004) Cosuppression of limonene-3-hydroxylase in peppermint promotes accumulation of limonene in the essential oil. *Phytochemistry* 65, 547–554.

Matsuo, K. and Matsumura, T. (2011) Deletion of fucose residues in plant N-glycans by repression of the GDP-mannose 4,6-dehydratase gene using virus-induced gene silencing and RNA interference. *Plant Biotechnology Journal* 9, 264–281.

Meli, V.S., Ghosh, S., Prabha, T.N., Chakraborty, N., Chakraborty, S. and Datta, A. (2010) Enhancement of fruit shelf life by suppressing *N*-glycan processing enzymes. *Proceedings of the National Academy of Sciences of the United States of America* 107, 2413–2418.

Moellering, E.R. and Benning, C. (2010) RNA interference silencing of a major lipid droplet protein affects lipid droplet size in *Chlamydomonas reinhardtii*. *Eukaryotic Cell* 9, 97–106.

Molina-Hidalgo, F.J., Franco, A.R., Villatoro, C., Medina-Puche, L., Mercado, J.A., Hidalgo, M.A., Monfort, A., Caballero, J.L., Muñoz-Blanco, J. and Blanco-Portales, R. (2013) The strawberry (*Fragaria x ananassa*) fruit-specific rhamnogalacturonate lyase 1 (*FaRGLyase1*) gene encodes an enzyme involved in the degradation of cell-wall middle lamellae. *Journal of Experimental Botany* 64, 1471–1483.

Mroczka, A., Roberts, P.D., Fillatti, J.J., Wiggins, B.E., Ulmasov, T. and Voelker, T. (2010) An intron sense suppression construct targeting soybean FAD2-1 requires a double-stranded RNA-producing inverted repeat T-DNA insert. *Plant Physiology* 153, 882–891.

Mussgnug, J.H., Thomas-Hall, S., Rupprecht, J., Foo, A., Klassen, V., McDowall, A., Schenk, P.M., Kruse, O. and Hankamer, B. (2007) Engineering photosynthetic light capture: impacts on improved solar energy to biomass conversion. *Plant Biotechnology Journal* 5, 802–814.

Nagels, B., Van Damme, E.J., Pabst, M., Callewaert, N. and Weterings, K. (2011) Production of complex multiantennary N-glycans in *Nicotiana benthamiana* plants. *Plant Physiology* 155, 1103–1112.

Nagels, B., Weterings, K., Callewaert, N. and Van Damme, E.J.M. (2012) Production of plant made pharmaceuticals: from plant host to functional protein. *Critical Reviews in Plant Sciences* 31, 148–180.

Nawaz-ul-Rehman, M.S., Mansoor, S., Khan, A.A., Zafar, Y. and Briddon, R.W. (2007) RNAi-mediated male sterility of tobacco by silencing TA29. *Molecular Biotechnology* 36, 159–165.

Ogita, S., Uefuji, H., Yamaguchi, Y., Koizumi, N. and Sano, H. (2003) Producing decaffeinated coffee plants. *Nature* 423, 823.

Otani, M., Hamada, T., Katayama, K., Kitahara, K., Kim, S.H., Takahata, Y., Suganuma, T. and Shimada, T. (2007) Inhibition of the gene expression for granule-bound starch synthase I by RNA interference in sweet potato plants. *Plant Cell Reports* 26, 1801–1807.

Pidkowich, M.S., Nguyen, H.T., Heilmann, I., Ischebeck, T. and Shanklin, J. (2007) Modulating seed beta-ketoacyl-acyl carrier protein synthase II level converts the composition of a temperate seed oil to that of a palm-like tropical oil. *Proceedings of the National Academy of Sciences of the United States of America* 104, 4742–4747.

Pilate, G., Guiney, E., Holt, K., Petit-Conil, M., Lapierre, C., Leplé, J.C., Pollet, B., Mila, I., Webster, E.A., Marstorp, H.G., Hopkins, D.W., Jouanin, L., Boerjan, W., Schuch, W., Cornu, D. and Halpin, C. (2002) Field and pulping performances of transgenic trees with altered lignification. *Nature Biotechnology* 20, 607–612.

Preisner, M., Kulma, A., Zebrowski, J., Dymińska, L., Hanuza, J., Arendt, M., Starzycki, M. and Szopa, J. (2014) Manipulating cinnamyl alcohol dehydrogenase (CAD) expression in flax affects fibre composition and properties. *BMC Plant Biology* 14, 50.

Ralph, J., Akiyama, T., Kim, H., Lu, F., Schatz, P.F., Marita, J.M., Ralph, S.A., Reddy, M.S., Chen, F. and Dixon, R.A. (2006) Effects of coumarate 3-hydroxylase down-regulation on lignin structure. *The Journal of Biological Chemistry* 281, 8843–8853.

Regina, A., Bird, A., Topping, D., Bowden, S., Freeman, J., Barsby, T., Kosar-Hashemi, B., Li, Z.Y., Rahman, S. and Morell, M. (2006) High-amylose wheat generated by RNA interference improves indices of large-bowel health in rats. *Proceedings of the National Academy of Sciences of the United States of America* 103, 3546–3551.

Regina, A., Kosar-Hashemi, B., Ling, S., Li, Z., Rahman, S. and Morell, M. (2010) Control of starch branching in barley defined through differential RNAi suppression of starch branching enzyme IIa and IIb. *Journal of Experimental Botany* 61, 1469–1482.

Rubin, E.M. (2008) Genomics of cellulosic biofuels. *Nature* 454, 841–845.

Saathoff, A.J., Sarath, G., Chow, E.K., Dien, B.S. and Tobias, C.M. (2011) Downregulation of cinnamyl-alcohol dehydrogenase in switchgrass by RNA silencing results in enhanced glucose release after cellulase treatment. *PLoS ONE* 6, e16416.

Sandhu, A.P., Abdelnoor, R.V. and Mackenzie, S.A. (2007) Transgenic induction of mitochondrial rearrangements for cytoplasmic male sterility in crop plants. *Proceedings of the National Academy of Sciences of the United States of America* 104, 1766–1770.

Sestili, F., Janni, M., Doherty, A., Botticella, E., D'Ovidio, R., Masci, S., Jones, H.D. and Lafiandra, D. (2010) Increasing the amylose content of durum wheat through silencing of the SBEIIa genes. *BMC Plant Biology* 10, 144.

Shigemitsu, T., Ozaki, S., Saito, Y., Kuroda, M., Morita, S., Satoh, S. and Masumura, T. (2012) Production of human growth hormone in transgenic rice seeds: co-introduction of RNA interference cassette for suppressing the gene expression of endogenous storage proteins. *Plant Cell Reports* 31, 539–549.

Shimajiri, Y., Oonishi, T., Ozaki, K., Kainou, K. and Akama, K. (2013) Genetic manipulation of the γ-aminobutyric acid (GABA) shunt in rice: overexpression of truncated glutamate decarboxylase (GAD2) and knockdown of γ-aminobutyric acid transaminase (GABA-T) lead to sustained and high levels of GABA accumulation in rice kernels. *Plant Biotechnology Journal* 11, 594–604.

Shin, Y.J., Chong, Y.J., Yang, M.S. and Kwon, T.H. (2011) Production of recombinant human granulocyte macrophage-colony stimulating factor in rice cell suspension culture with a human-like N-glycan structure. *Plant Biotechnology Journal* 9, 1109–1119.

Sourrouille, C., Marquet-Blouin, E., D'Aoust, M.A., Kiefer-Meyer, M.C., Seveno, M., Pagny-Salehabadi, S., Bardor, M., Durambur, G., Lerouge, P., Vezina, L. and Gomord, V. (2008) Down-regulated expression of plant-specific glycoepitopes in alfalfa. *Plant Biotechnology Journal* 6, 702–721.

Stoutjesdijk, P.A., Hurlestone, C., Singh, S.P. and Green, A.G. (2000) High-oleic acid Australian *Brassica napus* and *B. juncea* varieties produced by co-suppression of endogenous Delta12-desaturases. *Biochemical Society Transactions* 28, 938–940.

Strasser, R. (2012) Challenges in O-glycan engineering of plants. *Frontiers in Plant Science* 3, 218.

Strasser, R., Stadlmann, J., Schähs, M., Stiegler, G., Quendler, H., Mach, L., Glössl, J., Weterings, K., Pabst, M. and Steinkellner, H. (2008) Generation of glyco-engineered *Nicotiana benthamiana* for the production of monoclonal antibodies with ahomogeneous human-like N-glycan structure. *Plant Biotechnology Journal* 6, 392–402.

Sun, L., Sun, Y., Zhang, M., Wang, L., Ren, J., Cui, M., Wang, Y., Ji, K., Li, P., Li, Q., Chen, P., Dai, S., Duan, C., Wu, Y. and Leng, P. (2012) Suppression of 9-cis-epoxycarotenoid dioxygenase, which encodes a key enzyme in abscisic acid biosynthesis, alters fruit texture in transgenic tomato. *Plant Physiology* 158, 283–298.

Trentacoste, E.M., Shrestha, R.P., Smith, S.R., Glé, C., Hartmann, A.C., Hildebrand, M. and Gerwick, W.H. (2013) Metabolic engineering of lipid catabolism increases microalgal lipid accumulation without compromising growth. *Proceedings of the National Academy of Sciences of the United States of America* 110, 19748–19753.

Trumbo, J.L., Zhang, B. and Stewart, C.N. Jr (2015) Manipulating microRNAs for improved biomass and biofuels from plant feedstocks. *Plant Biotechnology Journal* 13, 337–354.

Vezina, L.P., Faye, L., Lerouge, P., D'Aoust, M.A., Marquet-Blouin, E., Burel, C., Lavoie, P.O., Bardor, M. and Gomord, V. (2009) Transient co-expression for fast and high-yield production of antibodies with human-like N-glycans in plants. *Plant Biotechnology Journal* 7, 442–455.

Wagner, A., Ralph, J., Akiyama, T., Flint, H., Phillips, L., Torr, K., Nanayakkara, B. and Te Kiri, L. (2007) Exploring lignification in conifers by silencing hydroxycinnamoyl-CoA:shikimate hydroxycinnamoyltransferase in *Pinus radiata*. *Proceedings of the National Academy of Sciences of the United States of America* 104, 11856–11861.

Wagner, N., Mroczka, A., Roberts, P.D., Schreckengost, W. and Voelker, T. (2011) RNAi trigger fragment truncation attenuates soybean FAD2-1 transcript suppression and yields intermediate oil phenotypes. *Plant Biotechnology Journal* 9, 723–728.

Weng, J.K. and Chapple, C. (2010) The origin and evolution of lignin biosynthesis. *New Phytologist* 187, 273–285.

Xu, B., Escamilla-Treviño, L.L., Sathitsuksanoh, N., Shen, Z., Shen, H., Zhang, Y.H., Dixon, R.A. and Zhao, B. (2011) Silencing of 4-coumarate:coenzyme A ligase in switchgrass leads to reduced lignin content and improved fermentable sugar yields for biofuel production. *New Phytologist* 192, 611–625.

Ye, J., Shakya, R., Shrestha, P. and Rommens, C.M. (2010) Tuber-specific silencing of the acid invertase gene substantially lowers the acrylamide forming potential of potato. *Journal of Agricultural and Food Chemistry* 58, 12162–12167.

Zhao, S.S., Dufour, D., Sánchez, T., Ceballos, H. and Zhang, P. (2011) Development of waxy cassava with different biological and physico-chemical characteristics of starches for industrial applications. *Biotechnology and Bioengineering* 108, 1925–1935.

# 8 Increasing Nutritional Value by RNA Silencing

Elsa Pons[1,2] and Leandro Peña[1,2]*

[1]*Instituto de Biología Molecular y Celular de Plantas (Consejo Superior de Investigaciones Científicas Universidad Politécnica de Valencia), Valencia, Spain;* [2]*Fundo de Defensa da Citricultura, Araraquara, Sao Paulo, Brazil*

## 8.1 Introduction

The first generation of plant biotechnology products commercialized was crops focused largely on input agronomic traits whose value was usually opaque to consumers. Present and future trends are intended to continue improvement of agronomic traits; and, as value-added output characteristics such as improved nutrition and food functionality are of great interest to consumers, so biotechnology efforts go increasingly in those directions (Cressey, 2013).

Molecules in plants with dietary relevance can be grouped into four main sets: macronutrients (proteins, carbohydrates, lipids (oils), fibre), micronutrients (vitamins, minerals, functional metabolites), antinutrients (substances such as phytate that limit the bioavailability of nutrients) and allergens (intolerances and toxins); the levels of the first two would need to be increased and the latter two need to be reduced or removed. Many macro- and micronutrients are produced by plants; therefore, approaches aiming to improve the availability of these nutrients involve the adjustment of endogenous plant metabolism (Capell and Christou, 2004). Several strategies can be applied to achieve this. One of the most widely used is to reduce a metabolic bottleneck by producing more of a rate-limiting enzyme. There are other ways to increase the level of beneficial compounds such as: (i) directing biosynthetic pathways towards certain directions by repressing competitive pathways; or (ii) reducing feedback inhibition by increasing or generating a metabolic sink that would lead to a higher level of the required product (reviewed in Zhu *et al*., 2013). In this regard, RNAi emerged as the most suitable strategy to enhance the nutritional composition of important crop plants. Moreover, RNAi can also be used to diminish the amount of antinutrients in the plant tissues we consume by suppressing the expression of genes producing such compounds (Tang and Galili, 2004; Tang *et al*., 2007; Katoch and Thakur, 2013a, b). We do not intend here to do an extensive review of all publications in the area but rather to illustrate with relevant examples (summarized

* Corresponding author e-mail: lpenya@fundecitrus.com.br

in Table 8.1) how the nutritional and health value of plant foods can be improved by means of RNAi technology, resulting in products with either higher levels of nutrients or lower levels of antinutrients.

## 8.2 Modifying Macro-nutrient Content

### 8.2.1 Essential amino acids

Maize is a very important crop and its productivity in terms of yield per hectare is one of the best among cereals. However, cereal grain protein composition is a critical constraint restricting its nutritional value. Essential amino acids such as tryptophan and lysine are very low in maize endosperm due to the very high levels of major storage proteins known as zeins. In particular, the abundance of α-zein proteins that do not contain lysin is the main reason for the low nutritional value. This can result in malnutrition, a serious problem in populations where maize is the main protein source, and there is increasing demand for high-lysine maize in these regions. High-lysine maize has been generated by mutating the transcriptional activator of the α-zein gene Opaque 2 (O2) causing an increased level of non-zein proteins. These plants contain twice as much lysine as normal maize. Unfortunately, the *o2* mutant is not suitable for cultivation because it is more susceptible to infections, has a low yield and a soft chalky endosperm. Breeders at the Centro Internacional de Mejoramiento de Maíz y Trigo, Mexico (CYMMIT) have, therefore, introduced QTLs known as *o2* modifiers, to select vitreous kernels to overcome the inferior properties of *o2* maize lines. The resulting modified maize is called quality protein maize (QPM). As QPM is vitreous and lysine-rich, it is cultivated in many developing countries in Africa, South America and Asia. However, conversion of QPM into local germplasms to develop hybrid maize is laborious, because maintaining o2 homozygosity through visual scoring is problematic and heterozygosity may result in loss of the opaque phenotype in QPM. Moreover, once wild-type pollen blows onto the QPM ears, progeny immediately lose their high-lysine quality. To simplify QPM breeding, lysine-rich maize lines were created through RNAi where silencing of the α-zein mRNA is a dominant trait (reviewed in Wu and Messing, 2011; Wu and Messing, 2012; Wu and Messing, 2014). The first report of this applied an inverted-repeat constructed from the sequence of a 22-kDa zein gene, which led to a dominant opaque phenotype (Segal *et al.*, 2003). The inverted-repeat dramatically reduced the expression of 22-kDa zeins, while other zein proteins were unaffected, and the trait segregated following the Mendelian rules. The transgenic plants contained increased levels of lysine and lower levels of alanine, leucine and glutamine. This illustrates the success of the strategy for nutritional enhancement in crop plants. Later, seed-specific RNAi approaches have also been used successfully to generate dominant lysine-enriched maize by silencing both 19- and 22-kD α-zeins, which resulted in higher lysine and tryptophan contents than previously reported for kernels with reduced zein levels (Huang *et al.*, 2006). Another approach for increasing lysine content in maize is suppressing lysine catabolism. This has been achieved by RNAi-induced silencing

Table 8.1. Gene suppression approaches for nutritional enhancement of crop plants.

| Trait (nutrient target) | Crop | Strategy: Silenced gene(s) | Strategy: Mechanism | Results (phenotypic description/trait details) | Reference |
|---|---|---|---|---|---|
| Protein and amino acids | Maize | 22 kDa α-Zein | Constitutive RNAi | Modification of aa balance: ↑lysine, ↓(leucine, alanine and lutamine) + dominant opaque phenotype | Segal *et al.*, 2003 |
| | Maize | 22-kDa and 19-kDa α-Zeins | Endosperm-specific RNAi | Modification of aa balance: ↑(lysine, tryptophan), ↑free amino acids (asparagine, asparate and glutamate) + dominant opaque phenotype | Huang *et al.*, 2006 |
| | Maize | *LKR/SDH* | Endosperm-specific RNAi | ↑Lysine / producing a high-lysine kernel | Houmard *et al.*, 2007 |
| | Maize | *LKR/SDH* (+ *CordapA*-overexpression) | Constitutive RNAi | ↑Lysine / producing a higher-lysine kernel (4000 ppm versus 100 ppm) | Frizzi *et al.*, 2008 |
| Starch/dietary fibre | Maize | *SBE* | Constitutive RNAi | Modification of starch composition: ↑amylose | Chai *et al.*, 2005 |
| | Sweet potato | *SBEII* | Constitutive RNAi | Modification of starch composition: ↑amylose | Shimada *et al.*, 2006 |
| | Potato | *SBEI* and *SBEII* | Constitutive AS | Modification of starch composition: ↑amylose + ↑phosphorus content of starch | Schwall *et al.*, 2000 |
| | Potato | *SBEI* and *SBEII* | Tuber-specific AS | Modification of starch composition: ↑amylose | Hofvander *et al.*, 2004 |
| | Potato | *SBEI* and *SBEII* | RNAi driven by *GBSS* promoter from potato | Modification of starch composition: ↑amylose | Andersson *et al.*, 2006 |

*Continued*

**Table 8.1.** Continued.

| Trait (nutrient target) | Crop | Strategy: Silenced gene(s) | Strategy: Mechanism | Results (phenotypic description/trait details) | Reference |
|---|---|---|---|---|---|
| | Wheat | *SBEIIa* and *SBEIIb* | Endosperm-specific RNAi | Modification of starch composition: ↑amylose + the modified wheat improved bowel functions in rats | Regina *et al.*, 2006 |
| Oils and fatty acids/ essential fatty acids | Soybean | *FAD3* | Cosuppression | Improved cooking oils: ↑oleic acid | Kinney, 1994; Kinney and Knowlton, 1998 |
| | *Brassica napus* and *B. juncea* | *FAD2* | Cosuppression driven by napin seed-storage promoter (FP1) | Improved cooking oils: ↑oleic acid | Stoutjesdijk *et al.*, 2000 |
| | *B. juncea* (Indian mustard) | *FAD2* | AS driven by a truncated napin promoter | Improved cooking oils: ↑oleic acid, ↓α-linoleic acid | Sivaraman *et al.*, 2004 |
| | Cotton | *SAD-1* and *FAD2-1* | Lectin-promoter RNAi and AS | Improved cooking oils: ↑oleic and stearic acids, ↓palmitic acid | Liu *et al.*, 2002 |
| | Soybean | *FAD3* | Seed-specific RNAi | Improved cooking oils: ↓α-linolenic acid | Flores *et al.*, 2008 |
| | *B. napus* | *FAD2* and *FAE1* | Seed-specific RNAi | Improved cooking oils: ↑oleic acid, ↓PUFAs, erucic acid eliminated | Peng *et al.*, 2010 |
| | *Crambe abyssinica* | *FAD2* and *FAE1* | Seed-specific RNAi | Improved cooking oils: ↑oleic acid, ↓erucic acid | Li *et al.*, 2016 |
| | *B. napus* | *FAE1* | Seed-specific RNAi | Improved cooking oils: ↑oleic acid, ↓erucic acid, slightly ↑PUFAs | Shi *et al.*, 2015 |

| | | | | | |
|---|---|---|---|---|---|
| Phytonutrients/ functional metabolites | Tomato | *LYCB* | Fruit-specific AS | Enrichment in carotenoids: ↑lycopene | Rosati *et al.*, 2000 |
| | Potato | *LYCE* | Tuber-specific AS | Enrichment in carotenoids: ↑β-carotene | Diretto *et al.*, 2006 |
| | Canola | *LYCE* | Constitutive RNAi | Enrichment in carotenoids: ↑(β-carotene, zeaxanthin, violaxanthin, lutein) + ↓total fatty acids content with minor changes in the proportions in seeds | Yu *et al.*, 2008 |
| | Potato | *CHYB* | Tuber-specific AS | Enrichment in carotenoids: ↑(β-carotene, lutein), ↓zeaxanthin | Diretto *et al.*, 2007 |
| | Sweet orange | *CHYB* | Constitutive RNAi | Enrichment in carotenoids: ↑(β-carotene, α-carotene), ↓xanthophylls + increased *in vivo* antioxidant properties of modified oranges | Pons *et al.*, 2014 |
| | Potato | *ZEP* | Tuber-specific AS and co-suppression (S) | Enrichment in carotenoids: ↑(zeaxanthin, β-carotene, lutein) ↓violaxanthin + ↑α-tocopherol | Romer *et al.*, 2002 |
| | Tomato | *SlNCED1* | Fruit-specific RNAi | Enrichment in carotenoids: ↑(lycopene, β-carotene) + ↓ABA | Sun *et al.*, 2012 |
| | Tomato | *DET1* | Fruit-specific RNAi | Enrichment in carotenoids (mainly lycopene and β-carotene) and flavonoids + no pleiotropic effects on fruit yield or quality | Davuluri *et al.*, 2005 |

*Continued*

**Table 8.1.** Continued.

| Trait (nutrient target) | Crop | Strategy: Silenced gene(s) | Strategy: Mechanism | Results (phenotypic description/trait details) | Reference |
|---|---|---|---|---|---|
| | Canola | *DET1* | Constitutive and seed-specific RNAi | Enrichment in carotenoids: ↑(lutein, β-carotene, zeaxanthin) (greater increase with constitutive RNAi) + ↓sinapate esters (antinutrients) in seeds | Wei *et al.*, 2009 |
| | Rapeseed | *SGT1* (+ *VST1*-overexpression) | Seed-specific RNAi | ↑Resveratrol glucoside (piceid) in seeds | Hüsken *et al.*, 2005 |
| Antinutrients/ toxins | Cotton | δ-Cadinene synthase | Tissue-specific RNAi | ↓Toxic gossypol specifically in seeds | Sunilkumar *et al.*, 2006 |
| | Coffee bean | *CaMxMt 1* | Constitutive RNAi | Decaffeinated coffee plants: ↓caffeine content in the seeds | Ogita *et al.*, 2003; Ogita *et al.*, 2004 |
| | Wheat | All the groups of gliadins (α-, γ- and ω-) | Endosperm-specific RNAi | Producing reduced-gliadin wheat: ↓toxicity for coeliac disease patients + resulting bread with unchanged baking, sensory properties and overall acceptance; ↑lysine content | Gil-Humanes *et al.*, 2008; Gil-Humanes *et al.*, 2010; Gil-Humanes *et al.*, 2014 |
| | Groundnut | Five aflatoxin-synthesis genes | Constitutive RNAi | ↓Aflatoxin accumulation in seeds following inoculation with *Aspergillus flavus* | Arias *et al.*, 2015 |

| | | | | | |
|---|---|---|---|---|---|
| Allergens | Rice | 16 kDa allergen | Seed-specific AS | Reduced allergens: ↓14-16 kDa proteins in rice seeds | Tada *et al.*, 1996 |
| | Soybean | P34 (Gly m Bd 30 K) protein | RNAi driven by β-conglycinin promoter | Reduced allergens: ↓Gly m Bd 30 K protein<br>+ no pleiotropic effects on composition, development, phenotype, proteome | Herman *et al.*, 2003 |
| | Groundnut | Ara h 2 | Constitutive RNAi | Reduced allergens: ↓Ara h 2; alleviating groundnut allergenicity | Dodo *et al.*, 2008 |
| | Groundnut | Ara h 2 and Ara h 6 | Constitutive RNAi | Reduced allergens: ↓(Ara h 2 and Ara h 6); alleviating groundnut allergenicity | Chu *et al.*, 2008; |
| | Carrot | Dau c 1.01 and Dau c 1.02 (two PR10 isoforms) | Constitutive RNAi | Reduced allergens: ↓Dau c 1, ↓PR10 accumulation , ↓*in vivo* allergenicity, proved by skin prick test | Peters *et al.*, 2011 |
| | Apple | Mal d 1 | Constitutive RNAi | Reduced allergens: ↓Mal d 1; ↓*in vivo* allergenicity, proved by skin prick test and oral challenges | Gilissen *et al.*, 2005;<br>Dubois *et al.*, 2015 |
| | Tomato | Lyc e 1.01 and Lyc e 1.02 (two profilin isoforms) | Constitutive RNAi | Reduced allergens: ↓Lyc e 1, ↓profilin allergenic protein, ↓*in vivo* allergenicity, proved by skin prick test | Le *et al.*, 2006a |
| | Tomato | Lyc e 3 | Constitutive RNAi | Reduced allergens: ↓Lyc e 3, alleviating tomato allergenicity | Le *et al.*, 2006b |

of a bifunctional lysine degradation enzyme, the lysine-ketoglutarate reductase/saccharophine dehydrogenase (LKR/SDH), either alone (Houmard *et al.*, 2007), or in combination with the overexpression of a deregulated lysine biosynthetic enzyme, CordapA (Frizzi *et al.*, 2008).

Lysine-rich maize has huge economic and humanitarian importance worldwide. It is expected that similar RNAi-mediated approaches will be applied to improve the lysine content in other staple crops.

### 8.2.2 Carbohydrates: starch

The main form of carbohydrate in plants is starch. Starch in plants is made up of two components: amylopectin and amylose, which are produced by two competitive pathways. High caloric food is required in regions where food supply is scarce but, as obesity is becoming a huge problem in developed countries, it is important that calorie intake is reduced (Maskarinec *et al.*, 2006). Because the long linear chains of amylose crystallize more readily than amylopectin (which has short, highly branched chains), high-amylose starch is more resistant to digestion. Unlike amylopectin, amylose is not soluble in cold water. It also reduces the crystallinity of amylopectin and how easily water can infiltrate the starch (Crowe *et al.*, 2000). Resistant starch is part of our dietary fibre, and fibre-rich food, where most of the carbohydrates are not broken down to simple sugars, are considered healthy in developed countries as they help the fight against obesity (Williams, 1995).

Starch is synthesized from adenosine diphosphoglucose (ADP-glucose), which is catalysed by several enzymes. The granule-bound starch synthase I enzyme is required to produce the linear polysaccharide amylose. Amylopectin, which differs from amylose by having a highly branched structure with α-1,4 glucan chains connected by α-1,6 glucosidic linkages, is synthesized by a complex pathway involving, among others, starch-branching enzymes (SBE) (Ball and Morell, 2003). The strategy carried out in numerous crops to rebalance the components of the starch towards the indigestible fraction is to silence gene(s) encoding SBE. In a first study, Chai and colleagues (Chai *et al.*, 2005) used RNAi to suppress the expression of an *SBE* gene in maize, leading to an increase in the level of amylose of starch compared to control maize, without changing the overall starch content. In the same way, Shimada *et al.* (2006) transformed sweet potato with a construct encoding a dsRNA targeting sweet potato *SBEII* expression, which resulted in an increase in amylose content in the starch. Similar approaches have been used in other studies, but using tissue-specific promoters. For example, production of high-amylose potato lines can be achieved by RNAi-induced inhibition of two genes coding for starch-branching enzymes (*SBEI* and *SBEII*) in the tuber (Andersson *et al.*, 2006). The findings of this work were consistent with a previous report in which simultaneous inhibition of *SBEI* and *SBEII* expression by antisense RNAs of both genes resulted in potato starch with a very high amylose content (Schwall *et al.*, 2000). However, the RNAi method was by far more efficient than the antisense approach for inhibition of *SBE* gene expression in potato (Schwall *et al.*, 2000; Hofvander *et al.*, 2004). RNAi has also been used to silence the two different isoforms of *SBEII* (*SBEIIa* and *SBEIIb*) in wheat endosperm to raise its amylose content. Silencing *SBEIIb* alone was not sufficient to alter amylose content, but the silencing of both *SBEIIa* and *SBEIIb* led to a strong enrichment of amylose (more than 70%). Rats fed with this amylose-rich wheat grain showed improved bowel function compared to rats eating standard wholemeal wheat (Regina *et al.*, 2006). This study revealed the potential of amylose-rich wheat to improve human health through enriching its digestion-resistant starch content.

### 8.2.3 Oils and fatty acids

The nutritional quality of seed oil is determined by its fatty acid (FA) profiles, which are made up of several FA species with distinct carbon chain lengths and levels of desaturation. Seed oil in general is composed of saturated and unsaturated FAs, the latter consisting of monounsaturated FAs (MUFAs, mainly oleic acid, C18:1) and polyunsaturated FAs (PUFAs, mainly linoleic acid, C18:2, and linolenic acid, C18:3). Generally, vegetable oil with high levels of oleic acid and low levels of erucic acid (C22:1; a toxic MUFA that has been linked to cardiac muscle damage) is considered to be of high nutritional quality (Wilson, 2012). This is due to the high heat and oxidative stability of oleic acid that makes high-oleic oils suitable for use directly in long-life deep-frying applications for the food industry without the need for hydrogenation. Hydrogenation could lead to the conversion of naturally occurring cis-FAs to trans-FAs, which are increasingly recognized as having cholesterol-raising properties and thus providing increased risk of cardiovascular diseases (Mozaffarian *et al.*, 2006). However, it is not sustainable to replace hydrogenated oil ingredients by the production of currently cultivated high-oleic oil crops. Even if the processing technology were to develop further, a more sustainable strategy to increase the supply of plant-derived oil is to change seed composition. This has been achieved in nearly every oilseed crop by modulating the expression of enzymes that control the synthesis of the major seed oil FAs (Wilson, 2012). *FAD2* is a key gene encoding an ER membrane-bound FA desaturase 2 (FAD2), which catalyses the conversion of oleic acid to linoleic acid (Okuley *et al.*, 1994). *FAD3* is another key gene for the enzyme converting 18:2 to 18:3 (Browse *et al.*, 1993), while *FAE1* encodes an enzyme responsible for the carbon chain elongation for synthesis of very long-chain FAs, such as erucic acid (Millar and Kunst, 1997). These genes are considered to be priority targets for genetic modification towards lowering the PUFA content and increasing the oleic acid levels. To date, a number of high-oleic oilseed crops have been developed through downregulating FA desaturase genes (mainly *FAD2*). Some authors tried to downregulate the expression of these genes by antisense and/or cosuppression approaches in different oilseed crops such as soybean, *Brassica napus* and *B. juncea* (Kinney, 1994; Stoutjesdijk *et al.*, 2000; Sivaraman *et al.*, 2004), but better results were achieved by silencing through RNAi. For example, Liu *et al.* (2002) reported important changes in the composition of cottonseed oils by silencing two FA desaturase genes from cotton (*FAD2-1* and *SAD-1*) using RNAi. Such changes consisted in increased accumulation of oleic and stearic acids, together with a reduction in the palmitic acid levels. Another example is that in which the RNAi-mediated silencing of a soybean *FAD3* resulted in a decrease in the α-linolenic acid in the seed, relative to the other PUFAs in soybean (Flores *et al.*, 2008). Since this FA contributes mostly to the instability of soybean and other seed oils, a significant reduction in its content may enhance the seed value. In other studies, the simultaneous silencing of *FAD2* and *FAE1* was achieved: in *B. napus* (Peng *et al.*, 2010) and in crambe (Li *et al.*, 2016), with the aim of increasing oleic acid and additionally reducing erucic acid contents. Finally, Shi *et al.* (2015) achieved high-oleic oil with very low erucic acid by silencing only *FAE1* in *B. napus*.

All this shows that genetic enhancement of oleic acid concentration is feasible in most oilseed crops. Private companies have already launched production of genetically enhanced oleic acid cultivars such as Nexera™ Omega-9 canola (Dow AgroSciences LLC, Indianapolis, Indiana, USA) and Omega-9 sunflower oils, and more are expected to be commercially available in coming years (Wilson, 2012; Ricroch and Henard-Damave, 2015).

## 8.3 Enrichment with Phytonutrients/Functional Metabolites

Food is viewed as a source of nutrition to meet minimum daily requirements in order to survive and, with increasing attention, to thrive. Epidemiological studies show that increased consumption of fruits and vegetables is correlated with a reduced risk of several diseases, including cancer and cardiovascular disease. The protective effects of plant-based food have been mainly attributed to the high concentrations of bioactive compounds which have antioxidant properties, such as vitamin C, phenolic compounds and carotenoids. These substances, known as phytonutrients, are necessary for proper functioning of the body but some may confer additional protection against chronic disease over and above basic nutrition. There is a great interest in new food products that contain high level of phytonutrients because it is generally accepted that they are more beneficial to human health than dietary supplements. In principle, traditional breeding techniques could be used to generate new cultivars; however, the relatively low level of genetic diversity within sexually compatible species of crops seriously limits the scope of potential improvement. Enhancing phytonutrient levels of plant-based foods through genetic modification is an attractive contribution to tackling the twin global health burdens of micronutrient deficiencies and diet-related non-communicable diseases (reviewed by Martin *et al.*, 2011).

Carotenoids are a primary class of phytonutrients found in many fruits and vegetables, providing nutritional benefits both as precursors to essential vitamins, and as antioxidants. Lycopene has proven nutritional value as an antioxidant. Higher levels of lycopene consumption are associated with lower risks of heart attack, and it is also a promising cancer chemopreventive, particularly against prostate cancer (Heber and Lu, 2002). β-carotene, another potent antioxidant carotenoid, is the most efficient precursor of vitamin A, deficiency in which is the most common dietary problem affecting children worldwide (Humphrey *et al.*, 1992). Flavonoids, a class of phenolic compounds, are shown to have many biological functions including antioxidative, anticarcinogenic, cardiovascular and anti-inflammatory activities (Benavente-García and Castillo, 2008; Huang and Ho, 2010). Health benefits of these secondary metabolites are probably potentiated by combinations with other phytochemicals occurring in plant foods, particularly carotenoids (Tanaka *et al.*, 2000; Kohno *et al.*, 2001).

Carotenoid levels have been raised in many crops by modulating biosynthetic steps (reviewed in Botella-Pavía and Rodríguez-Concepción, 2006; DellaPenna and Pogson, 2006). Carotenoids are produced in plants through a complex pathway catalysed by several enzymes. The first step requires phytoene synthase (PSY) and the downstream steps, including desaturation and isomerization, are catalysed by different enzymes belonging to the carotene desaturase family. There is a step in the pathway, where beta and epsilon cyclases (LYCB and LYCE) compete for the same substrate (lycopene) and they produce either β-carotene (pro-vitamin A) or α-carotene. Carotene hydroxylases (CHYB, CHYE) then process these compounds into lutein or zeaxanthin. Finally, the epoxidation reaction of the xanthophyll cycle consists of the conversion of zeaxanthin to antheraxanthin and violaxanthin, and this is carried out by the zeaxanthin epoxidase (ZEP). This metabolic pathway has been modulated in numerous crops by inhibiting the synthesis of endogenous carotenogenic enzymes (Fig. 8.1). For example, Rosati *et al.* (2000) used an antisense strategy to downregulate the expression of the LCY-B gene in tomato in a fruit-specific manner, achieving moderate increases in the lycopene content. Similarly, β-carotene levels were increased by up to 14-fold and total carotenoid levels by up to 2.5-fold in potato by silencing LYCE in the tuber (Diretto *et al.*, 2006). Silencing LYCE in canola seeds also increased the level of β-carotene and total carotenoid levels (Yu *et al.*, 2008). The silencing of CHYB expression blocked the conversion of β-carotene into zeaxanthin. This approach achieved a 38-fold increase in β-carotene levels in potatoes, and also led to a 3.7-fold increase in lutein levels and a 0.5-fold reduction in zeaxanthin levels (Diretto *et al.*, 2007). Likewise, Pons *et al.* (2014) succeeded in enhancing the β-carotene level of oranges by silencing the CHYB gene. This resulted in oranges with significant increases in β-carotene (up to 36-fold) and α-carotene content in the pulp, accompanied by a general mild decrease in accumulation of downstream xanthophylls. Furthermore, the β-carotene-enriched oranges induced a 20% higher antioxidant effect than the isogenic control, as was demonstrated by performing *in vivo* assays using *Caenorhabditis elegans*. Silencing the next enzyme in the pathway (ZEP) reduces conversion of zeaxanthin into downstream products, and this approach increased carotenoid levels 5.7-fold, β-carotene levels 3.4-fold, lutein levels 1.9-fold and zeaxanthin levels 133-fold in potato (Romer *et al.*, 2002). Finally, Sun *et al.* (2012) demonstrated that it is possible to engineer higher carotenoid levels by altering carotenoid catabolism. In this work, an inverted-repeat containing the sequence of the *SlNCED1* gene encoding 9-cis-epoxycarotenoid dioxygenase (NCED) was expressed from the fruit-specific E8 promoter in tomato. NCED is a key enzyme in the ABA biosynthesis pathway and silencing this gene caused a higher level accumulation of lycopene and β-carotene, which are the upstream compounds in the pathway.

However, while it would be more beneficial to increase the general level of carotenoids, the methods discussed so far have enriched only one or a few compounds. One approach to obtain a more general enhancement in a group of metabolites could be to interfere with the expression of regulatory genes that control flux through several biosynthetic pathways. For example, RNAi-mediated silencing of a photomorphogenesis regulatory gene (*DET1*) in tomato fruit caused flavonoid and carotenoid enrichment in the fruit but did not affect the level of other compounds (Davuluri *et al.*, 2005). In another study, *DET1* was suppressed both constitutively

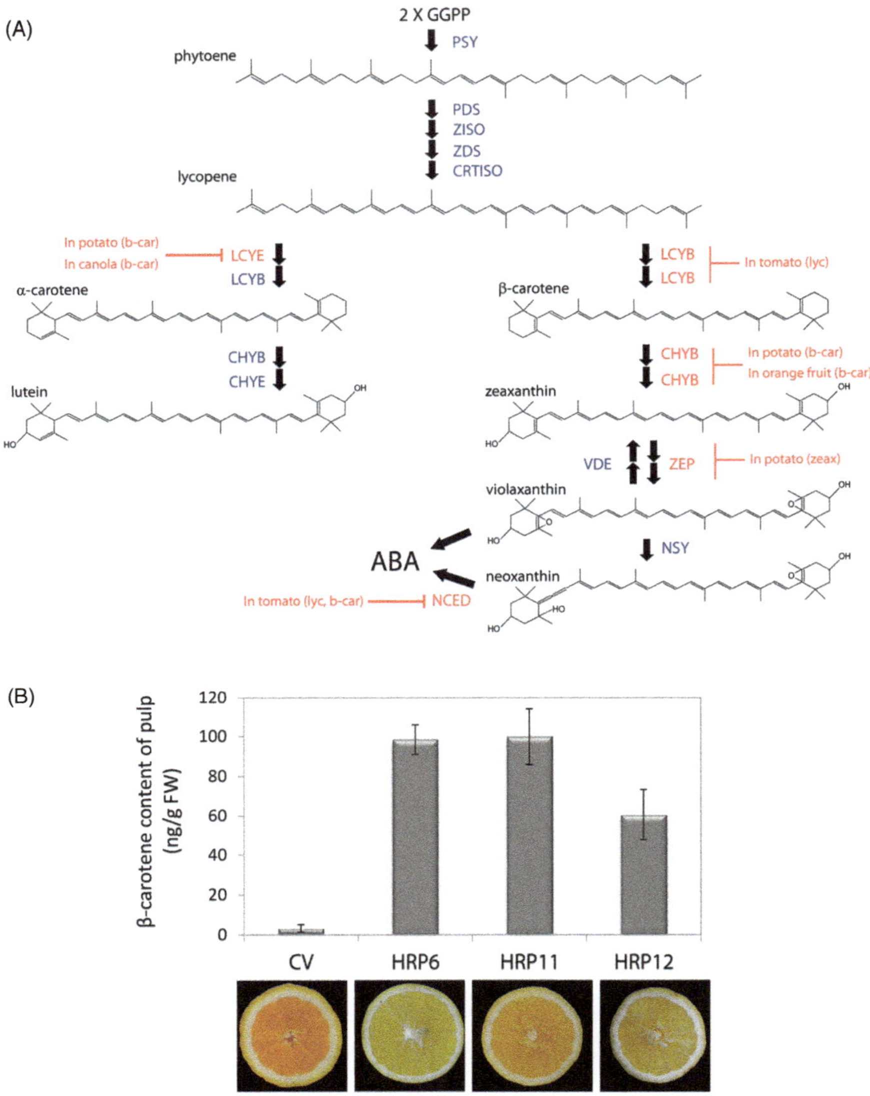


**Fig. 8.1.** A) Summary of nutritional improvement approaches based on the RNAi-mediated suppression of key carotenogenic genes. Schematic diagram of the carotenoid biosynthesis pathway in plants. In red are represented the different enzymes targeted, the crop(s) engineered with each strategy and the main enhanced carotenoid as a result of the modification (shown in parenthesis). GGPP: geranylgeranyl diphosphate; ABA: abscisic acid. Enzymes: PSY: phytoene synthase; PDS: phytoene desaturase; ZISO: ζ-carotene isomerase; ZDS: ζ-carotene desaturase; CRTISO: carotenoid isomerase; LCYB: lycopene β-cyclase; LCYE: lycopene ε-cyclase; CHYB: carotenoid β-ring hydroxylase; CHYE: carotenoid ε-ring hydroxylase; ZEP: zeaxanthin epoxidase; VDE: violaxanthin de-epoxidase; NSY: neoxanthin synthase; NCED: 9-cis-epoxycarotenoid dioxygenase. B) Levels of β-carotene content achieved in pulp of sweet orange fruits from three independent transgenic lines carrying an RNAi-cassette designed for silencing a *CHYB* gene (named HRP6, 11 and 12). CV: control fruits, transformed with the empty vector (Pons *et al.*, 2014.)

and in a seed-specific manner in canola by RNAi. Silencing of *DET1* resulted in transgenic seeds with substantially elevated levels of lutein, β-carotene and zeaxanthin relative to non-transgenic seeds, especially when silencing was constitutive. Moreover, the levels of sinapate esters (antinutritive compounds present in *B. napus*) in seeds in both sets of transgenic plants were lower compared to non-transgenic counterparts. These results revealed that *DET1* suppression in *B. napus* can increase the levels of carotenoids and reduce the levels of sinapate esters simultaneously in the seeds, thus enhancing their overall nutritional value (Wei *et al.*, 2009).

Another interesting target for metabolic engineering in crops is resveratrol. Resveratrol is a phytoalexin produced in plants as varied as grapevine, groundnut and pine in response to fungal infection or UV irradiation, but it is absent in members of the *Brassicaceae*. Resveratrol and its glucoside (piceid) are considered to have beneficial effects on human health; they are known to reduce heart disease, atherosclerosis and cancer mortality (Khan *et al.*, 2016; Xia *et al.*, 2016). With the aim of improving the quality of rapeseed products, the stilbene synthase gene isolated from grapevine (*Vitis vinifera*) was cloned under the control of the seed-specific napin promoter and introduced into rapeseed, together with an RNAi construct deduced from the gene sequence encoding the key enzyme for sinapate ester biosynthesis, UDP-glucose:sinapate glucosyltransferase (SGT1). The authors of this work assumed that the suppression of sinapate ester biosynthesis might increase resveratrol production in seeds through the increased availability of the precursor 4-coumarate. Resveratrol glucoside (piceid) was, in fact, produced at levels up to 361 μg/g in the seeds of the primary transformants (Hüsken *et al.*, 2005).

## 8.4 Reduction in Antinutrients, Toxins and Allergens

Cottonseed is a clear example of an underused oilseed crop. It contains a high level of oil (21%) and high-quality protein (23%) and is the third largest field crop in the world that produces edible oilseed. However, the seeds contain gossypol, produced from (+)-δ-cadinene by the enzyme δ-cadinene synthase. Unfortunately this is toxic, so we cannot eat cottonseed. All parts of cotton plants accumulate gossypol and related terpenoids, which play a protective role against both insects and pathogens. Gossypol-free seeds were generated by tissue-specific silencing of δ-cadinene synthase. Since silencing was tissue specific, the transgenic plants were not more susceptible to diseases and insects because the levels of gossypol and related terpenoids in other parts of the plant were not reduced (Sunilkumar *et al.*, 2006). This is a good example of a tissue-specific RNAi-based approach making a byproduct into a potential main product which could help in feeding the increasing population of the world.

Coffee is extremely popular worldwide and many people like drinking it even when their physiology changes and caffeine causes problems for them (e.g. increased blood pressure and insomnia). There is, therefore, an increased demand for decaffeinated coffee, but decaffeination is expensive and affects the flavour (Ashihara and Crozier, 2001). Caffeine is produced from xanthosine by three *N*-methyltransferase enzymes in coffee plants by adding methyl groups. These are CaXMT1 (xanthosine methyltransferase), CaMXMT1 (theobromine synthase) and CaDXMT1 (caffeine synthase) (Uefuji *et al.*, 2003). Transgenic coffee plants of *Coffea canephora* and *C. arabica* were generated, where silencing of the *CaMXMT1* gene led to a reduction in caffeine by up to 70% (Ogita *et al.*, 2003; Ogita *et al.*, 2004). Since there is enough sequence similarity between the three *N*-methyltransferase enzymes, the construct designed against CaMXMT1 also silenced the other two genes to some extent. The RNAi system developed in these studies to produce genetically modified coffee plants with a low level of caffeine appears commercially promising.

Despite aflatoxin being one of the most dangerous natural toxins, approximately 4.5 billion people are exposed to it. These carcinogen mycotoxins are produced by the fungi *Aspergillus flavus* and *A. parasiticus* and are present in one-quarter of edible crops, including such important crops as rice, maize, cereals, cassava and nuts. Although some procedures can reduce the effect of aflatoxins, there is no effective scheme that could be used worldwide, and all cultivated crop species are susceptible to these fungi (Williams *et al.*, 2004). The level of aflatoxin induced by *A. flavus* infection was successfully reduced in transgenic groundnut plants expressing an RNAi construct against five aflatoxin-synthesis genes (Arias *et al.*, 2015). In fact, in some lines the level of aflatoxins B1 and B2 was undetectable after *A. flavus* inoculation, while it was high in the control plants. This work illustrates how application of RNAi technology in cultivated groundnut and other crops would be effective against aflatoxins, and probably against other toxins produced by fungi.

Wheat is one of the main components of most people's diets because it was one of the first cultivated plants as a result of the relative ease of its cultivation, its calorie content and its nutritional value. However, wheat products contain gluten, and a substantial number of people are allergic or intolerant to gluten. There are different types of gluten allergy/intolerance. Coeliac disease (CD) is the best understood condition, but still the only option for CD patients is to avoid eating products containing gluten; this is complicated to follow and detrimental to gut health. Gluten is encoded by gliadin genes that belong to a fairly large gene family. Since these genes are not closely linked on the wheat genome, it is difficult to remove them all by traditional breeding techniques while preserving all the good traits of the cultivated cultivars. Hence RNAi technology has been used for downregulating gliadin genes in the endosperm of cultivated wheat. First, γ-gliadin genes were silenced in bread wheat (Gil-Humanes *et al.*, 2008); later, the same research group achieved downregulation of α-, γ- and ω-gliadins in transgenic lines by using a RNAi construct containing a fragment of 361 nucleotides highly conserved among α-, ω- and γ-gliadins (Gil-Humanes *et al.*, 2010). The resulting transgenic wheat showed less ability to stimulate gliadin-reactive T-cells from CD patients. The reduced-gliadin bread (in fact it was almost free of gliadin) made with flour from transgenic wheat showed very similar properties to that of regular flour. In addition, the bread made from low-gliadin flour contained higher levels of lysine than bread made from normal flour, and so was even healthier (Gil-Humanes *et al.*, 2014). The development of this reduced-gliadin bread offers a promising prospect to improve the life quality for millions of people who suffer from gluten allergy/intolerance.

Food allergy can be a serious health problem, and any food that contains protein has the potential to elicit an allergic reaction in a percentage of the human population. Avoidance of the food is the only treatment available, thus severely limiting dietary choices and the quality of life of food-allergic individuals. Current proposed alternative solutions focus primarily on ways to alter the immune system of allergic patients. However, with the advent of genetic engineering, novel strategies can be proposed to solve the problem of allergy from the source. Groundnut allergy is one of the most life-threatening food allergies and a serious challenge facing the groundnut and food industries. Some authors addressed this by knocking down expression of the major groundnut allergens, Ara h 2 and Ara h 6, through RNAi (Chu *et al.*, 2008; Dodo *et al.*, 2008). Similarly, this approach has been applied to reduce allergen concentrations in several plants, including soybean allergen P34 (Herman *et al.*, 2003), a 14-16 kDa rice allergen (Tada *et al.*, 1996), the apple allergen Mal d 1 (Gilissen *et al.*, 2005), tomato allergens Lyc e 1 and Lyc e 3 (Le, *et al.*, 2006a; Le *et al.*, 2006b) and carrot allergens Dau c 1.01 and Dau c 1.02 (Peters *et al.*, 2011). All these studies demonstrate that RNAi constitutes a reasonable strategy for allergen avoidance.

## Acknowledgements

We thank the Fundação de Amparo à Pesquisa do Estado de São Paulo (FAPESP) for partial funding of our research on increasing nutritional value of sweet orange fruit and juice through project no. 2014/12616-9.

## References

Andersson, M., Melander, M., Pojmark, P., Larsson, H., Bülow, L. and Hofvander, P. (2006) Targeted gene suppression by RNA interference: an efficient method for production of high-amylose potato lines. *Journal of Biotechnology* 123(2), 137–148.

Arias, R.S., Dang, P.M. and Sobolev, V.S. (2015) RNAi-mediated control of aflatoxins in peanut: method to analyze mycotoxin production and transgene expression in the peanut/*Aspergillus* pathosystem. *Journal of Visualized Experiments* 106, e53398, DOI:10.3791/53398

Ashihara, H. and Crozier, A. (2001) Caffeine: a well known but little mentioned compound in plant science. *Trends in Plant Science* 6(9), 407–413.

Ball, S.G. and Morell, M.K. (2003) From bacterial glycogen to starch: understanding the biogenesis of the plant starch granule. *Annual Review of Plant Biology* 54, 207–233.

Benavente-García, O. and Castillo, J. (2008) Update on uses and properties of citrus flavonoids: new finding in anticancer, cardiovascular, and anti-inflammatory activity. *Journal of Agricultural and Food Chemistry* 56(15), 6185–6205.

Botella-Pavía, P. and Rodríguez-Concepción, M. (2006) Carotenoid biotechnology in plants for nutritionally improved foods. *Physiologia Plantarum* 126(3), 369–381.

Browse, J., McConn, M., James, D. and Miquel, M. (1993) Mutants of *Arabidopsis* deficient in the synthesis of alpha-linolenate. Biochemical and genetic characterization of the endoplasmic reticulum linoleoyl desaturase. *The Journal of Biological Chemistry* 268(22), 16345–16351.

Capell, T. and Christou, P. (2004) Progress in plant metabolic engineering. *Current Opinion in Biotechnology* 15(2), 148–154.

Chai, X.J., Wang, P.W., Guan, S.Y. and Xu, Y.W. (2005) Reducing the maize amylopectin content through RNA interference manipulation. *Journal of Plant Physiology and Molecular Biology* 31(6), 625–630.

Chu, Y., Faustinelli, P., Ramos, M.L., Hajduch, M., Stevenson, S., Thelen, J.J., Maleki, S.J., Cheng, H. and Ozias-Akins, P. (2008) Reduction of IgE binding and nonpromotion of

*Aspergillus flavus* fungal growth by simultaneously silencing Ara h 2 and Ara h 6 in peanut. *Journal of Agricultural and Food Chemistry* 56(23), 11225–11233.

Cressey, D. (2013) Transgenics: a new breed. *Nature* 497(7447), 27–29.

Crowe, T.C., Seligman, S.A. and Copeland, L. (2000) Inhibition of enzymic digestion of amylose by free fatty acids *in vitro* contributes to resistant starch formation. *Journal of Nutrition* 130(8), 2006–2008.

Davuluri, G.R., van Tuinen, A., Fraser, P.D., Manfredonia, A., Newman, R., Burgess, D., Brummell, D.A., King, S.R., Palys, J., Uhlig, J., Bramley, P.M., Pennings, H.M. and Bowler, C. (2005) Fruit-specific RNAi-mediated suppression of *DET1* enhances carotenoid and flavonoid content in tomatoes. *Nature Biotechnology* 23(7), 890–895.

DellaPenna, D. and Pogson, B.J. (2006) Vitamin synthesis in plants: tocopherols and carotenoids. *Annual Review of Plant Biology* 57, 711–738.

Diretto, G., Tavazza, R., Welsch, R., Pizzichini, D., Mourgues, F., Papacchioli, V., Beyer, P. and Giuliano, G. (2006) Metabolic engineering of potato tuber carotenoids through tuber-specific silencing of lycopene epsilon cyclase. *BMC Plant Biology* 6, 13.

Diretto, G., Welsch, R., Tavazza, R., Mourgues, F., Pizzichini, D., Beyer, P. and Giuliano, G. (2007) Silencing of beta-carotene hydroxylase increases total carotenoid and beta-carotene levels in potato tubers. *BMC Plant Biology* 7, 11.

Dodo, H.W., Konan, K.N., Chen, F.C., Egnin, M. and Viquez, O.M. (2008) Alleviating peanut allergy using genetic engineering: the silencing of the immunodominant allergen Ara h 2 leads to its significant reduction and a decrease in peanut allergenicity. *Plant Biotechnology Journal* 6(2), 135–145.

Dubois, A.E.J., Pagliarani, G., Brouwer, R.M., Kollen, B.J., Dragsted, L.O., Eriksen, F.D., Callesen, O., Gilissen, L.J., Krens, F.A., Visser, R.G., Smulders, M.J., Vlieg-Boerstra, B.J., Flokstra-de Blok, B.J. and van de Weg, W.E. (2015) First successful reduction of clinical allergenicity of food by genetic modification: Mal d 1-silenced apples cause fewer allergy symptoms than the wild-type cultivar. *Allergy* 70(11), 1406–1412.

Flores, T., Karpova, O., Su, X., Zeng, P., Bilyeu, K., Sleper, D.A., Nguyen, H.T. and Zhang, Z.J. (2008) Silencing of *GmFAD3* gene by siRNA leads to low alpha-linolenic acids (18:3) of *fad3*-mutant phenotype in soybean (*Glycine max* (Merr.)). *Transgenic Research* 17(5), 839–850.

Frizzi, A., Huang, S., Gilbertson, L.A., Armstrong, T.A., Luethy, M.H. and Malvar, T.M. (2008) Modifying lysine biosynthesis and catabolism in corn with a single bifunctional expression/silencing transgene cassette. *Plant Biotechnology Journal* 6(1), 13–21.

Gil-Humanes, J., Pistón, F., Hernando, A., Alvarez, J.B., Shewry, P.R. and Barro, F. (2008) Silencing of γ-gliadins by RNA interference (RNAi) in bread wheat. *Journal of Cereal Science* 48(3), 565–568.

Gil-Humanes, J., Piston, F., Tollefsen, S., Sollid, L.M. and Barro, F. (2010) Effective shutdown in the expression of celiac disease-related wheat gliadin T-cell epitopes by RNA interference. *Proceedings of the National Academy of Sciences of the United States of America* 107(39), 17023–17028.

Gil-Humanes, J., Piston, F., Altamirano-Fortoul, R., Real, A., Comino, I., Sousa, C., Rosell, C.M. and Barro, F. (2014) Reduced-gliadin wheat bread: an alternative to the gluten-free diet for consumers suffering gluten-related pathologies. *PloS One* 9(3), e90898.

Gilissen, L.J.W.J., Bolhaar, S.T.H.P., Matos, C.I., Rouwendal, G.J.A., Boone, M.J., Krens, F.A., Zuidmeer, L., Van Leeuwen, A., Akkerdaas, J., Hoffmann-Sommergruber, K., Knulst, A.C., Bosch, D., Van de Weg, W.E. and Van Ree, R. (2005) Silencing the major apple allergen Mal d 1 by using the RNA interference approach. *The Journal of Allergy and Clinical Immunology* 115(2), 364–369.

Heber, D. and Lu, Q.-Y. (2002) Overview of mechanisms of action of lycopene. *Experimental Biology and Medicine (Maywood, N.J.)* 227(10), 920–923.

Herman, E.M., Helm, R.M., Jung, R. and Kinney, A.J. (2003) Genetic modification removes an immunodominant allergen from soybean. *Plant Physiology* 132(1), 36–43.

Hofvander, P., Andersson, M., Larsson, C.-T. and Larsson, H. (2004) Field performance and starch characteristics of high-amylose potatoes obtained by antisense gene targeting of two branching enzymes. *Plant Biotechnology Journal* 2(4), 311–320.

Houmard, N.M., Mainville, J.L., Bonin, C.P., Huang, S., Luethy, M.H. and Malvar, T.M. (2007) High-lysine corn generated by endosperm-specific suppression of lysine catabolism using RNAi. *Plant Biotechnology Journal* 5(5), 605–614.

Huang, S., Frizzi, A., Florida, C.A., Kruger, D.E. and Luethy, M.H. (2006) High lysine and high tryptophan transgenic maize resulting from the reduction of both 19- and 22-kD alpha-zeins. *Plant Molecular Biology* 61(3), 525–535.

Huang, Y.S. and Ho, S.C. (2010) Polymethoxy flavones are responsible for the anti-inflammatory activity of citrus fruit peel. *Food Chemistry* 119(3), 868–873.

Humphrey, J.H., West, K.P. and Sommer, A. (1992) Vitamin A deficiency and attributable mortality among under-5-year-olds. *Bulletin of the World Health Organization* 70(2), 225–232.

Hüsken, A., Baumert, A., Milkowski, C., Becker, H., Strack, D. and Möllers, C. (2005) Resveratrol glucoside (Piceid) synthesis in seeds of transgenic oilseed rape (*Brassica napus* L.) *Theoretical and Applied Genetics* 111(8), 1553–1562.

Katoch, R. and Thakur, N. (2013a) Advances in RNA interference technology and its impact on nutritional improvement, disease and insect control in plants. *Applied Biochemistry and Biotechnology* 169(5), 1579–1605.

Katoch, R. and Thakur, N. (2013b) RNA interference: a promising technique for the improvement of traditional crops. *International Journal of Food Sciences and Nutrition* 64(2), 248–259.

Khan, O.S., Bhat, A.A., Krishnankutty, R., Mohammad, R.M. and Uddin, S. (2016) Therapeutic potential of resveratrol in lymphoid malignancies. *Nutrition and Cancer* 68(3), 365–373.

Kinney, A.J. (1994) Genetic modification of the storage lipids of plants. *Current Opinion in Biotechnology* 5(2), 144–151.

Kinney, A.J. and Knowlton, S. (1998) Designer oils: the high oleic acid soybean. In: Roller, S. and Harlander, S. (eds) *Genetic Modification in the Food Industry: A Strategy for Food Quality Improvement*. Chapman & Hall, London, pp. 192–213.

Kohno, H., Taima, M., Sumida, T., Azuma, Y., Ogawa, H. and Tanaka, T. (2001) Inhibitory effect of mandarin juice rich in beta-cryptoxanthin and hesperidin on 4-(methylnitrosamino)-1-(3-pyridyl)-1-butanone-induced pulmonary tumorigenesis in mice. *Cancer Letters* 174(2), 141–150.

Le, L.Q., Lorenz, Y., Scheurer, S., Fötisch, K., Enrique, E., Bartra, J., Biemelt, S., Vieths, S. and Sonnewald, U. (2006a) Design of tomato fruits with reduced allergenicity by dsRNAi-mediated inhibition of ns-LTP (Lyc e 3) expression. *Plant Biotechnology Journal* 4(2), 231–242.

Le, L.Q., Mahler, V., Lorenz, Y., Scheurer, S., Biemelt, S., Vieths, S. and Sonnewald, U. (2006b) Reduced allergenicity of tomato fruits harvested from Lyc e 1-silenced transgenic tomato plants. *The Journal of Allergy and Clinical Immunology* 118(5), 1176–1183.

Li, X., Mei, D., Liu, Q., Fan, J., Singh, S., Green, A., Zhou, X.R. and Zhu, L.H. (2016) Down-regulation of crambe fatty acid desaturase and elongase in *Arabidopsis* and crambe resulted in significantly increased oleic acid content in seed oil. *Plant Biotechnology Journal* 14(1), 323–331.

Liu, Q., Singh, S.P. and Green, A.G. (2002) High-stearic and High-oleic cottonseed oils produced by hairpin RNA-mediated post-transcriptional gene silencing. *Plant Physiology* 129(4), 1732–1743.

Martin, C., Butelli, E., Petroni, K. and Tonelli, C. (2011) How can research on plants contribute to promoting human health? *The Plant Cell* 23(5), 1685–1699.

Maskarinec, G., Takata, Y., Pagano, I., Carlin, L., Goodman, M.T., Le Marchand, L., Nomura, A.M., Wilkens, L.R. and Kolonel, L.N. (2006) Trends and dietary determinants of overweight and obesity in a multiethnic population. *Obesity (Silver Spring, Md.)* 14(4), 717–726.

Millar, A.A. and Kunst, L. (1997) Very-long-chain fatty acid biosynthesis is controlled through the expression and specificity of the condensing enzyme. *The Plant Journal* 12(1), 121–131.

Mozaffarian, D., Katan, M.B., Ascherio, A., Stampfer, M.J. and Willett, W.C. (2006) Trans fatty acids and cardiovascular disease. *New England Journal of Medicine* 354(15), 1601–1613.

Ogita, S., Uefuji, H., Yamaguchi, Y., Koizumi, N. and Sano, H. (2003) Producing decaffeinated coffee plants. *Nature* 423(6942), 823.

Ogita, S., Uefuji, H., Morimoto, M. and Sano, H. (2004) Application of RNAi to confirm theobromine as the major intermediate for caffeine biosynthesis in coffee plants with potential for construction of decaffeinated varieties. *Plant Molecular Biology* 54(6), 931–941.

Okuley, J., Lightner, J., Feldmann, K., Yadav, N., Lark, E. and Browse, J. (1994) *Arabidopsis FAD2* gene encodes the enzyme that is essential for polyunsaturated lipid synthesis. *The Plant Cell* 6(1), 147–158.

Peng, Q., Hu, Y., Wei, R., Zhang, Y., Guan, C., Ruan, Y. and Liu, C. (2010) Simultaneous silencing of *FAD2* and *FAE1* genes affects both oleic acid and erucic acid contents in *Brassica napus* seeds. *Plant Cell Reports* 29(4), 317–325.

Peters, S., Imani, J., Mahler, V., Foetisch, K., Kaul, S., Paulus, K.E., Scheurer, S., Vieths, S. and Kogel, K.H. (2011) Dau c 1.01 and Dau c 1.02-silenced transgenic carrot plants show reduced allergenicity to patients with carrot allergy. *Transgenic Research* 20(3), 547–556.

Pons, E., Alquezar, B., Rodriguez, A., Martorell, P., Genoves, S., Ramón, D., Rodrigo, M.J., Zacarías, L. and Peña, L. (2014) Metabolic engineering of beta-carotene in orange fruit increases its *in vivo* antioxidant properties. *Plant Biotechnology Journal* 12(1), 17–27.

Regina, A., Bird, A., Topping, D., Bowden, S., Freeman, J., Barsby, T., Kosar-Hashemi, B., Li, Z., Rahman, S. and Morell, M. (2006) High-amylose wheat generated by RNA interference improves indices of large-bowel health in rats. *Proceedings of the National Academy of Sciences of the United States of America* 103(10), 3546–3551.

Ricroch, A.E. and Henard-Damave, M.C. (2015) Next biotech plants: new traits, crops, developers and technologies for addressing global challenges. *Critical Reviews in Biotechnology* 36, 675–690.

Romer, S., Lubeck, J., Kauder, F., Steiger, S., Adomat, C. and Sandmann, G. (2002) Genetic engineering of a zeaxanthin-rich potato by antisense inactivation and co-suppression of carotenoid epoxidation. *Metabolic Engineering* 4(4), 263–272.

Rosati, C., Aquilani, R., Dharmapuri, S., Pallara, P., Marusic, C., Tavazza, R., Bouvier, F., Camara, B. and Giuliano, G. (2000) Metabolic engineering of beta-carotene and lycopene content in tomato fruit. *The Plant Journal* 24(3), 413–419.

Schwall, G.P., Safford, R., Westcott, R.J., Jeffcoat, R., Tayal, A., Shi, Y.C., Gidley, M.J. and Jobling, S.A. (2000) Production of very-high-amylose potato starch by inhibition of *SBE* A and B. *Nature Biotechnology* 18(5), 551–554.

Segal, G., Song, R. and Messing, J. (2003) A new opaque variant of maize by a single dominant RNA-interference-inducing transgene. *Genetics* 165(1), 387–397.

Shi, J., Lang, C., Wu, X., Liu, R., Zheng, T., Zhang, D., Chen, J. and Wu, G. (2015) RNAi knockdown of fatty acid elongase1 alters fatty acid composition in *Brassica napus*. *Biochemical and Biophysical Research Communications* 466(3), 518–522.

Shimada, T., Otani, M., Hamada, T. and Kim, S.-H. (2006) Increase of amylose content of sweetpotato starch by RNA interference of the starch branching enzyme II gene (*IbSBEII*) *Plant Biotechnology* 23(1), 85–90.

Sivaraman, I., Arumugam, N., Sodhi, Y.S., Gupta, V., Mukhopadhyay, A., Pradhan, A.K., Burma, P.K. and Pental, D. (2004) Development of high oleic and low linoleic acid transgenics in a zero erucic acid *Brassica juncea* L. (Indian mustard) line by antisense suppression of the *fad2* gene. *Molecular Breeding* 13(4), 365–375.

Stoutjesdijk, P.A., Hurlestone, C., Singh, S.P. and Green, A.G. (2000) High-oleic acid Australian *Brassica napus* and *B. juncea* varieties produced by co-suppression of endogenous Δ12-desaturases. *Biochemical Society Transactions* 28(6), 938–940.

Sun, L., Yuan, B., Zhang, M., Wang, L., Cui, M., Wang, Q. and Leng, P. (2012) Fruit-specific RNAi-mediated suppression of *SlNCED1* increases both lycopene and beta-carotene contents in tomato fruit. *Journal of Experimental Botany* 63(8), 3097–3108.

Sunilkumar, G., Campbell, L.M., Puckhaber, L., Stipanovic, R.D. and Rathore, K.S. (2006) Engineering cottonseed for use in human nutrition by tissue-specific reduction of toxic gossypol. *Proceedings of the National Academy of Sciences* 103(48), 18054–18059.

Tada, Y., Nakase, M., Adachi, T., Nakamura, R., Shimada, H., Takahashi, M., Fujimura, T. and Matsuda, T. (1996) Reduction of 14-16 kDa allergenic proteins in transgenic rice plants by antisense gene. *FEBS Letters* 391(3), 341–345.

Tanaka, T., Kohno, H., Murakami, M., Shimada, R., Kagami, S., Sumida, T., Azuma, Y. and Ogawa, H. (2000) Suppression of azoxymethane-induced colon carcinogenesis in male F344 rats by mandarin juices rich in beta-cryptoxanthin and hesperidin. *International Journal of Cancer* 88(1), 146–150.

Tang, G. and Galili, G. (2004) Using RNAi to improve plant nutritional value: from mechanism to application. *Trends in Biotechnology* 22(9), 463–469.

Tang, G., Galili, G. and Zhuang, X. (2007) RNAi and microRNA: breakthrough technologies for the improvement of plant nutritional value and metabolic engineering. *Metabolomics* 3(3), 357–369.

Uefuji, H., Ogita, S., Yamaguchi, Y., Koizumi, N. and Sano, H. (2003) Molecular cloning and functional characterization of three distinct N-methyltransferases involved in the caffeine biosynthetic pathway in coffee plants. *Plant Physiology* 132(1), 372–380.

Wei, S., Li, X., Gruber, M.Y., Li, R., Zhou, R., Zebarjadi, A. and Hannoufa, A. (2009) RNAi-mediated suppression of *DET1* alters the levels of carotenoids and sinapate esters in seeds of *Brassica napus*. *Journal of Agricultural and Food Chemistry* 57(12), 5326–5333.

Williams, C.L. (1995) Importance of dietary fiber in childhood. *Journal of the American Dietetic Association* 95(10), 1140–1146.

Williams, J.H., Phillips, T.D., Jolly, P.E., Stiles, J.K., Jolly, C.M. and Aggarwal, D. (2004) Human aflatoxicosis in developing countries: a review of toxicology, exposure, potential health consequences, and interventions. *The American Journal of Clinical Nutrition* 80(5), 1106–1122.

Wilson, R.F. (2012) The role of genomics and biotechnology in achieving global food security for high-oleic vegetable oil. *Journal of Oleo Science* 61(7), 357–367.

Wu, Y. and Messing, J. (2011) Novel genetic selection system for quantitative trait loci of quality protein maize. *Genetics* 188(4), 1019–1022.

Wu, Y. and Messing, J. (2012) RNA interference can rebalance the nitrogen sink of maize seeds without losing hard endosperm. *PloS One* 7(2), e32850.

Wu, Y. and Messing, J. (2014) Proteome balancing of the maize seed for higher nutritional value. *Frontiers in Plant Science* 5, 240.

Xia, N., Daiber, A., Förstermann, U. and Li, H. (2016) Antioxidant effects of resveratrol in the cardiovascular system. *British Journal of Pharmacology*. DOI: 10.1111/bph.13492

Yu, B., Lydiate, D.J., Young, L.W., Schäfer, U.A. and Hannoufa, A. (2008) Enhancing the carotenoid content of *Brassica napus* seeds by downregulating lycopene epsilon cyclase. *Transgenic Research* 17(4), 573–585.

Zhu, C., Sanahuja, G., Yuan, D., Farré, G., Arjó, G., Berman, J., Zorrilla-López, U., Banakar, R., Bai, C., Pérez-Massot, E., Bassie, L., Capell, T. and Christou, P. (2013) Biofortification of plants with altered antioxidant content and composition: genetic engineering strategies. *Plant Biotechnology Journal* 11(2), 129–141.

# 9 RNA-based Control of Plant Diseases: A Case Study with *Fusarium graminearum*

**Aline Koch and Karl-Heinz Kogel***

*Justus Liebig University, Giessen, Germany*

## 9.1 Introduction

Current strategies of plant production measures, including conventional and organic farming, are still characterized by high usage of pesticides such as copper compounds. Moreover resistance breeding, though largely successful over the years, lacks solutions where monogenic resistance traits are unspecified. In consequence, a growing population will require developing groundbreaking strategies that promote sustainable plant production. RNA interference has emerged as a powerful genetic tool for scientific research. Demonstration that agricultural pests and microbial pathogens are killed by exogenously supplied RNA targeting their essential genes has raised the possibility that plant health can be controlled by lethal RNA signals. Here we discuss recent work conducted to assess the potential of double-stranded (ds)RNA targeting essential fungal ergosterol biosynthesis genes to control the infection of barley by the necrotrophic pathogenic fungus *Fusarium graminearum*. Regardless of how dsRNA is applied – by transgene expression (host-induced gene silencing, HIGS) or spray application (spray-induced gene silencing, SIGS) – the use of target-specific inhibitory RNA to mediate protection against fungal infection is a potential alternative to conventional chemicals, though many questions concerning field application of the novel technique remain unresolved.

## 9.2 Fusarium is a Cereal Killer Which Requires More Efficient Strategies for Disease Control

Current types of plant production measures are still characterized by high use of pesticides. While modern pesticides are largely safe, there are still several issues that demand further optimization of their use, including: (i) the emergence of compound resistance due to wrong management of pesticide use; (ii) soil/water pollution (e.g. excessive use of copper in organic farming); and (iii) largely inefficient

* Corresponding author e-mail: karl-heinz.kogel@agrar.uni-giessen.de

 

pesticide activities against certain ear and root diseases. In consequence there is a never-ending 'arms race' between food producers (humans) and pathogens/pests for efficient crop yield, and we are in constant need of novel production strategies.

More than half of the world's harvested area is allotted to cereals such as rice, maize and wheat (circa 2.3 billion tonnes in 2010; FAO, 2013). Diseases of cereal crops such as Fusarium head blight (FHB) and Fusarium seedling blight (FSB), caused by necrotrophic fungi of the genus *Fusarium*, have a particularly severe economic and agronomic impact on global grain production and the grain industry (Bai *et al.*, 2004; Broekaert *et al.*, 2015). Food safety can be compromised by contamination of agricultural products with mycotoxins, which are produced during FHB and FSB development (Jansen *et al.*, 2005; Merhej *et al.*, 2011) and represent a serious threat to human and animal health. One of the predominant mycotoxins, the trichothecene zearalenone, has a high acute toxicity ($LD_{50}$ = 7 mg $kg^{-1}$ body weight, mouse oral) that is much higher than that of a modern fungicide such as strobilurins ($LD_{50}$ = >5000 mg $kg^{-1}$ body weight) and azoles ($LD_{50}$ = between 600 to >2000 mg $kg^{-1}$ body weight). Currently, the major strategies to control Fusarium diseases include resistance breeding, crop rotation and ploughing, along with the application of demethylation inhibitor (DMI) fungicides (Kazan *et al.*, 2012). These chemicals are the most successful fungicides worldwide. DMI fungicides, such as tebuconazole, fenbuconazole and propiconazole, inhibit ergosterol biosynthesis by binding to cytochrome P450 lanosterol C-14 $\alpha$-demethylase (CYP51), thereby disrupting fungal membrane integrity (Kuck, 2012). However, heavy reliance on DMI fungicides since their discovery in the mid-1970s risks the emergence of DMI-tolerant strains of plant pathogens (Gsaller *et al.*, 2016). Greater compound tolerance is at least partly because of the enhanced ability of fungi to detoxify the chemicals. Even worse, the quantitative nature of FHB and FSB resistance and the lack of true resistance (R) genes do not allow straightforward breeding programmes (Jansen *et al.*, 2005). Together, these problems reveal *Fusarium* species as some of the most problematic cereal pathogens worldwide.

Our group has a research focus on the study of double-stranded (ds)RNA applications as sustainable alternatives to control pests and diseases in plant production. Here we discuss current knowledge of the mechanism by which dsRNA affects target organisms. We will focus our discussion on recent work on the interaction of barley (*Hordeum vulgare*) and *Fusarium graminearum*.

## 9.3 Application Opportunities for RNAi in Agriculture

Exogenous dsRNA triggers suppression of gene activity in a homology-dependent manner (Fire *et al.*, 1998). Since this discovery and the identification of small RNAs (sRNAs) as a new class of regulatory molecules (Hamilton and Baulcombe, 1999) that functions via RNA interference (RNAi), our understanding of the essential cellular function of gene silencing has increased considerably (Vaucheret and Fagard, 2001; Castel and Martienssen, 2013). Mobile RNA silencing signals are capable of translocating from the host to its interacting organism, and vice versa (Tomilov *et al.*, 2008; for reviews see Knip *et al.*, 2014; Koch and Kogel, 2014; Baulcombe, 2015). Consistent with this, a recent work showed the significant role that small RNAs may play in the communication between plants and a pathogenic fungus (Weiberg *et al.*, 2013; Weiberg *et al.*, 2015). Exploiting this mechanism in plants also has strong potential for agricultural applications. Supportively, transgenic expression of inhibitory dsRNAs in the corresponding host plant resulted in protection from predation or infection by targeted gene silencing (Price and Gatehouse, 2008; Nowara *et al.*, 2010; Koch *et al.*, 2013; Ghag *et al.*, 2014; Cheng *et al.*, 2015).

### 9.3.1 Host-induced gene silencing against *Fusarium graminearum*

In *Arabidopsis* (*A. thaliana*) and barley, transgenic expression of *CYP3*-dsRNA, a 791-nt long dsRNA targeting the three fungal *Cytochrome P450 lanosterol C-14α-demethylase* genes *CYP51A*, *CYP51B* and *CYP51C*, mediates resistance to infection with *F. graminearum* (Fig. 9.1). HIGS of these essential ergosterol biosynthesis genes is a highly efficient strategy for controlling the growth and development of the phytopathogenic fungus (Koch *et al.*, 2013). Initially, *CYP51* genes were selected as potential HIGS targets because of the previous demonstration that CYP51 enzymes are well-known targets for DMI fungicides, whose inhibition leads to fungal arrest and disease control. Specifically, dysfunction of CYP51 enzymes results in depletion of ergosterol and the accumulation of sterol precursors (including the 14α-demethylated sterols, 4,14-dimethylzymosterol and 24 methylenedihydrolanosterol) in the plasma membrane, and concurrent decrease in CYP51 products such as 14-methylated sterols (Ghannoum and Rice, 1999). This imbalance alters plasma membrane structure and function, as elevated levels of ergosterol precursors induce permeability changes, membrane leakiness, changes in membrane-bound enzymes and inhibition of fungal growth. In addition, the 14α-demethylsterols cannot replace ergosterol to stimulate cell proliferation (Fernández-Ortuño *et al.*, 2010; Liu *et al.*, 2011; Fan *et al.*, 2013). Consistent with the predicted function of the three *F. graminearum* CYP51 paralogues in ensuring membrane integrity and fungal virulence, silencing their encoding genes altered fungal growth and development. Treatment of fungal axenic cultures with *CYP3*-dsRNA resulted in an increase in hyphal branching and inhibition of hyphal growth (see Fig. 9.1 and Koch *et al.*, 2013).

In transgenic *Arabidopsis* expressing *CYP3*-dsRNA, fungal growth was restricted to nearly 100%, with a small amount of growth occurring at the wounded area immediately surrounding the inoculation sites. Consistent with this, ears from barley plants expressing the inhibitory dsRNA were virtually free of fungal infections as compared to wild-type plants, which showed heavily infected roots, leaves and spikelets. Analysis of *CYP51* expression at infection sites showed that all three fungal genes were partially silenced. Thus, the altered growth and morphology of the fungus appears to be provoked by a reduction in fungal *Cytochrome P450 lanosterol C-14α-demethylase* expression.

The mechanism through which *in planta* expression of a nuclear-integrated *CYP3*-dsRNA construct silences gene expression in plant-colonizing fungi is still unresolved. One can speculate that siRNAs generated by the plant's silencing machinery from the long *CYP3*-dsRNA are transferred and secreted via vesicles. Consistent with this scenario, small interfering 21–24 nt (si)RNAs corresponding to the targeted sequences were detected in *CYP3*-dsRNA-expressing *Arabidopsis*,

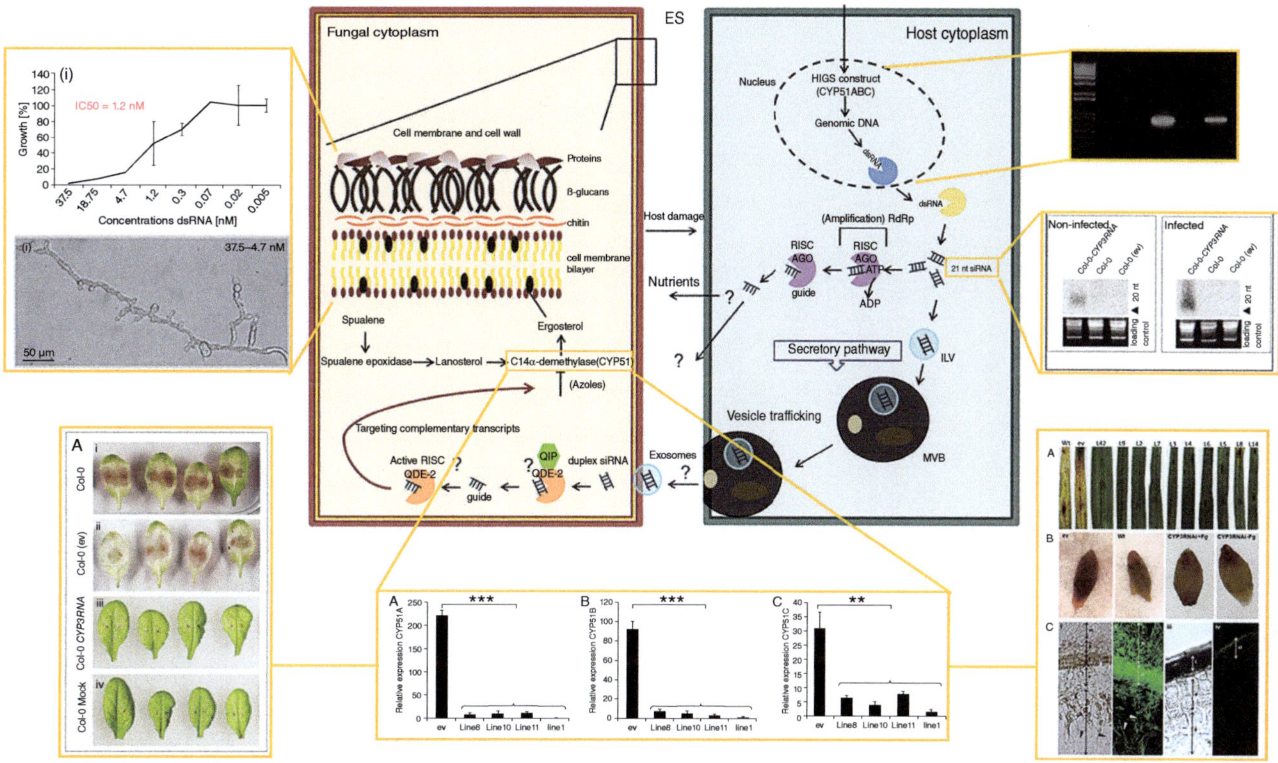


**Fig. 9.1.** Host-induced gene silencing (HIGS) of fungal ergosterol biosynthesis genes leads to growth inhibition of the necrotrophic fungus *Fusarium graminearum*. Hypothetical HIGS mechanism of dsRNA processing, sRNA transport, silencing of fungal target genes, and effect on fungal membranes (*central part of the drawing*). Treatment of axenic fungal cultures with *CYP3*-dsRNA, a 791 nt dsRNA complementary to the *Cytochrome P450 lanosterol C-14α-demethylase* genes *CYP51A*, *CYP51B*, and *CYP51C* resulted in growth inhibition and altered fungal morphology (*upper left*). Transgenic expression of the same dsRNA in *Arabidopsis* and barley (*gel blot upper right*) rendered susceptible plants highly resistant to fungal infection (*bottom left:* Arabidopsis/*bottom right: barley*). Microscopic analysis revealed that mycelium formation on *CYP3*-dsRNA-expressing leaves was restricted to the inoculation sites, and that inoculated barley caryopses were virtually free of fungal hyphae (*bottom right*). The inhibition of fungal growth correlated with *in planta* accumulation of sRNAs corresponding to the targeted *CYP51* sequences (*middle right*), as well as highly efficient silencing of the fungal genes *CYP51A*, *CYP51B*, and *CYP51C* (*bottom*).

independent of fungal infection. However, further studies are required to elucidate whether *CYP51* gene silencing is mediated by fungal uptake of siRNAs generated by the plant's RNAi machinery, or by uptake of the unprocessed precursor *CYP3*-dsRNA.

### 9.3.2 Spray-induced gene silencing against *Fusarium graminearum*

While the HIGS-mediated control of *F. graminearum* provided proof-of-concept that RNA-based plant protection is an effective strategy against diseases caused by devastating necrotrophic pathogens, the broad applicability of this transgenic method remains questionable due to the persisting weak acceptance of GMO strategies for food and feed production in many countries. More important, a broad application of this transgenic approach is hampered by the lack of transformability of various crop plants and the missing genetic stability of the silencing trait. Owing to these political and scientific constraints, we also investigated the potential of a protection strategy using SIGS to apply dsRNA and target *F. graminearum* (Koch *et al*., 2016). Spraying *CYP3*-dsRNA (20 ng $\mu L^{-1}$) strongly reduces the development of the pathogen on barley leaves, supporting the idea that dsRNA could be used as a fungicide to control plant diseases (Fig. 9.2).

### 9.3.3 Mechanistic considerations

Elucidating the molecular mechanisms of SIGS is key to successful future implementation. While HIGS is probably based on the plant's silencing machinery, the mechanism of gene silencing by exogenously delivered dsRNA constitutes a more complex situation, because of: (i) the possible involvement of silencing machineries from both the plant and the fungus; (ii) the requirement for local and remote transport of channelled dsRNA molecules; and (iii) the problem of dsRNA transport at the apoplast–symplast interface.

Gene annotation of *F. graminearum*'s genome (http://www.broadinstitute.org) predicted genes coding for two ARGONAUT-like proteins, two DICER-like proteins and five RNA-dependent RNA polymerases (RDR; Chen *et al*., 2015). Consistent with these findings, RNAseq analysis of axenically grown *F. graminearum*, treated with *CYP3*-dsRNA, showed high numbers of reads of *CYP3*-dsRNA-derived siRNAs, together showing that *F. graminearum* has a functional gene silencing system, which is a prerequisite for disease control by SIGS.

To test the antifungal activity of *CYP3*-dsRNA and their siRNA derivatives, a detached leaf assay was used; this enabled us to assess fungal growth in local (directly sprayed) and distal (semi-systemic, non-sprayed) leaf segments (Koch *et al*., 2016). Using confocal laser scanning microscopy, inhibitory fluorescent-labelled *CYP3*-$dsRNA_{A488}$ was detected in the plant vascular system and was absorbed eventually by the pathogen from (systemic) leaf tissue. The profile of *CYP3*-dsRNA accumulation, as demonstrated by northern blot analysis and RNAseq, showed that both long *CYP3*-dsRNA and plant-processed *CYP3*-dsRNA-derived siRNA accumulate in the plant vascular system. Translocation of siRNA, however, seems to be

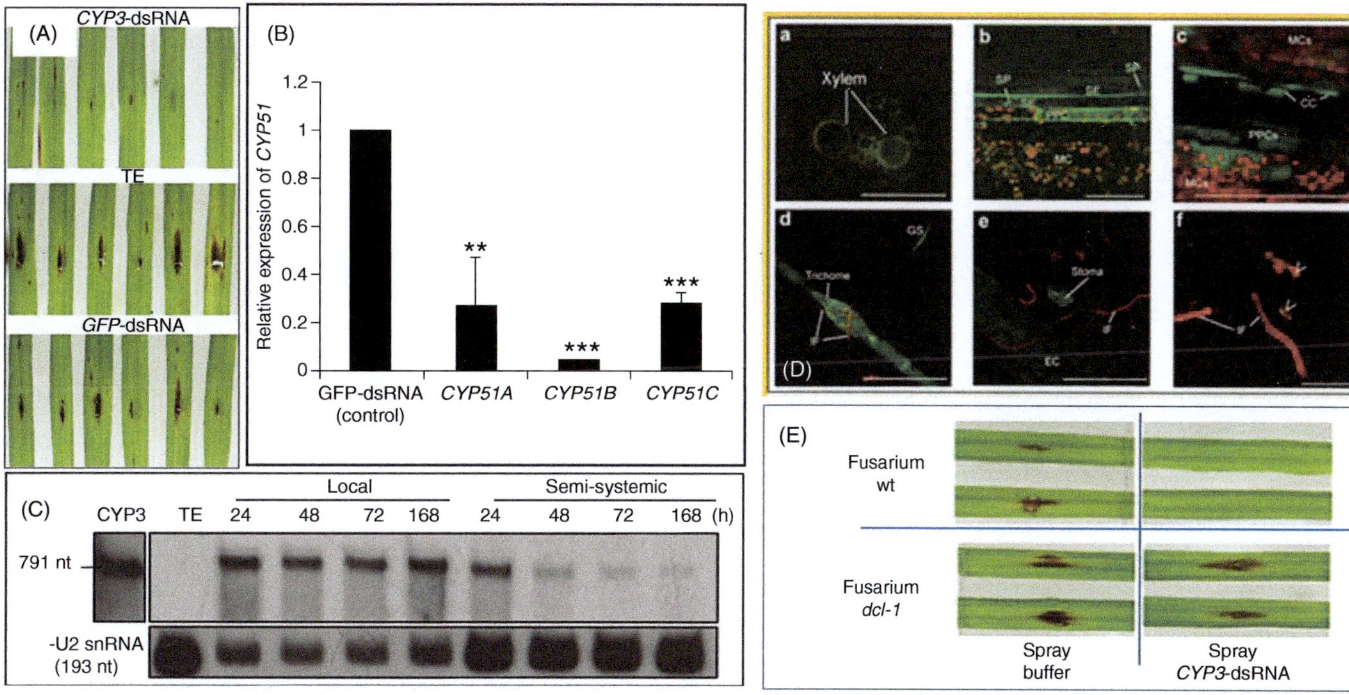


**Fig. 9.2.** Spray-induced gene silencing (SIGS) of fungal ergosterol biosynthesis genes leads to growth inhibition of the necrotrophic fungus *Fusarium graminearum*. A,B: Upon *CYP3*-dsRNA spraying to barley leaf segments and inoculation of non-sprayed distal parts of the leaf with conidia, fungal growth is inhibited; while dsRNA derived from green fluorescent protein (GFP) or mock spray with buffer (TE) do not affect fungal growth. Fungal arrest is correlated with reduced expression of *CYP51A*, *CYP51B* and *CYP51C* (B). C: Northern blot analysis shows that the double-stranded (ds)RNA is translocated to and accumulated in non-sprayed leaf areas. D: Upon spraying, ATTO488-labelled dsRNA (*CYP3*-dsRNAA488) is taken up by the plant and transferred via the vascular system to fungal infection sites. CC, companion cells; EC, epidermal cells; GS, germinating spore; IF, infection hyphae; MC, mesophyll cells; PPC, phloem parenchyma cells; SE, sieve elements; SP, sieve plate. RNA signals in germinated conidia are marked by arrowheads. Scale bars 100 µm. E: *CYP3*-dsRNA is translocated to and processed by the fungus at distal leaf areas that are not directly sprayed, as suggested by the finding that the fungal DICER mutant *Fg-dcl1-1*, unlike the wild type, is compromised in SIGS.

less efficient and thus siRNA concentration at the remote infection sites was not high enough to induced SIGS. Nevertheless, spraying high concentrations of *CYP3*-dsRNA-derived siRNA (20 ng $\mu L^{-1}$) induced the SIGS process, demonstrating that the fungus is able to absorb siRNAs from barley leaves.

Importantly, the transport and translocation of the unprocessed 791-nt long *CYP3*-dsRNA has a critical role in the SIGS process (Koch *et al.*, 2016). First, *CYP3*-dsRNA was translocated systemically via the vascular system as evidenced by northern blot detection of the dsRNA at remote sites. Second, the dsRNA was detected in pure phloem saps that were collected at remote sites by stylectomy using aphids (unpublished). Third and most convincing, when inoculated to remote sites, the fungal dicer-1 mutant $IFA65_{\Delta dcl\text{-}1}$ showed a compromised SIGS phenotype, suggesting that processing of *CYP3*-dsRNA by the fungus is critical in this case. Compromised DICER activity resulted in the fungus' inability to cleave *CYP3*-dsRNA into siRNA, thus interrupting the RNA interference mechanism. That unprocessed *CYP3*-dsRNA could be absorbed from remote leaf tissue has further implications for future disease control strategies using dsRNA. Application of longer dsRNAs might be more efficient than application of siRNAs given their efficient translocation. It is anticipated that long dsRNA is processed into many different inhibitory siRNAs by the fungus. This view is supported by the fact that RNAseq analysis of either axenic fungal cultures or infected leaf material showed a number of different *CYP3*-dsRNA-derived siRNA species.

Silencing signals in plants are mobile (Molnar *et al.*, 2010), just as viroids (Ding, 2009), preferably moving via the vascular system in the source-to-sink direction although some reports discussed transport in the opposite route (for review see Melnyk *et al.*, 2011). Source-to-sink movement is one reason why the phloem rather than the xylem is generally considered as the conduit for movement of the silencing signal. This hypothesis is supported by the finding that the xylem sap, which transports water and ions, commonly is free of RNA (Buhtz *et al.*, 2008). However, spray application of dsRNA onto detached leaves cannot be compared with the situation in an intact leaf. Exogenously applied dsRNA first reached the apoplast, including the xylem, and subsequently translocated into the symplast by a yet unknown mechanism. Apoplastic movement of RNA has been proposed, e.g. to explain how maternally expressed siRNAs could be transferred from the endosperm of developing seeds into the symplastically isolated embryo (Martienssen, 2010).

## 9.4 Questions to Be Resolved in the Future

In mammalian cells, perception of certain dsRNAs via toll-like receptors triggers an inflammation response (Karpala *et al.*, 2005; Gantier and Williams, 2007). In contrast, spraying *CYP3*-dsRNA onto barley leaves did not trigger an innate immune response (Koch *et al.*, 2016). While expression of salicylate-responsive *Pathogenesis-related 1* (*HvPR1*) and Jasmonate-responsive *S-adenosyl-l-methionine:jasmonic acid carboxyl methyltransferase* (*HvJMT*) in TRIS-EDTA-treated leaves was strongly induced upon infection by *F. graminearum*, *CYP3*-dsRNA treatment failed to induce these genes. Furthermore, fungus-induced expression of either gene was much lower in *CYP3*-dsRNA-treated leaves as compared with buffer-treated leaves. This result showed that *CYP3*-dsRNA does not induce immune responses in barley, and that the SIGS mechanism does not rely on activation of canonical defence pathways. Since the plant's immune system is not triggered by dsRNA, these results also suggest that efficient SIGS does not require a relevant fitness cost, and so may not negatively affect yield performance under field conditions.

Further research also is required to establish rules for optimal dsRNA structures, including dsRNA lengths, combinatorial order of gene fragments, target sites in a given gene target and the number of genes targeted by one dsRNA. Supporting the requirement for more information on the design of dsRNA constructs, RNAseq analysis revealed that most of the *CYP3*-dsRNA-derived siRNAs that accumulated in the axenic fungal mycelium treated with the dsRNA were not equally distributed at the *CYP3*-dsRNA scaffold, but could be assigned to the fragment of the *CYP51A* gene. As the sequence of the fragments in *CYP3*-dsRNA is BAC, corresponding to the *CYP51B*, *CYP51A* and *CYP51C* gene, further analysis is required to explain this bias in the production of siRNAs from *CYP3*-dsRNA.

Many more questions have to be addressed in the future to eventually judge the agronomical potential of SIGS, including the costs of RNA applications and their stability under field conditions. More research also is required into dsRNA uptake by compound design and chemical formulations. Another yet unassessed issue is the risk that microbial strains become insensitive to a commercial dsRNA product. Such scenario could probably be resolved by application of dsRNA that targets different regions in one gene or different genes. Most importantly, a commercial dsRNA product should be designed not to have off-target effects in other organisms that might be relevant in the respective agroecosystem, including beneficial fungi and bacteria. When considering the regulatory issue of RNA-based plant protection it is crucial to emphasize that the principles of SIGS and HIGS rely on the mechanisms found for trans-kingdom communication in mutualistic and parasitic interactions, and thus rely on natural phenomena (Weiberg *et al.*, 2013).

Regardless of how target-specific inhibitory RNAs are applied – by transgene expression (HIGS) or spray application (SIGS) – the use of target-specific inhibitory dsRNA to mediate protection against pathogens and pests is a potential alternative to conventional chemicals because dsRNAs are (i) highly specific and depend solely on their nucleotide sequence; and (ii) can be developed against an unlimited range of pathogens provided that the RNAi machinery is in place. Given the accumulation of dsRNA in the plant phloem, sucking insects also are realistic SIGS targets, as their efficient control by HIGS has been largely demonstrated (Eamens *et al.*, 2008; Abdellatef *et al.*, 2015).

Apart from the prospects for dsRNAs in future plant protection strategies, there is an additional technological potential in developing new pesticides. The simple phenotyping adopted by the SIGS screens renders them a powerful tool for genetic studies to assess compound targets with high efficiency and low cost.

## Acknowledgements

We thank P. Ganz, C. Birkenstock, U. Schnepp and V. Weisel for excellent technical assistance.

## References

Abdellatef, E., Will, T., Koch, A., Imani, J., Vilcinskas, A. and Kogel, K.H. (2015) Silencing the expression of the salivary sheath protein causes transgenerational feeding suppression in the aphid *Sitobion avenae*. *Plant Biotechnology Journal* 13, 849–857.

Bai, G. and Shaner, G. (2004) Management and resistance in wheat and barley to Fusarium head blight. *Annual Review of Phytopathology* 42, 135–161.

Baulcombe, D.C. (2015) VIGS, HIGS and FIGS: small RNA silencing in the interactions of viruses or filamentous organisms with their plant hosts. *Current Opinion in Plant Biology* 26, 141–146.

Broekaert, N., Devreese, M., De Baere, S., De Backer, P. and Croubels, S. (2015) Modified Fusarium mycotoxins unmasked: from occurrence in cereals to animal and human excretion. *Food Chemical Toxicology* 80, 17–31.

Buhtz, A., Springer, F., Chappell, L., Baulcombe, D.C. and Kehr, J. (2008) Identification and characterization of small RNAs from the phloem of *Brassica napus*. *The Plant Journal* 53, 739–749.

Castel, S.E. and Martienssen, R.A. (2013) RNA interference in the nucleus: roles for small RNAs in transcription, epigenetics and beyond. *Nature Reviews Genetics* 14, 100–112.

Chen, Y., Gao, Q., Huang, M., Liu, Y., Liu, Z., Liu, X. and Ma, Z. (2015) Characterization of RNA silencing components in the plant pathogenic fungus *Fusarium graminearum*. *Scientific Reports* 5, 12500.

Cheng, W., Song, X.S., Li, H.P., Cao, L.-H., Sun, K., Qiu, X.L., Xu, Y.B., Yang, P., Huang, T., Zhang, J.B., Qu, B. and Liao, Y.C. (2015) Host-induced gene silencing of an essential chitin synthase gene confers durable resistance to Fusarium head blight and seedling blight in wheat. *Plant Biotechnology Journal* 13, 1335–1345.

Ding, B. (2009) The biology of viroid-host interactions. *Annual Review of Phytopathology* 47, 105–131.

Eamens, A., Wang, M.B., Smith, N.A. and Waterhouse, P.M. (2008) RNA silencing in plants: yesterday, today, and tomorrow. *Plant Physiology* 147, 456–468.

Fan, J., Urban, M., Parker, J.E., Brewer, H.C., Kelly, S.L., Hammond-Kosack, K.E., Fraaije, B.A., Liu, X. and Cools, H.J. (2013) Characterization of the sterol 14α-demethylases of *Fusarium graminearum* identifies a novel genus-specific CYP51 function. *New Phytologist* 198, 821–835.

FAO (2013) *FAO Statistical Yearbook. World Food and Agriculture. Part 3: Feeding the World*. FAO, Rome. Available at: http://www.fao.org, accessed 15 December 2016.

Fernández-Ortuño, D., Loza-Reyes, E., Atkins, S.L. and Fraaije, B.A. (2010) The CYP51C gene, a reliable marker to resolve interspecific phylogenetic relationships within the Fusarium species complex and a novel target for species-specific PCR. *International Journal of Food Microbiology* 144, 301–309.

Fire, A., Xu, S., Montgomery, M.K., Kostas, S.A., Driver, S.E. and Mello, C.C. (1998) Potent and specific genetic interference by double-stranded RNA in *Caenorhabditis elegans*. *Nature* 391, 806–811.

Gantier, M.P. and Williams, B.R.G. (2007) The response of mammalian cells to double-stranded RNA. *Cytokine and Growth Factor Reviews* 18, 363–371.

Ghag, S.B., Shekhawat, U.K.S. and Ganapathi, T.R. (2014) Host-induced post-transcriptional hairpin RNA-mediated gene silencing of vital fungal genes confers efficient resistance against Fusarium wilt in banana. *Plant Biotechnology Journal* 12, 541–553.

Ghannoum, M.A. and Rice, L.B. (1999) Antifungal agents: mode of action, mechanisms of resistance, and correlation of these mechanisms with bacterial resistance. *Clinical Microbiology Reviews* 12, 501–517.

Gsaller, F., Hortschansky, P., Furukawa, T., Carr, P.D., Rash, B., Capilla, J., Müller, C., Bracher, F., Bowyer, P., Haas, H., Brakhage, A. and Bromley, M.J. (2016) Sterol biosynthesis and azole tolerance is governed by the opposing actions of SrbA and the CCAAT binding complex. *PLOS Pathogens* 12, e1005775.

Hamilton, A.J. and Baulcombe, D.C. (1999) A species of small antisense RNA in posttranscriptional gene silencing in plants. *Science* 286, 950–952.

Jansen, C., von Wettstein, D., Schäfer, W., Kogel, K.H., Felk, A. and Maier, F.J. (2005) Infection patterns in barley and wheat spikes inoculated with wild-type and trichodiene synthase gene disrupted *Fusarium graminearum*. *Proceedings of the National Academy of Sciences of the United States of America* 102, 16892–16897.

Karpala, A.J., Dorant, T.J. and Bean, A.G.D. (2005) Immune responses to dsRNA: implications for gene silencing technologies. *Immunology and Cell Biology* 83, 211–216.

Kazan, K., Gardiner, D.M. and Manners, J.M. (2012) On the trail of a cereal killer: recent advances in *Fusarium graminearum* pathogenomics and host resistance. *Molecular Plant Pathology* 13, 399–413.

Knip, M., Constantin, M.E. and Thordal-Christensen, H. (2014) Trans-kingdom cross-talk: small RNAs on the move. *PLoS Genetics* 10, e1004602.

Koch, A. and Kogel, K.H. (2014) New wind in the sails: improving the agronomic value of crop plants through RNAi-mediated gene silencing. *Plant Biotechnology Journal* 12, 821–831.

Koch, A., Kumar, N., Weber, L., Keller, H., Imani, J. and Kogel, K.H. (2013) Host-induced gene silencing of cytochrome P450 lanosterol C14α-demethylase-encoding genes confers strong resistance to Fusarium species. *Proceedings of the National Academy of Sciences of the United States of America* 110, 19324–19329.

Koch, A., Biedenkopf, D., Furch, A., Weber, L., Rossbach, O., Abdellatef, E., Linicus, L., Johannsmeier, J., Jelonek, L., Goesmann, A., Cardoza, V., McMillan, J., Mentzel, T. and Kogel, K.H. (2016) An RNAi-based control of *Fusarium graminearum* infections through spraying of long dsRNAs involves a plant passage and is controlled by the fungal silencing machinery. *PLOS Pathogens*. Available at: http://dx.doi.org/10.1371/journal.ppat.1005901, accessed 15 December 2016.

Kuck, K.H., Stenzel, K. and Vors, J.P. (2012) *Sterol Biosynthesis Inhibitors: Modern Crop Protection Compounds*. Wiley-VCH Verlag, Weinberg, Germany, pp. 761–805.

Liu, X., Yu, F., Schnabel, G., Wu, J., Wang, Z. and Ma, Z. (2011) Paralogous cyp51 genes in *Fusarium graminearum* mediate differential sensitivity to sterol demethylation inhibitors. *Fungal Genetics and Biology* 48, 113–123.

Martienssen, R.A. (2010) Heterochromatin, small RNA and post-fertilization dysgenesis in allopolyploid and interploid hybrids of *Arabidopsis*. *New Phytologist* 186, 46–53.

Melnyk, C.W., Molnar, A. and Baulcombe, D.C. (2011) Intercellular and systemic movement of RNA silencing signals. *The EMBO Journal* 30, 3553–3563.

Merhej, J., Richard-Forget, F. and Barreau, C. (2011) Regulation of trichothecene biosynthesis in *Fusarium*: recent advances and new insights. *Applied Microbiology and Biotechnology* 91, 519–528.

Molnar, A., Molnar, A., Melnyk, C.W., Bassett, A., Hardcastle, T.J., Dunn, R. and Baulcombe, D.C. (2010) Small silencing RNAs in plants are mobile and direct epigenetic modification in recipient cells. *Science* 328, 872–875.

Nowara, D., Gay, A., Lacomme, C., Shaw, J., Ridout, C., Douchkov, D., Hensel, G., Kumlehn, J. and Schweizer, P. (2010) HIGS: host-induced gene silencing in the obligate biotrophic fungal pathogen *Blumeria graminis*. *Plant Cell* 22, 3130–3141.

Price, D.R.G. and Gatehouse, J.A. (2008) RNAi-mediated crop protection against insects. *Trends in Biotechnology* 26, 393–400.

Tomilov, A.A., Tomilova, N.B., Wroblewski, T., Michelmore, R. and Yoder, J.I. (2008) Trans-specific gene silencing between host and parasitic plants. *The Plant Journal* 56, 389–397.

Vaucheret, H. and Fagard, M. (2001) Transcriptional gene silencing in plants: targets, inducers and regulators. *Trends in Genetics* 17, 29–35.

Weiberg, A., Wang, M., Lin, F.M., Zhao, H., Zhang, Z., Kaloshian, I., Huang, H.D. and Jin, H. (2013) Fungal small RNAs suppress plant immunity by hijacking host RNA interference pathways. *Science* 342, 118–123.

Weiberg, A., Bellinger, M. and Jin, H. (2015) Conversations between kingdoms: small RNAs. *Current Opinion in Biotechnology* 32, 207–215.

# 10 Targeting Nematode Genes by RNA Silencing

**John Fosu-Nyarko*, Sadia Iqbal and Michael G.K. Jones**
*Murdoch University, Perth, Australia*

## 10.1 Introduction

The free-living nematode *Caenorhabditis elegans* is arguably the best model for understanding RNA silencing: its anatomy and physiology allow many aspects of this natural gene regulation mechanism in eukaryotes, including the genes and the pathways involved, to be studied in detail. The successful development of this mechanism into a functional analysis tool has facilitated characterization of almost all of its genes. Following the required modifications this tool has been invaluable in assessing gene function of parasitic nematodes, and in particular in assessing the role of essential and nematode genes required for plant parasitism. Demonstration of effective silencing of plant-parasitic nematode genes after soaking infective stages in buffered dsRNA *in vitro*, and the ability of transgenic plant-processed small RNA triggers to effect silencing of nematodes, has opened new avenues for safe, environmentally friendly and sustainable approaches to nematode control based on RNA silencing. Current data indicate that this approach will deliver nematode resistance in crops of economic importance.

Although RNA silencing as a subject can be traced to discoveries in genetic manipulation of plants and virus–plant interactions (Napoli *et al.*, 1990; Hamilton and Baulcombe, 1999; Voinnet, 2001) as described in Chapters 1–4, much of the progress in this field can arguably be attributed to research on the free-living Rhabditid nematode *C. elegans*. *C. elegans* research continues to influence discoveries in other species including other *Caenorhabditis* species (e.g. *C. brenneri*, *C. briggsae*, *C. japonica*), and also animal and plant-parasitic nematodes (PPNs). In particular, development of RNA silencing mechanisms and their applications for functional genomics of parasitic nematodes has been directed by progress in the field using the model nematode *C. elegans*.

*C. elegans* research, in particular into RNA silencing, has also advanced understanding in the current genomic era, highlighted by next-generation sequencing of whole genomes. Sequencing and analyses of the *C. elegans* genome (Sulston *et al.*, 1992; *C. elegans* Sequencing Consortium, 1998) has contributed immensely to

* Corresponding author e-mail: j.fosu-nyarko@murdoch.edu.au

 

annotation of nematode genomes including the omnivorous *Pristionchus pacificus* (Dieterich *et al.*, 2008), the animal parasitic nematodes *Ascaris suum* (Jex *et al.*, 2011) and *Brugia malayi* (Ghedin *et al.*, 2007), the PPNs *Meloidogyne incognita* (Abad *et al.*, 2008), *M. hapla* (Opperman *et al.*, 2008), *Globodera pallida* (Cotton *et al.*, 2014) and *Pratylenchus coffeae* (Burke *et al.*, 2015). These advances have identified conserved biological processes as well as defined evolutionary differences and adaptations, and functional divergence of some genes between obligate parasitic and free-living nematodes.

Despite prominent differences in lifestyles of these nematodes, the intensive research in characterizing genes and proteins of *C. elegans* using RNA silencing has inspired research and technological applications for parasitic nematodes. This is exemplified in dsRNA-mediated silencing of PPN genes to study their function, and demonstration of the existence of similar functional mechanisms of RNA silencing in these species. Armed with this knowledge it then became possible to explore and identify appropriate genes of PPNs as targets for new strategies of nematode control: the methods and applications of RNA silencing to PPNs are the focus of this chapter.

## 10.2 Nematodes: *Caenorhabditis* Species

Nematodes (Phylum Nematoda) are the most abundant and diverse of multicellular organisms: the >25,000 species described to date are adapted to almost all forms of ecosystems (Blaxter, 2003). Free-living nematodes feed on bacteria or exist as scavengers, of which the best-studied are *Caenorhabditis* species, although more than half of known nematodes are parasites of animals and plants.

Since Sydney Brenner's choice of *C. elegans* to study the molecular biology of a simple nervous system (Brenner, 1974), it has become a powerful laboratory model organism for in-depth study of gene function and important molecular processes such as metabolism, chemoreception and neurobiology. Its transparent body, relatively short and simple 3-day life cycle normally with a 2–3 week lifespan, high reproduction rate and feeding habit (on laboratory strains of *E. coli*) allow easy culture, maintenance and use for a range of experimental purposes. In addition it has a relatively small genome (100Mb), and was the first multicellular eukaryote to be sequenced. The dedication of research scientists to functionally characterize its genome in detail, and the work to make resources available, has reinforced *C. elegans* as the model for studies of many facets of biology in which the mechanism of RNA silencing and its application to study gene function is widely employed.

## 10.3 Plant-parasitic Nematodes of Economic Importance

PPNs of economic importance are mostly the ubiquitous sedentary endoparasitic cyst (*Heterodera* spp., *Globodera* spp.) and root knot nematodes (*Meloidogyne* spp.), followed closely by the migratory endoparasitic root lesion nematodes (*Pratylenchus* spp.). The latter are major economic pests in cereal growing areas of Australia and the Pacific North-west of the USA (Jones *et al.*, 2013; Jones and Fosu-Nyarko, 2014). More than a decade ago, a study by Chitwood (2003) reported that these and related PPNs were reducing the yield and value of major agricultural and industrial crops worldwide to an estimated $US125 billion. Some nematodes cause economic losses to many crops, whereas others are more specific; for example, infestation of bananas and plantains in subtropical and tropical climates of Africa, Asia and the Americas by *Radopholus similis* (http:/www.cabi.org/isc/datasheet/46685).

PPNs are obligate, biotrophic pests which feed from the cytoplasmic contents of cells of host plants. They undergo six developmental stages: the egg, four juvenile (J1–J4) and the adult. Their lifestyles are characterized by their feeding habits and interactions with host cells after entering host tissues. In appropriate hosts and under suitable conditions the egg hatches and develops into the J1 stage, which then moults to the infective J2 stage. After leaving the eggshell the J2 then locates and enters a host root (or other tissue) with the aid of mechanical probing with a sclerotized stylet, accompanied by secretions (mainly of cell-wall modifying enzymes) from the subventral gland cells (Perry and Moens, 2011). For sedentary endoparasitic forms, such as cyst and root knot nematodes, it is the beginning of an intimate relationship with the host in which the J2, after entering the root and migrating from cell to cell, selects suitable 'feeding cells' and becomes sedentary. The feeding cells (syncytia or giant cells) develop in response to a combination of secretions from the dorsal gland cells and removal of contents by the nematode, which induce the differentiation of specific host cells into metabolically active sinks with features of transfer cells (Jones, 1981). The J2 nematodes feed from the syncytia or giant cells which continue to expand for about 3 weeks; moult three more times during this period; and develop to the adult stage by 4–5 weeks. The mechanisms by which sedentary nematodes induce and maintain feeding cells have been studied in increasing detail: they are able to induce changes in plant cell structures and modulate host metabolism using secretions from their gland cells so they can feed without being recognized by the host, and so evade host defences (Jones and Goto, 2011; Kyndt *et al.*, 2013).

In contrast, the life cycle of *Pratylenchus* spp. lasts for 3–9 weeks depending on host and conditions (Jones and Fosu-Nyarko, 2014). Unlike their sedentary counterparts, all stages (except the egg and J1) of *Pratylenchus* species are infective and can enter and leave the host root during feeding. This nomadic habit and mode of feeding causes characteristic necrotic brown lesions on roots: hence the name 'root lesion nematodes' (Fosu-Nyarko and Jones, 2016). Pratylenchid nematodes of the genera *Radopholus*, *Hirschmanniella* and *Scutellonema* (family Haplolaimidae) are also economic pests of particular crop species and are also sometimes referred to as lesion nematodes. The migratory pine wilt nematode *Bursaphelenchus xylophilus*, indigenous to North America and also present in Europe, similarly feeds from host cells, and causes major damage to managed and natural pine forests in Japan and elsewhere. For most PPNs, RNA silencing is now the tool of choice for functional analysis of their genes as a means of understanding their biology and to develop new control strategies.

## 10.4 RNA Silencing in Nematodes

The general concept of RNA silencing is discussed in Chapters 1–4 of this volume: regulation of the process in different nematodes and transitive RNA interference

(RNAi) are therefore only discussed briefly here. Specific processes of RNA silencing in nematodes, as in most eukaryotes, are regulated by proteins or effectors via the small interfering RNA (siRNA) and microRNA (miRNA) pathways. To date, circa 90 genes of *C. elegans* encoding such effectors have been identified. These include genes required for transporting silencing triggers (i.e. dsRNA, siRNA and miRNA), components of the Dicer and the RNA interference silencing complexes (RISC), those required for amplifying or inhibiting silencing signals and/or processes, nuclear RNA silencing effectors and argonaute proteins (Dalzell *et al.*, 2011; Maule *et al.*, 2011). Bioinformatic analyses of genomes and transcriptomes indicate that similar but fewer orthologous genes of *C. elegans* RNA silencing effectors are encoded by other nematodes (Dalzell *et al.*, 2011; Fosu-Nyarko *et al.*, 2016; Iqbal *et al.*, 2016). For animal and plant parasitic nematodes it appears that only about half the number of these genes encoding RNA silencing effectors (reported for *C. elegans*) are present, although orthologous genes for every essential process of the RNA silencing pathway appear to be present (Fig. 10.1). Possible explanations are that the parasitic genomes are more highly specialized; or that the gene products have diverged but still contain the key components required to sustain a functional RNA silencing mechanism; or

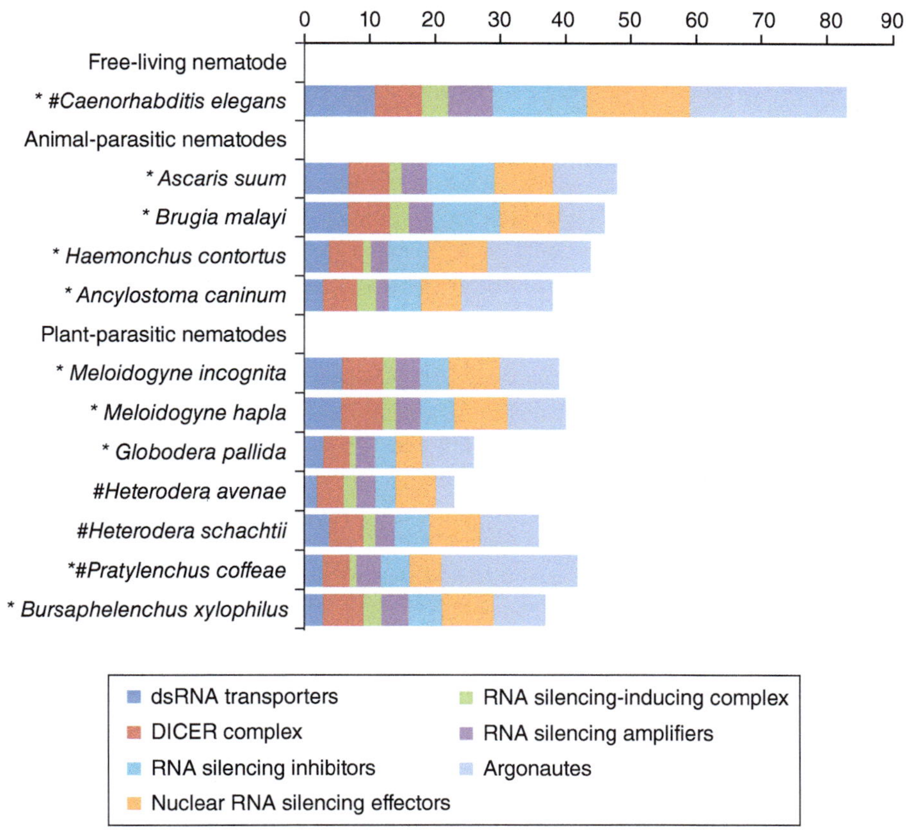


**Fig. 10.1.** Number of effectors of distinct process of the RNA silencing pathways in nematodes of different lifestyles. *, data sourced from genomic sequence; #, sourced from transcriptome data.

indeed that the pathways differ slightly in these nematodes (Dalzell *et al.*, 2011). Even if the effector repertoire is reduced, the existence of a functional RNA silencing pathway is clearly supported by the successful silencing of genes with different functions in parasitic nematodes of different genera and lifestyles (Knox *et al.*, 2007; Lilley *et al.*, 2007; Viney and Thompson, 2008; Li *et al.*, 2011).

A distinctive feature of the RNA silencing pathway in nematodes, as for fungi and plants, is an amplification step that ensures siRNAs are continuously present to perpetuate silencing of a target, referred to as transitive RNAi (Vaistij *et al.*, 2002; Alder *et al.*, 2003; Fernandez *et al.*, 2012). In this process, once RNAi is triggered, the sense strand of siRNAs not incorporated into the RISC becomes a primer for 5′–3′ synthesis of more dsRNA by RNA-dependent RNA polymerases using the target mRNA as a template (Sijen *et al.*, 2001; Alder *et al.*, 2003). The secondary siRNAs produced via transitive RNAi have di- and tri-phosphates at the 5′ ends compared to monophosphates for primary siRNAs, and are also exclusively associated with secondary argonautes (SAGOs) or worm-specific argonautes (WAGOs), identified only in nematodes (Faehnle and Joshua-Tor, 2007). With the aid of proteins for spread of dsRNA/siRNA within and between cells, the RNAi amplification mechanism perpetuates silencing after the original dsRNA or siRNA trigger is consumed. These characteristic features of the RNA silencing mechanisms in both plants and nematodes have been exploited to great effect in developing RNA silencing as a control strategy for PPNs, as discussed later in this chapter.

## 10.5 *C. elegans* as an Invaluable Model for RNA Silencing

Double-stranded RNA as a trigger for RNA silencing in animals was first demonstrated in *C. elegans* (Fire *et al.*, 1998). Since then, exogenous long (100–250 bp) and short (up to 25 nucleotides) dsRNA delivered by microinjection or through ingestion (where nematodes of different stages are soaked in buffered solutions with dsRNA, or by feeding on *E. coli* engineered to produce dsRNA corresponding to target genes) have become standard in nematode research (Timmons and Fire, 1998; Timmons *et al.*, 2001). What makes RNA silencing in *C. elegans* even more appealing is its anatomy and physiology; for example, the transparent body, which allows efficient observation of most structures and detection of changes in biological processes *in situ* after RNAi. Screening systems that allow single or multiple genes to be knocked down to study their roles in essential biological processes have since been developed (Kamath and Ahringer, 2003; Simmer *et al.*, 2003). For example, large-scale, genome-wide and high-throughput techniques have identified multiple genes of *C. elegans* required for embryogenesis (Sönnichsen *et al.*, 2005), moulting (Frand *et al.*, 2005), ageing (Hamilton *et al.*, 2005), fat regulation (Ashrafi *et al.*, 2003) and the nervous system (Schmitz *et al.*, 2007). Currently, most of the predicted genes (circa 20,000) of *C. elegans* have been knocked down to study their roles in biological processes.

The importance of RNA silencing research on *C. elegans* is emphasized by the availability of associated resources to the scientific community. Arguably the most comprehensive among the many electronic-based repositories for genetic information on *C. elegans* is the free, user-friendly and well-organized database, Wormbase (http://www.wormbase.org). Wormbase has evolved since it was first established and currently integrates readily accessible sequence data, information on molecular characterization of genes, tools for comparative sequence analysis and an impressive compilation of functional data generated via RNA silencing of almost every gene of *C. elegans*. Wormbase has recently been expanded to include similar data, not only for other *Caenorhabditis* species (e.g. *C. briggssae and C. brenneri*), but also tools which have been made available for comparative structural and *in silico* functional analysis of orthologous sequences of parasitic nematodes. In particular, the comprehensive description of RNAi phenotypes of *C. elegans* genes, detailed methods and associated published literature, and the integration of gene ontology to observed phenotypes, has become an invaluable tool for studies of orthologous genes in other organisms. For most genes, the RNAi data complement and integrate functional data such as spatial and temporal expression and compare these to orthologues of other organisms from other public databases such as the National Center for Biotechnology Information and SWISSPROT, which also happens to curate *C. elegans* resources.

There is currently no such comprehensive resource on RNA silencing for parasitic nematodes, and so *C. elegans* resources have been the main source and driver of advances of the current knowledge of RNA silencing in parasitic nematodes. However, since parasitic nematodes are not as amenable to some experimental manipulations that are possible with *C. elegans*, some initial challenges and scepticism to studying RNA silencing in the parasitic species had to be overcome. The following sections describe recent developments in RNA silencing of PPN genes.

## 10.6 RNA Silencing and Functional Analysis of PPN Genes

Despite sharing many physiological and anatomical features that make *Caenorhabditis* species pliable for RNA silencing research, the methods are not transferable for PPN research without modifications. PPNs are largely obligate parasites which live, feed from and develop within a host; in sedentary endoparasites, for example, only the pre-infective J2s are found outside the host for the brief period before they enter host tissues. This makes it impractical to deliver dsRNA triggers to most stages via microinjection or through genetically modified bacteria. Also, because PPNs do not feed outside the host, it was initially thought that soaking in dsRNA with buffered solutions would not result in ingestion of enough dsRNA to trigger RNA silencing. Addition of neurostimulants such as octopamine and resorcinol, capable of increasing pharyngeal pumping of PPNs in soaking media, resulted in 'forced uptake' of dsRNA by *Heterodera glycines* (Urwin *et al*., 2002), but subsequent experiments with *M. incognita* indicate such treatments are not always necessary (Bakhetia *et al*., 2005). These discoveries provided early proof that a functional RNA silencing mechanism existed in PPNs; and that soaking J2s in buffer with or without neurostimulants, and with 3 mM spermidine, 0.05% gelatine and 1–2 mM of fluorescein isothiocyante to trace uptake, could trigger silencing. The procedure has since become a widely used strategy for RNA silencing in PPNs.

To date, soaking of J2s (or, for migratory nematodes, mixed vermiform stages) with dsRNA has been used successfully to downregulate more than 50 essential and parasitism genes of PPNs of economic importance. These include the cyst nematodes *H. glycines, H. schachtii, G. pallida and G. rostochiensis*; the root knot nematodes *M. incognita, M. javanica, M. arenaria and M. artiellia*; the Pratylenchid root lesion nematodes *Pratylenchus zeae, P. thornei, P. coffeae, P. vulnus, P. penetrans*; and *Radopholus similis* and *B. xylophilus* (Lilley *et al.*, 2007; Li *et al.*, 2011; Fosu-Nyarko and Jones, 2016). Generally, the nematodes are soaked in buffered solutions with dsRNAs (42–750 bp) at 1–2 mg/mL for 4–24 h, after which the effects of gene knockdown are assessed phenotypically and quantitatively: this treatment leads to varying degrees of gene knockdown in different nematodes. To avoid possible off-target effects of silencing when using long dsRNAs (especially when genomic sequences of several nematode species were not available and it was not possible to analyse such potential effects *in silico a priori*), it was suggested that using double-stranded synthetic siRNAs as triggers, or separately as sense and antisense forms in a soaking method would result in specific silencing. An example is the specific knockdown of transcripts of FMRF amide-like peptides in J2s of both *G. pallida* and *M. incognita* using discrete 21 bp siRNAs (Dalzell *et al.*, 2010).

Despite the possibility of functional divergence of genes of free-living and parasitic nematodes, RNA silencing effects of orthologues of *C. elegans* have been important guides for silencing PPN genes. So far, RNA silencing of some genes indicates there is functional gene conservation in these nematodes (Tan *et al.*, 2013). Targeted and successfully silenced essential nematode genes are expressed in a range of tissue and cell types including muscle (Tan *et al.*, 2013), reproductive systems (Urwin *et al.*, 2002), hypodermis (Urwin *et al.*, 2002), neurons (Kimber *et al.*, 2007) and gland cells, and are involved in several vital molecular and biological processes such as RNA processing, moulting, reproduction, movement and feeding (Li *et al.*, 2011; Fosu-Nyarko *et al.*, 2015). In addition, genes encoding RNA silencing effectors of *M. incognita* have been silenced successfully, resulting in severe biological consequences to the nematodes, including abnormal development and reduction in infectivity of appropriate hosts (Iqbal *et al.*, 2016).

It is perhaps not surprising that there are variations in the levels of susceptibility of PPNs to RNA silencing and to silencing of specific genes. For example, in dsRNA-mediated silencing of *pat-10* and *unc-87* using equal amounts of dsRNA, Tan *et al.* (2013) reported that for both genes there was a greater degree of paralysis and incoordination and a greater reduction in transcript abundance for *P. thornei* than for *P. zeae*, suggesting that *P. thornei* was more susceptible to silencing than *P. zeae*. Also, differences in the extent of knockdown of different genes in the same organism are apparent for several cyst, root knot and root lesion nematodes. Such differences, which include the levels of gene knockdown induced by different lengths and sequences of dsRNA of the same gene (Sukno *et al.*, 2007), have been attributed to several factors. These include the amount, length and type of dsRNA triggers, the length of time nematodes are soaked in dsRNA, possible differences in the mode of dsRNA uptake and mechanisms of RNA silencing, the specific functions and turnover of particular genes, their copy number and their temporal expression patterns (Sukno *et al.*, 2007; Tan *et al.*, 2013).

Differences between the mechanisms of RNA silencing of particular genes of *C. elegans* and PPNs are also emerging. Whereas neuronally expressed genes of *C. elegans* are refractory to RNA silencing, similar genes – particularly FMRF amide-like (*flp*) neuropeptide genes of *G. pallida* and *M. incognita* – have been silenced successfully (Dalzell *et al.*, 2010; Papolu *et al.*, 2013). A possible explanation is that dsRNA uptake may follow a different pathway in the different nematodes.

Over the last few years, the application of RNA silencing to study the role of parasitism genes in nematode infection processes has provided major advances in PPN research, and this has been coupled to studying how secreted gene products interact with components of host plants. This has provided new understanding of the function of PPN parasitism gene products, including proteins that modify plant cell walls, and peptides that act to protect nematodes from plant defences, aid digestion of host cell contents or interact with host machinery to modulate plant processes. These proteins, commonly referred to as 'nematode effectors', are secreted principally by the dorsal and subventral pharyngeal glands cells, but some are also secreted by the amphids and hypodermis (Haegeman *et al.*, 2012). For sedentary PPNs, the repertoire of effectors includes specific peptides/proteins from the dorsal glands required for initiation, establishment and maintenance of permanent feeding sites which provide a source of nutrients for development and growth of the associated nematode after it has become sedentary (Haegeman *et al.*, 2012). Parasitism genes have been major targets of RNA silencing in almost all PPNs. The premise is that, if such determinants of the nematode infection process can be disrupted, this enables study of mechanisms involved in the infection process. In turn the knowledge generated could be applied as a nematode control strategy to knockdown genes vital for successful nematode parasitism. PPN parasitism genes are predominantly absent in the free-living *C. elegans*, except for a few such as gluthatione-S-transferase, which in fact may have diverged functionally in PPNs to play a role in host interaction.

The effectiveness of RNA silencing in PPNs can be assessed from measuring target transcript abundance and the nematode infectivity on appropriate hosts after dsRNA treatment. This approach has successfully complemented traditional tools for classifying putative effectors of nematode parasitism identified from gland cell secretions. Phenotypic abnormalities in nematodes, such as paralysis or abnormal behaviour, are not usually observed immediately after soaking in dsRNA of target genes, but interference of expression of known and putative nematode effectors usually results in some level of reduction in infectivity, as measured by the number of nematodes able to establish in appropriate hosts (Jones *et al.*, 2011). Actual reduction in the amount of a secreted nematode effector peptide still remains to be demonstrated; however, silencing of genes essential for parasitism or other vital functions have been used *in planta* to silence PPN genes, in which the host plant delivers the dsRNA silencing triggers.

Nevertheless, the development of RNA silencing of PPN genes has lagged behind that of *C. elegans*. The reasons are that genomic sequence information of PPNs has only recently become available; PPNs are more difficult experimental organisms because of their longer life cycle; and it is not yet possible to generate transgenic PPNs or to generate mutants to enable reverse genetics studies. In addition, apart from soaking in dsRNA there is no convenient synthetic feeding system that allows these obligate parasites to feed outside a host. The application of RNA silencing to essential PPN genes is nevertheless an exciting prospect, both for functional genomics studies, and to develop viable nematode control options.

## 10.7 Plant-derived RNA Silencing of Nematode Genes: Applications

Since the first demonstration of RNA silencing of splicing factor and integrase genes of *M. incognita* triggered by plant-processed dsRNA/siRNA 10 years ago (Yadav *et al.*, 2006), host-induced gene silencing (HIGS) has been explored for two purposes: (i) to study how nematode effectors interact with host plants; and (ii) as an alternative strategy of conferring genetic resistance to nematodes (Fosu-Nyarko and Jones, 2015). The principle of HIGS is that a plant is transformed to transcribe a nematode gene in the form of hairpin RNA, which is then processed by plant RNA-silencing machinery, such that a feeding nematode will ingest the processed dsRNA and/or siRNAs. The latter will then trigger RNA silencing of the targeted genes in the nematode. For genes vital for life processes or nematode parasitism, initially identified by soaking in dsRNA, plant expression via HIGS then provides a continuous source of the silencing trigger when nematodes feed. HIGS is then also an option to silence vital genes for which effects of silencing by soaking do not appear to persist after the trigger is removed (Tan *et al.*, 2013).

HIGS complements other molecular approaches used to assess the role of PPN parasitism genes and transcripts of gland cell secretions, conveniently using the model plants *Arabidopsis thaliana* and *Nicotiana* spp., and crop plants such as tomato, potato and grapes. The results of successful HIGS of essential genes depends on the gene targeted, and is manifested by a failure of nematodes to initiate successful infection, inability to migrate through or between cells, reduced feeding and poor development, resulting in no or low feeding cell establishment in an otherwise suitable host. The overall result is a significant reduction in numbers of adult nematodes that develop; a reduction in the number of eggs produced by females; malformed adults; and smaller and fewer galls per plant in the case of root knot nematodes (Table 10.1). Significant reductions in nematode reproduction can be achieved; for example, reductions in the number of females (up to 91%) and eggs (up to 95%) of cyst nematodes, provides evidence that HIGS is a viable strategy for control of major PPNs when essential genes are targeted (Fosu-Nyarko and Jones, 2015; Fosu-Nyarko and Jones, 2016).

## 10.8 Factors that Affect the Application and Efficacy of HIGS of Nematode Genes

Some detailed aspects of how HIGS of nematode genes works are yet to be understood, and so far it appears that a 100% control of PPNs using this approach alone is hard to achieve. Although there are differences in responses to HIGS by different PPNs, it now appears that other factors are also important. These include the target gene selected (the function, length and composition of dsRNA/hairpin), the specific transgenic events and the number studied, whether stable transgenic plants or hairy roots are used, and the vector constructs used to generate plants such as hairpin dsRNA expressed via constitutive promoters or tissue-specific expression. The levels of transgene expression can vary, although it is not clear whether any

**Table 10.1.** HIGS of parasitism genes of cyst and root knot nematodes: a summary of effects and reduced infectivity in model and crop plants.

| Nematode effector | Nematode putative or known effector | Nematode species | Plant/crop | RNAi phenotype and effect on nematode lifestyle | Reference |
|---|---|---|---|---|---|
| 16D10 | Secretory peptide | *Meloidogyne incognita*<br>*M. hapla*<br>*M. arenaria*<br>*M. javanica* | *Arabidopsis thaliana* | 39–83% reduction in number of eggs/g of root, 63–90% reduction in number of galls, general decrease in gall sizes | Huang *et al.*, 2006 |
| 16D10 | Secretory peptide | *M. incognita* | Grape | Significant reduction in number of eggs/g of hairy root | Yang *et al.*, 2013 |
| 16D10 | Secretory peptide | *M. chitwoodi* | Potato | Up to 68% reduction in number of egg masses/g of root | Dinh *et al.*, 2014b |
| 16D10 | Secretory peptide | *M. chitwoodi* | *Arabidopsis thaliana* | 57% and 67% reduction in number of egg masses and eggs, respectively | Dinh *et al.*, 2014a |
| | | | Potato | 71% and 63% reduction in number of egg masses and eggs, respectively | |
| Mi-crt | Calreticulin | *M. incognita* | *N. benthamiana* | J2s of silenced progeny induced 84% less galls on plants. dsRNA delivered via virus-induced gene silencing (VIGS) | Dubreuil *et al.*, 2009 |
| Mi-crt | Calreticulin | *M. incognita* | *Arabidopsis thaliana* | Reduced gene expression in nematodes by up to 75%; 62% reduction in galls/plant | Jaouannet *et al.*, 2013 |
| Mi8D05 | Secreted effector | *M. incognita* | *Arabidopsis thaliana* | Up to 90% reduction in gall formation on plant roots | Xue *et al.*, 2013 |
| NULG1a | Expressed in dorsal glands | *M. javanica* | *Arabidopsis thaliana* | 88% reduction in nematodes in roots | Lin *et al.*, 2013 |

*Continued*

**Table 10.1.** Continued.

| Nematode effector | Nematode putative or known effector | Nematode species | Plant/crop | RNAi phenotype and effect on nematode lifestyle | Reference |
|---|---|---|---|---|---|
| Mj-far-1 | Fatty acid and retinol binding protein secreted during sedentary stages | *M. javanica* | Tomato | Arrested nematode development, 80% reduction in far-1 transcripts in nematodes | Iberkleid *et al.*, 2013 |
| 8H07 | SKP-1 like protein | *Heterodera schachtii* | *Arabidopsis thaliana* | Up to 64% reduction in developing females | Sindhu *et al.*, 2009 |
| 10A06 | Zinc finger protein | | | Up to 42% reduction in developing females | |
| 3B05 | Cellulose binding protein | | | 12–47% reduction in developing females | |
| 4G06 | Ubiquitin-like protein | | | 23–64% reduction in developing females | |
| Hssyv46 | Nematode secreted peptides | *H. schachtii* | *Arabidopsis thaliana* | 36% reduced cyst formation | Patel *et al.*, 2008 |
| Hs5d08 | | | | 20% reduced cyst formation | |
| Hs4e02 | | | | 20% reduced cyst formation | |
| Hs4F01 | | | | 55% reduced cyst formation | |
| Hg30C02 | Parasitism effector | *H. schachtii* | *Arabidopsis thaliana* | 92% reduced cyst formation | Hamamouch *et al.*, 2012 |
| Gp-hyp-1 | Effector expressed in amphids | *Globodera pallida* | Potato | 50–60% reduction in infection | Eves-van den Akker *et al.*, 2014 |
| MiMsp40 | Effector expressed in subventral pharyngeal gland | *M. incognita* | *Arabidopsis thaliana* | Up to 51% reduction in number of galls; up to 18% reduction in egg masses/g of root | Niu *et al.*, 2016 |

such level correlates with the level of effective siRNAs produced by transgenic plants, or with the numbers of such triggers ingested by a feeding nematode.

Despite initial concerns that the success of HIGS could be limited by the presence of feeding tubes in giant cells and syncytia of root knot and cyst nematodes, respectively, current data indicate that silencing triggers are ingested from host cells by PPNs, and that this leads to silencing of target genes. The challenge is knowing the threshold amount and type of triggers ingested, be they long, unprocessed dsRNA or siRNAs such as one of four discrete sizes processed by DICERS of *Arabidopsis* (Margis *et al.*, 2006) or combinations of these (Fig. 10.2). This is an area where we do not know how these factors contribute to the variations in silencing phenomena observed with HIGS of nematode genes. The importance of understanding and optimizing all the parameters is that specific target sequences, sizes or conformations can be chosen to maximize the efficiency of a trigger of silencing, to increase its efficacy and stability.

One intriguing possibility is that, by careful selection of target sequences from different nematode species, it should be possible to develop resistance to a range of nematode species, either of the same genera or to include different genera. This requires more genomic sequencing, combined with bioinformatic tools that enable identification of such 'super targets'. There are examples of highly efficient HIGS, such as using the 16D10 parasitism effector for *Meloidogyne* species, and evidence for cross-species dsRNA-soaking-mediated silencing of *pat-10* and *unc-87* genes of *P. thornei* and *P. zeae* (Huang *et al.*, 2006; Tan *et al.*, 2013). Silencing two or more nematode genes using the same transgenic plant via HIGS would be both

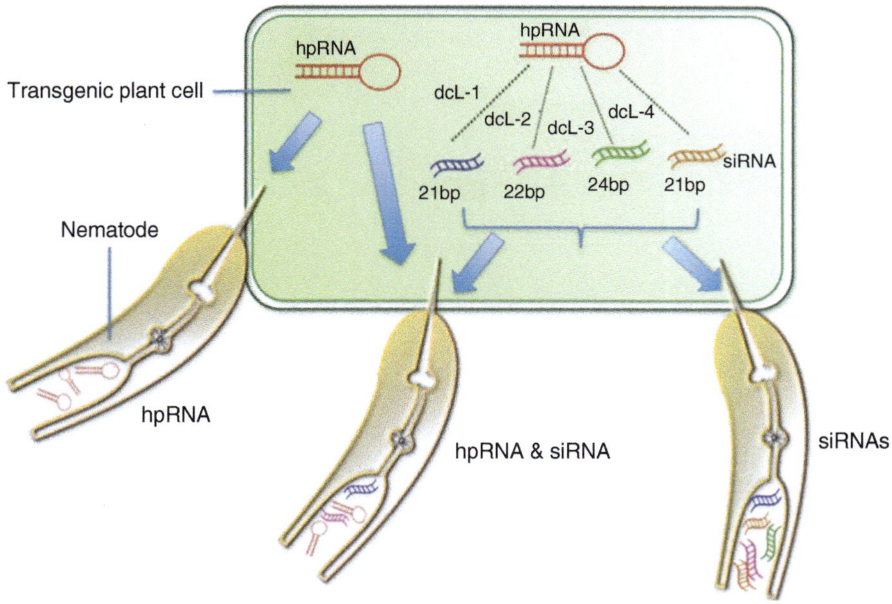


Fig. 10.2. Diagrammatic representation of plant-parasitic nematodes ingesting hairpin RNA (hpRNA) and/or DICER-like (dcl-1, dcl-2, dcl-3 and dcl-4) processed small interfering RNA (siRNA) from cell of transgenic host plant. (Diagram courtesy of Ms Fareeha Naz.)

cost-effective and efficient. Current data indicate the effect of multiple gene silencing of nematode genes either from crosses of plants with different hairpins, or for plants transformed with a single construct, may not be additive; but this situation can be improved with better selection of genes and transgenic events (Charlton *et al.*, 2010).

HIGS of nematode genes has also inspired development of alternative technologies to deliver dsRNA, such as ectopic delivery of dsRNA targeting essential nematode genes to mature plants or via seed dressings. The principle is that suitably formulated dsRNA can be taken up by plants and can move systemically, such that feeding nematodes will ingest dsRNA molecules to trigger silencing of vital genes. Such potential developments are attractive in jurisdictions where genetically modified plants cannot be grown.

In conclusion, there is no doubt that commercial implementation of biotechnology-based control strategies for nematode pests via HIGS, for example, is now closer to reality. This advance is based on RNA silencing technology, in which research into the free-living nematode *C. elegans* has played a major role. Intensive efforts to address practical issues of implementation will make gene silencing technology capable of delivering effective nematode resistance for crops, either as a stand-alone approach, or in combination with alternative nematode control options.

## References

Abad, P., Gouzy, J., Aury, J.-M., Castagnone-Sereno, P., Danchin, E.G., Deleury, E., Perfus-Barbeoch, L., Anthouard, V., Artiguenave, F., Blok, V.C., Caillaud, M.C., Coutinho, P.M., Dasilva, C., De Luca, F., Deau, F., Esquibet, M., Flutre, T., Goldstone, J.V., Hamamouch, N., Hewezi, T., Jaillon, O., Jubin, C., Leonetti, P., Magliano, M., Maier, T.R., Markov, G.V., McVeigh, P., Pesole, G., Poulain, J., Robinson-Rechavi, M., Sallet, E., Ségurens, B., Steinbach, D., Tytgat, T., Ugarte, E., van Ghelder, C., Veronico, P., Baum, T.J., Blaxter, M., Bleve-Zacheo, T., Davis, E.L., Ewbank, J.J., Favery, B., Grenier, E., Henrissat, B., Jones, J.T., Laudet, V., Maule, A.G., Quesneville, H., Rosso, M.N., Schiex, T., Smant, G., Weissenbach, J. and Wincker, P. (2008) Genome sequence of the metazoan plant-parasitic nematode *Meloidogyne incognita*. *Nature Biotechnology* 26, 909–915.

Alder, M.N., Dames, S., Gaudet, J. and Mango, S.E. (2003) Gene silencing in *Caenorhabditis elegans* by transitive RNA interference. *RNA* 9, 25–32.

Ashrafi, K., Chang, F.Y., Watts, J.L., Fraser, A.G., Kamath, R.S., Ahringer, J. and Ruvkun, G. (2003) Genome-wide RNAi analysis of *Caenorhabditis elegans* fat regulatory genes. *Nature* 421, 268–272.

Bakhetia, M., Charlton, W., Atkinson, H.J. and Mcpherson, M.J. (2005) RNA interference of dual oxidase in the plant nematode *Meloidogyne incognita*. *Molecular Plant-Microbe Interactions* 18, 1099–1106.

Blaxter, M.L. (2003) Nematoda: genes, genomes and the evolution of parasitism. *Advances in Parasitology* 54, 101–195.

Brenner, S. (1974) The genetics of *Caenorhabditis elegans*. *Genetics* 77, 71–94.

Burke, M., Scholl, E.H., Bird, D.M., Schaff, J.E., Coleman, S., Crowell, R., Diener, S., Gordon, O., Graham, S., Wang, X., Windham, E., Wright, G.M. and Opperman, C.H. (2015) The plant parasite *Pratylenchus coffeae* carries a minimal nematode genome. *Nematology* 17, 621–637.

*C. elegans* Sequencing Consortium (1998) Genome sequence of the nematode *C. elegans*: a platform for investigating biology. *Science* 282, 2012–2018.

Charlton, W.L., Harel, H.Y.M., Bakhetia, M., Hibbard, J.K., Atkinson, H.J. and McPherson, M.J. (2010) Additive effects of plant expressed double-stranded RNAs on root-knot nematode development. *International Journal for Parasitology* 40, 855–864.

Chitwood, D.J. (2003) Research on plant-parasitic nematode biology conducted by the United States Department of Agriculture–Agricultural Research Service. *Pest Management Science* 59, 748–753.

Cotton, J.A., Lilley, C.J., Jones, L.M., Kikuchi, T., Reid, A.J., Thorpe, P., Tsai, I.J., Beasley, H., Blok, V., Cock, P.J., Eves-van den Akker, S., Holroyd, N., Hunt, M., Mantelin, S., Naghra, H., Pain, A., Palomares-Rius, J.E., Zarowiecki, M., Berriman, M., Jones, J.T. and Urwin, P.E. (2014) The genome and life-stage specific transcriptomes of *Globodera pallida* elucidate key aspects of plant parasitism by a cyst nematode. *Genome Biology* 15, R43.

Dalzell, J.J., Mcmaster, S., Fleming, C.C. and Maule, A.G. (2010) Short interfering RNA-mediated gene silencing in *Globodera pallida* and *Meloidogyne incognita* infective stage juveniles. *International Journal for Parasitology* 40, 91–100.

Dalzell, J.J., Mcveigh, P., Warnock, N.D., Mitreva, M., Bird, D.M., Abad, P., Fleming, C.C., Day, T.A., Mousley, A., Marks, N.J. and Maule, A.G. (2011) RNAi effector diversity in nematodes. *PLOS Neglected Tropical Diseases* 5, e1176.

Dieterich, C., Clifton, S.W., Schuster, L.N., Chinwalla, A., Delehaunty, K., Dinkelacker, I., Fulton, L., Fulton, R., Godfrey, J., Minx, P., Mitreva, M., Roeseler, W., Tian, H., Witte, H., Yang, S.P., Wilson, R.K. and Sommer, R.J. (2008) The *Pristionchus pacificus* genome provides a unique perspective on nematode lifestyle and parasitism. *Nature Genetics* 40, 1193–1198.

Dinh, P.T., Brown, C.R. and Elling, A.A. (2014a) RNA interference of effector gene Mc16D10L confers resistance against *Meloidogyne chitwoodi* in *Arabidopsis* and potato. *Phytopathology* 104, 1098–1106.

Dinh, P.T., Zhang, L., Brown, C.R. and Elling, A.A. (2014b) Plant-mediated RNA interference of effector gene Mc16D10L confers resistance against *Meloidogyne chitwoodi* in diverse genetic backgrounds of potato and reduces pathogenicity of nematode offspring. *Nematology* 16, 669–682.

Dubreuil, G., Magliano, M., Dubrana, M., Lozano, J., Lecomte, P., Favery, B., Abad, P. and Rosso, M.N. (2009) Tobacco rattle virus mediates gene silencing in a plant parasitic root-knot nematode. *Journal of Experimental Botany* 60, 4041–4050.

Eves-Van Den Akker, S., Lilley, C.J., Jones, J.T. and Urwin, P.E. (2014) Identification and characterisation of a hyper-variable apoplastic effector gene family of the potato cyst nematodes. *PLOS Pathology* 10, e1004391.

Faehnle, C.R. and Joshua-Tor, L. (2007) Argonautes confront new small RNAs. *Current Opinion in Chemical Biology* 11, 569–577.

Fernandez, E.Q., Moyer, D.L., Maiyuran, S., Labaro, A. and Brody, H. (2012) Vector-initiated transitive RNA interference in the filamentous fungus *Aspergillus oryzae*. *Fungal Genetics and Biology* 49, 294–301.

Fire, A., Xu, S., Montgomery, M.K., Kostas, S.A., Driver, S.E. and Mello, C.C. (1998) Potent and specific genetic interference by double-stranded RNA in *Caenorhabditis elegans*. *Nature* 391, 806–811.

Fosu-Nyarko, J. and Jones, M.G.K. (2015) Chapter fourteen-application of biotechnology for nematode control in crop plants. *Advances in Botanical Research* 73, 339–376.

Fosu-Nyarko, J. and Jones, M.G.K. (2016) Advances in understanding the molecular mechanisms of root lesion nematode host interactions. *Annual Review of Phytopathology* 54, 253–278.

Fosu-Nyarko, J., Tan, J.C.H., Gill, R., Agrez, V.G., Rao, U. and Jones, M.G.K. (2015) *De novo* analysis of the transcriptome of *Pratylenchus zeae* to identify transcripts for proteins required for structural integrity, sensation, locomotion and parasitism. *Molecular Plant Pathology* 17, 532–552.

Fosu-Nyarko, J., Nicol, P., Naz, F., Gill, R. and Jones, M.G.K. (2016) Analysis of the transcriptome of the infective stage of the beet cyst nematode, *H. schachtii*. *PLOS One* 11, e0147511.

Frand, A.R., Russel, S. and Ruvkun, G. (2005) Functional genomic analysis of *C. elegans* molting. *PLOS Biology* 3, e312.

Ghedin, E., Wang, S., Spiro, D., Caler, E., Zhao, Q., Crabtree, J., Allen, J.E., Delcher, A.L., Guiliano, D.B., Miranda-Saavedra, D., Angiuoli, S.V., Creasy, T., Amedeo, P., Haas, B., El-Sayed, N.M., Wortman, J.R., Feldblyum, T., Tallon, L., Schatz, M., Shumway, M., Koo, H., Salzberg, S.L., Schobel, S., Pertea, M., Pop, M., White, O., Barton, G.J., Carlow, C.K., Crawford, M.J., Daub, J., Dimmic, M.W., Estes, C.F., Foster, J.M., Ganatra, M., Gregory, W.F., Johnson, N.M., Jin, J., Komuniecki, R., Korf, I., Kumar, S., Laney, S., Li, B.W., Li, W., Lindblom, T.H., Lustigman, S., Ma, D., Maina, C.V., Martin, D.M., McCarter, J.P., McReynolds, L., Mitreva, M., Nutman, T.B., Parkinson, J., Peregrín-Alvarez, J.M., Poole, C., Ren, Q., Saunders, L., Sluder, A.E., Smith, K., Stanke, M., Unnasch, T.R., Ware, J., Wei, A.D., Weil, G., Williams, D.J., Zhang, Y., Williams, S.A., Fraser-Liggett, C., Slatko, B., Blaxter, M.L. and Scott, A.L. (2007) Draft genome of the filarial nematode parasite *Brugia malayi*. *Science* 317, 1756–1760.

Haegeman, A., Mantelin, S., Jones, J.T. and Gheysen, G. (2012) Functional roles of effectors of plant-parasitic nematodes. *Gene* 492, 19–31.

Hamamouch, N., Li, C., Hewezi, T., Baum, T.J., Mitchum, M.G., Hussey, R.S., Vodkin, L.O. and Davis, E.L. (2012) The interaction of the novel 30C02 cyst nematode effector protein with a plant β-1, 3-endoglucanase may suppress host defence to promote parasitism. *Journal of Experimental Botany* 63, 3683–3695.

Hamilton, A.J. and Baulcombe, D.C. (1999) A species of small antisense RNA in posttranscriptional gene silencing in plants. *Science* 286, 950–952.

Hamilton, B., Dong, Y., Shindo, M., Liu, W., Odell, I., Ruvkun, G. and Lee, S.S. (2005) A systematic RNAi screen for longevity genes in *C. elegans*. *Genes and Development* 19, 1544–1555.

Huang, G., Allen, R., Davis, E.L., Baum, T.J. and Hussey, R.S. (2006) Engineering broad root-knot resistance in transgenic plants by RNAi silencing of a conserved and essential root-knot nematode parasitism gene. *Proceedings of the National Academy of Sciences* 103, 14302–14306.

Iberkleid, I., Vieira, P., De Almeida Engler, J., Firester, K., Spiegel, Y. and Horowitz, S.B. (2013) Fatty acid-and retinol-binding protein, Mj-FAR-1 induces tomato host susceptibility to root-knot nematodes. *PLOS One* 8, e64586.

Iqbal, S., Fosu-Nyarko, J. and Jones, M.G.K. (2016) Genomes of parasitic nematodes (*Meloidogyne hapla*, *Meloidogyne incognita*, *Ascaris suum* and *Brugia malayi*) have a reduced complement of small RNA interference pathway genes: knockdown can reduce host infectivity of *M. incognita*. *Functional and Integrative Genomics*, 16(4), 441–57.

Jaouannet, M., Magliano, M., Arguel, M., Gourgues, M., Evangelisti, E., Abad, P. and Rosso, M.N. (2013) The root-knot nematode calreticulin Mi-CRT is a key effector in plant defense suppression. *Molecular Plant-Microbe Interactions* 26, 97–105.

Jex, A.R., Liu, S., Li, B., Young, N.D., Hall, R.S., Li, Y., Yang, L., Zeng, N., Xu, X., Xiong, Z., Chen, F., Wu, X., Zhang, G., Fang, X., Kang, Y., Anderson, G.A., Harris, T.W., Campbell, B.E., Vlaminck, J., Wang, T., Cantacessi, C., Schwarz, E.M., Ranganathan, S., Geldhof, P., Nejsum, P., Sternberg, P.W., Yang, H., Wang, J., Wang, J. and Gasser, R.B. (2011) *Ascaris suum* draft genome. *Nature* 479, 529–533.

Jones, J.T., Haegeman, A., Danchin, E.G.J., Gaur, H.S., Helder, J., Jones, M.G.K., Kikuchi, T., Manzanilla-López, R., Palomares-Rius, J.E., Wesemael, W.M. and Perry, R.N. (2013) Top 10 plant-parasitic nematodes in molecular plant pathology. *Molecular Plant Pathology* 14, 946–961. DOI: 10.1111/mpp.12057

Jones, L.M., De Giorgi, C. and Urwin, P.E. (2011) *C. elegans* as a resource for studies on plant parasitic nematodes. In: Jones, J., Gheysen, G. and Fenoll, C. (eds) *Genomics and Molecular Genetics of Plant-nematode Interactions*. Springer, Dordrecht, The Netherlands, pp. 175–220.

Jones, M.G.K. (1981) Host cell responses to endoparasitic nematode attack: structure and function of giant cells and syncytia. *Annals of Applied Biology* 97, 353–372.

Jones, M.G.K. and Fosu-Nyarko, J. (2014) Molecular biology of root lesion nematodes (*Pratylenchus* spp.) and their interaction with host plants. *Annals of Applied Biology* 164, 163–181.

Jones, M.G.K. and Goto, D.B. (2011) Root-knot nematodes and giant cells. In: Jones, J., Gheysen, G. and Fenoll, C. (eds) *Genomics and Molecular Genetics of Plant-Nematode Interactions*. Springer, Dordrecht, The Netherlands, pp. 83–100.

Kamath, R.S. and Ahringer, J. (2003) Genome-wide RNAi screening in *Caenorhabditis elegans*. *Methods* 30, 313–321.

Kimber, M.J., Mckinney, S., Mcmaster, S., Day, T.A., Fleming, C.C. and Maule, A.G. (2007) *flp* gene disruption in a parasitic nematode reveals motor dysfunction and unusual neuronal sensitivity to RNA interference. *The FASEB Journal* 21, 1233–1243.

Knox, D.P., Geldhof, P., Visser, A. and Britton, C. (2007) RNA interference in parasitic nematodes of animals: a reality check? *Trends in Parasitology* 23, 105–107.

Kyndt, T., Vieira, P., Gheysen, G. and De Almeida-Engler, J. (2013) Nematode feeding sites: unique organs in plant roots. *Planta* 238, 807–818.

Li, J., Todd, T.C., Lee, J. and Trick, H.N. (2011) Biotechnological application of functional genomics towards plant parasitic nematode control. *Plant Biotechnology Journal* 9, 936–944.

Lilley, C.J., Bakhetia, M., Charlton, W.L. and Urwin, P.E. (2007) Recent progress in the development of RNA interference for plant parasitic nematodes. *Molecular Plant Pathology* 8, 701–711.

Lin, B., Zhuo, K., Wu, P., Cui, R., Zhang, L.-H. and Liao, J. (2013) A novel effector protein, MJ-NULG1a, targeted to giant cell nuclei plays a role in *Meloidogyne javanica* parasitism. *Molecular Plant-Microbe Interactions* 26, 55–66.

Margis, R., Fusaro, A.F., Smith, N.A., Curtin, S.J., Watson, J.M., Finnegan, E.J. and Waterhouse, P.M. (2006) The evolution and diversification of Dicers in plants. *FEBS Letters* 580, 2442–2450.

Maule, A.G., Mcveigh, P., Dalzell, J.J., Atkinson, L., Mousley, A. and Marks, N.J. (2011) An eye on RNAi in nematode parasites. *Trends in Parasitology* 27, 505–513.

Napoli, C., Lemieux, C. and Jorgensen, R. (1990) Introduction of a chimeric chalcone synthase gene into petunia results in reversible co-suppression of homologous genes in trans. *The Plant Cell* 2, 279–289.

Niu, J., Liu, P., Liu, Q., Chen, C., Guo, Q., Yin, J., Yang, G. and Jian, H. (2016) Msp40 effector of root-knot nematode manipulates plant immunity to facilitate parasitism. *Scientific Reports* 6, 19443.

Opperman, C.H., Bird, D.M., Williamson, V.M., Rokhsar, D.S., Burke, M., Cohn, J., Cromer, J., Diener, S., Gajan, J., Graham, S., Houfek, T.D., Liu, Q., Mitros, T., Schaff, J., Schaffer, R., Scholl, E., Sosinski, B.R., Thomas, V.P. and Windham, E. (2008) Sequence and genetic map of *Meloidogyne hapla*: a compact nematode genome for plant parasitism. *Proceedings of the National Academy of Sciences* 105, 14802–14807.

Papolu, P.K., Gantasala, N.P., Kamaraju, D., Banakar, P., Sreevathsa, R. and Rao, U. (2013) Utility of host delivered RNAi of two FMRF amide like peptides, flp-14 and flp-18, for the management of root knot nematode, *Meloidogyne incognita*. *PLOS One* 8, e80603.

Patel, N., Hamamouch, N., Chunying, L., Hussey, R., Mitchum, M., Baum, T., Xiaohong, W. and Davis, E.L. (2008) Similarity and functional analyses of expressed parasitism genes in *Heterodera schachtii* and *Heterodera glycines*. *Journal of Nematology* 40, 299–310.

Perry, R.N. and Moens, M. (2011) Introduction to plant-parasitic nematodes; modes of parasitism. In: Jones, J., Gheysen, G. and Fenoll, C. (eds) *Genomics and Molecular Genetics of Plant-Nematode Interactions*. Springer, Dordrecht, The Netherlands, pp. 3–20.

Schmitz, C., Kinge, P. and Hutter, H. (2007) Axon guidance genes identified in a large-scale RNAi screen using the RNAi-hypersensitive *Caenorhabditis elegans* strain nre-1 (hd20) lin-15b (hd126). *Proceedings of the National Academy of Sciences* 104, 834–839.

Sijen, T., Fleenor, J., Simmer, F., Thijssen, K.L., Parrish, S., Timmons, L., Plasterk, R.H. and Fire, A. (2001) On the role of RNA amplification in dsRNA-triggered gene silencing. *Cell* 107, 465–476.

Simmer, F., Moorman, C., Van Der Linden, A.M., Kuijk, E., Van Den Berghe, P.V., Kamath, R.S., Fraser, A.G., Ahringer, J. and Plasterk, R.H. (2003) Genome-wide RNAi of *C. elegans* using the hypersensitive rrf-3 strain reveals novel gene functions. *PLOS Biology* 1, e12.

Sindhu, A.S., Maier, T.R., Mitchum, M.G., Hussey, R.S., Davis, E.L. and Baum, T.J. (2009) Effective and specific *in planta* RNAi in cyst nematodes: expression interference of four parasitism genes reduces parasitic success. *Journal of Experimental Botany* 60, 315–324.

Sönnichsen, B., Koski, L., Walsh, A., Marschall, P., Neumann, B., Brehm, M., Alleaume, A.M., Artelt, J., Bettencourt, P., Cassin, E., Hewitson, M., Holz, C., Khan, M., Lazik, S., Martin, C., Nitzsche, B., Ruer, M., Stamford, J., Winzi, M., Heinkel, R., Röder, M., Finell, J., Häntsch, H., Jones, S.J., Jones, M., Piano, F., Gunsalus, K.C., Oegema, K., Gönczy, P., Coulson, A., Hyman, A.A. and Echeverri, C.J. (2005) Full-genome RNAi profiling of early embryogenesis in *Caenorhabditis elegans*. *Nature* 434, 462–469.

Sukno, S.A., Mccuiston, J., Wong, M.-Y., Wang, X., Thon, M.R., Hussey, R., Baum, T. and Davis, E. (2007) Quantitative detection of double-stranded RNA-mediated gene silencing of parasitism genes in *Heterodera glycines*. *Journal of Nematology* 39, 145–152.

Sulston, J., Du, Z., Thomas, K., Wilson, R., Hillier, L., Staden, R., Halloran, N., Green, P., Thierry-Mieg, J., Qiu, L., Dear, S., Coulson, A., Craxton, M., Durbin, R., Berks, M., Metzstein, M., Hawkins, T., Ainscough, R. and Waterston, R. (1992) The *C. elegans* genome sequencing project: a beginning. *Nature* 356, 37–41.

Tan, J.C.H., Jones, M.G.K. and Fosu-Nyarko, J. (2013) Gene silencing in root lesion nematodes (*Pratylenchus* spp.) significantly reduces reproduction in a plant host. *Experimental Parasitology* 133, 166–178.

Timmons, L. and Fire, A. (1998) Specific interference by ingested dsRNA. *Nature* 395, 854–854. DOI:10.1038/27579

Timmons, L., Court, D.L. and Fire, A. (2001) Ingestion of bacterially expressed dsRNAs can produce specific and potent genetic interference in *Caenorhabditis elegans*. *Gene* 263, 103–112.

Urwin, P., Lilley, C.J. and Atkinson, H.J. (2002) Ingestion of double-stranded RNA by preparasitic juvenile cyst nematodes leads to RNA interference. *Molecular Plant-Microbe Interactions* 15, 747–752.

Vaistij, F.E., Jones, L. and Baulcombe, D.C. (2002) Spreading of RNA targeting and DNA methylation in RNA silencing requires transcription of the target gene and a putative RNA-dependent RNA polymerase. *The Plant Cell* 14, 857–867.

Viney, M. and Thompson, F. (2008) Two hypotheses to explain why RNA interference does not work in animal parasitic nematodes. *International Journal for Parasitology* 38, 43–47.

Voinnet, O. (2001) RNA silencing as a plant immune system against viruses. *Trends in Genetics* 17, 449–459.

Xue, B., Hamamouch, N., Li, C., Huang, G., Hussey, R.S., Baum, T.J. and Davis, E.L. (2013) The 8D05 parasitism gene of *Meloidogyne incognita* is required for successful infection of host roots. *Phytopathology* 103, 175–181.

Yadav, B.C., Veluthambi, K. and Subramaniam, K. (2006) Host-generated double stranded RNA induces RNAi in plant-parasitic nematodes and protects the host from infection. *Molecular and Biochemical Parasitology* 148, 219–222.

Yang, Y., Jittayasothorn, Y., Chronis, D., Wang, X., Cousins, P. and Zhong, G.Y. (2013) Molecular characteristics and efficacy of 16D10 siRNAs in inhibiting root-knot nematode infection in transgenic grape hairy roots. *PLOS One* 8, e69463.

# 11 Gene Silencing Provides Efficient Protection against Plant Viruses

**Mario Tavazza,[1*] Alessandra Lucioli[1] and Vincenza Ilardi[2]**

*[1]Agenzia Nazionale per le Nuove Tecnologie, l'Energia e lo Sviluppo Economico Sostenibile, Rome, Italy; [2]Consiglio per la Ricerca in Agricoltura e l'Analisi dell'Economia Agraria – Centro di Ricerca per la Patologia Vegetale, Rome, Italy*

## 11.1 Introduction

One of the most significant social and economic challenges in the future will be to increase agricultural production while preserving the environment and human health (UN-DESA, n.d.). Viral diseases can affect both the yield and the quality of crop harvest. Although there are no exact estimates on a global scale, losses of up to 80–100% have been reported depending on the type of virus, crop, cultivar, year and region analysed (e.g. French and Stenger, 2003). These considerations underline the importance of developing robust strategies to reduce the impact exerted by viruses on crop productivity.

Plant viruses, which are obligate intracellular pathogens, are difficult to manage as they cannot be controlled directly by chemical applications on infected plants. Prophylactic measures consist of virus-free planting material, early detection and eradication of infected plants, crop rotation and control of virus vectors through chemicals. Extensive use of pesticides, however, induces the selection of chemical-resistant virus-vector populations (Elbert and Nauen, 2000), and impacts on health and the environment (WHO, n.d.).

The development of virus-resistant plants thus remains the most efficient and sustainable approach but, unfortunately, for several plant–virus combinations no useful natural sources of resistance have been identified. Also, in some cases, the resistance trait is too complex to be transferred to the desired cultivar without incurring agronomical deleterious effects due to linkage drag. Depending on the species, other constraints such as a high degree of heterozygosis, incompatibility barriers between species or challenging selection processes due to long generation times seriously limit breeding programmes. On top of that is the unique ability of viruses to evolve quickly through mutations, recombination and reassortment; often, under specific selective pressure, resistance-breaking viruses emerge.

In the early 1980s, the ability to transform and regenerate plants opened the way for developing biotechnological strategies to control virus infections. This

* Corresponding author e-mail: mario.tavazza@enea.it

 

chapter focuses on how knowledge of RNA silencing has been applied successfully to confer virus resistance in principal crops. We also illustrate how ongoing research in this field has allowed the development of new tools that, having been proved to be efficient in model systems, will be shortly applied to agriculture.

## 11.2 Pathogen-derived RNA Silencing-mediated Resistance: A Brief History

The most fertile biotechnological research area in the field of virus control stemmed from the concept of pathogen-derived resistance (PDR) proposed by Sanford and Johnston (1985). They suggested that the expression of a viral product outside of the normal context of infection should interfere with viral infection, potentially leading to resistance. From a mechanistic point of view, PDR falls into two broad classes: protein- and RNA silencing-mediated resistance.

PDR was initially conceived to be protein-mediated (Abel *et al.*, 1986). Thus, the underlying idea was to express a viral protein product in plants. In 1992–1993, Dougherty's group brought new life to the study of PDRs. They showed that the plants with the highest level of virus resistance had actively transcribed transgenes but low steady-state levels of transgene RNA. Importantly, this was also true for some plants harbouring untranslatable viral sequences, thus demonstrating the role of the RNA and not of the protein in these resistances (Lindbo *et al.*, 1993). The link between post-transcriptionally silenced transgenes and virus resistance was subsequently demonstrated using a recombinant virus harbouring exogenous sequences (English *et al.*, 1996). Two years later, a breakthrough in the nascent field of 'RNA silencing' was the discovery that transforming plants with virus or reporter gene constructs that produce RNA duplexes confer virus immunity or gene silencing on the plants (Waterhouse *et al.*, 1998). Shortly after the seminal work of Waterhouse *et al.* (1998) the same research group showed that transgenic expression of a viral sequence of interest (VSOI) arranged in an intron-hairpin RNA (hpRNA) conformation confers robust virus resistance (Smith *et al.*, 2000) (Fig. 11.1A).

RNA silencing refers to a family of sequence-specific phenomena downregulating gene expression (Bologna and Voinnet, 2014). The triggers of RNA silencing are partially or wholly double-stranded RNAs (dsRNAs), which are cut by Dicer-like enzymes into short molecules (small interfering RNAs, siRNAs) of 21–24 nucleotides in length. The siRNAs, protected from degradation by 2′-O-methylation, are loaded onto an RNA-induced silencing complex (RISC), which in the case of post-transcriptional gene silencing (PTGS), guides sequence-specific degradation of homologous RNAs. Also, RNA-dependent RNA polymerases, in particular RDR6 together with SGS3, amplify the RNA silencing response to convert aberrant RNAs into dsRNA. In plants, RNA silencing has, among other functions, a pivotal role in defending the host from viruses. To counterattack, viruses encode proteins called viral suppressors of RNA silencing (VSRs), which are able to interfere with different steps in the RNA silencing pathway (Csorba *et al.*, 2015).

Retrospectively, it appears that although most of the first transgenic plants were planned to express a viral protein product, in several cases resistance was the result of unpredictable activation of RNA silencing, stimulated by the integration

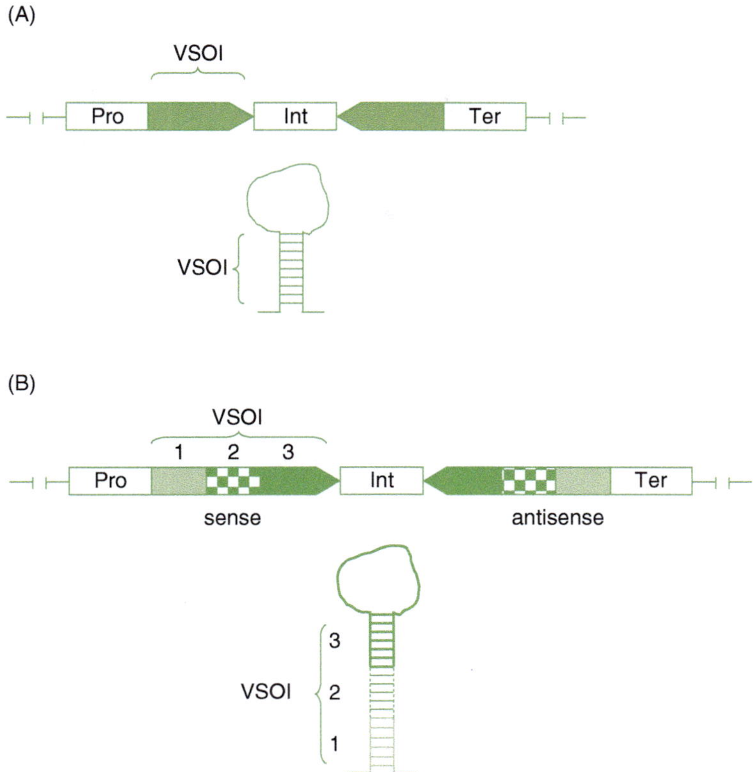


**Fig. 11.1.** Schematic representation of hpRNA constructs targeting one (A) or more (B) VSOI. VSOI: viral sequences of interest; Pro: promoter; Int: intron or spacer; Ter: terminator.

of multiple/rearranged transgene copies, which led to transcription of aberrant RNAs or dsRNAs. In this regard, two examples are the fruit tree clones approved for commercialization, papaya 55-1 and plum C5, resistant to the potyviruses *Papaya ring spot virus* (PRSV) and *Plum pox virus* (PPV), respectively.

## 11.3 Commercialized Crops Resistant to Viruses by Transgenic RNA Silencing Activation

Interestingly, all except one of the commercialized events reported in the disease resistance trait section of the 'GM approval database' (http://www.isaaa.org/gmapprovaldatabase/default.asp) are plants resistant to viruses.

Several of these transgenic plants were produced in the 1990s and were transformed with a translatable viral gene. Nevertheless, as mentioned above, in many cases the resistance was the result of activation of RNA silencing, as in the cases of the intensively studied papaya 55-1 and plum C5 events.

Papaya is an important fruit crop widely cultivated worldwide in lowland tropical and subtropical regions. PRSV is its most devastating virus, and in Brazil, Taiwan and Hawaii the papaya industry was forced to move to escape its destructive effects during the second half of the 20th century (Gonsalves, 2014). In 1989 the transgenic line papaya 55-1 was obtained, introducing the coat protein (*CP*) gene of the mild strain PRSV HA 5-1. Although the *CP* transgene was highly transcribed in line 55-1, CP mRNA accumulated at a low level, indicating that PRSV resistance was RNA-mediated. Two commercial cultivars, 'SunUp' and 'Rainbow' (homozygous and hemizygous for the *CP* gene, respectively), were subsequently derived from line 55-1. 'Rainbow' was resistant to Hawaiian PRSV isolates, while 'SunUp' was resistant to most PRSV isolates. Since 1997, papaya 55-1 and its derivatives have been approved for commercialization. Transgenic papaya acreage in the state of Hawaii increased from 39% of total acreage in 2000 to about 85% in 2012, witness to the deep social and economic impact of RNA silencing technology (Gonsalves, 2014).

The 'Stanley' plum C5 clone, renamed 'HoneySweet', was produced in 1994 and has been approved for commercialization in the USA since 2009 (Scorza *et al.*, 2013). Its resistance to PPV, the most devastating disease of stone fruits worldwide, derives from RNA silencing. In fact, C5 possesses multiple, rearranged and methylated *CP* transgene copies, expresses a low level of CP mRNA and does not produce a detectable amount of CP protein, while producing CP-specific siRNAs. Resistance of the C5 clone has been extensively evaluated for over 10 years in field trials in several countries, at sites characterized by endemic PPV infections (Scorza *et al.*, 2013). C5 plants were also graft-inoculated with different combinations of viruses affecting stone fruits. The analyses confirmed that C5 is highly resistant to PPV and that heterologous virus infections do not suppress PPV resistance.

The common bean line EMBRAPA 5.1, resistant to the *Begomovirus Bean golden mosaic virus* (BGMV), is the first commercialization event obtained through the hpRNA approach (Aragão *et al.*, 2013). BGMV is the most important virus infecting common beans in tropical and semi-tropical regions, and may account for yield losses up to 100%. The common bean line EMBRAPA 5.1, which harbours an hpRNA for BGMV *AC1* sequences, was approved for commercialization in Brazil in 2011. Homozygous R2 plants were immune when challenged with viruliferous whiteflies and the resistance was stable in field trials performed in distinct regions. Furthermore, resistance was maintained after eight self-pollinated generations and in backcrosses with non-transgenic commercial varieties.

## 11.4 Virus-derived Hairpin-RNA-mediated Resistance: A Robust Tool for Introducing New Virus Resistance Traits in Crops

As seen in the previous section, the first commercialized hpRNA-derived crop was EMBRAPA 5.1. However, many other virus-resistant crops have been obtained using hpRNA constructs, and it is expected that several of these events will shortly be approved for commercialization. Importantly, different VSOIs can be assembled as functional molecular building blocks in a single hpRNA transgene, thus conferring resistance to multiple viruses (Bucher *et al.*, 2006) (Fig. 11.1B). We describe some selected examples of hpRNA-mediated virus resistance for the top ten crops ranked by worldwide tonnage production. Seven of these: maize, rice, wheat, potato, soybean, cassava and barley are staple food crops (FAO, n.d.).

### 11.4.1 Sugarcane

Sugarcane accounts for 80% of world sugar production, and has recently been used as an important source for the production of ethanol. One of the most important viral diseases affecting sugarcane worldwide is the sugarcane mosaic disease, caused by two potyviruses, *Sugarcane mosaic virus* and *Sorghum mosaic virus* (SrMV) (Guo *et al.*, 2015).

Due to the high genetic complexity and low fertility of sugarcane, traditional breeding is difficult, and the transgenic approach seems to be the best route to achieve virus resistance. Accordingly, Guo and colleagues (2015) transformed 'ROC22' (the most popular cultivar in China) with an hpRNA construct containing SrMV *CP* sequences, and most of the regenerants were immune to SrMV.

### 11.4.2 Maize

The *Potyvirus Maize dwarf mosaic virus* (MDMV) causes yield losses up to 40% in maize. Although genetic resistances to MDMV have been reported, resistant germplasms such as the line 'H9-21' are of no use in breeding programmes because of their poor agronomic characteristics. Conversely, useful levels of MDMV resistance were introduced in the elite maize '18-599' through an hpRNA construct targeting the virus protease *P1* gene (Zhang *et al.*, 2013). Importantly, field trials showed that transgenic $T_2$ maize lines had MDMV resistance levels similar to that of 'H9-21'.

### 11.4.3 Rice

Rice supports nearly half of the world's population, and viral infections cause enormous yield losses worldwide. Viruses belonging to the genera *Oryzavirus*, *Phytoreovirus* and *Fijivirus* of the *Reoviridae* family, and the *Tenuivirus* genus, are those mostly impacting rice production. A few rice virus resistance genes are available, and resistance breaking is frequent. hpRNA technology has been successfully applied against most rice viruses (reviewed in Sasaya *et al.*, 2014). Two systematic studies, targeting each of the seven *Tenuivirus Rice stripe virus* (RSV) genes, and each of the twelve *Phytoreovirus Rice dwarf virus* (RDV) genes, identified tenuivirus genes encoding the CP or the movement protein, and *Reoviridae* genes encoding the viroplasm-associated protein as the best viral targets. These results were also applied to other rice viruses. Rice plants almost immune to the phytoreoviruses RDV and *Rice gall dwarf virus*, the *Fijivirus Southern rice black-streaked dwarf virus*, and to the tenuiviruses RSV and *Rice grassy stripe virus* were obtained. Importantly, this hpRNA technology was successfully applied to rice varieties used for human food and also to forage cultivars.

Finally, a detailed study of the $T_4$ generation of transgenic 'Daesanbyeo' rice plants highly resistant to RSV by targeting the *CP* gene showed that the agronomic traits of the transgenic rice plants were similar or even better than those of the untransformed control (Park *et al.*, 2012).

### 11.4.4 Wheat

One of the most important viral pathogens of wheat is the *Tritimovirus Wheat streak mosaic virus* (WSMV). WSMV, which is transmitted by the wheat curl mite *Aceria tosichella*, can cause yield loss up to 100% (French and Stenger, 2003). Of the three natural sources of WSMV resistance described, two are temperature sensitive. To overcome this limitation, Fahim and colleagues (2010) transformed wheat with an hpRNA containing WSMV *NIa* gene sequences. All plants of the line *hpws*2b were immune to WSMV.

### 11.4.5 Potato

Several viruses infect potato, and nine of them cause significant economic losses worldwide either alone or in mixed infections (Palukaitis, 2012). In particular, *Potato virus Y* (PVY) and *Potato leaf roll virus* (PLRV) can cause yield losses up to 80%. Several groups have obtained potato plants resistant or immune to one or more potato-infecting viruses such as the potyviruses PVY and *Potato virus A* (PVA), the *Luteovirus* PLRV and the *Potexvirus Potato virus X* (PVX). Immunity to PVY isolates of three distinct subgroups was achieved expressing a PVY CP-derived hpRNA (Missiou *et al.*, 2004). Marker-free transgenic potato plants concomitantly resistant to PVY and PVX under field conditions have also been obtained by combining sequences of both viruses in a single hpRNA construct (Bai *et al.*, 2009). A similar approach has been used to confer multiple resistances to PLRV, PVY and PVX (Arif *et al.*, 2012) and PLRV, $PVY^O$ and PVA (Chung *et al.*, 2013).

### 11.4.6 Soybean

The *Potyvirus Soybean mosaic virus* (SMV) is one of the most important viruses affecting soybean production and, depending on the cultivar analysed, yield losses up to 50% have been observed. Recently, Gao *et al.* (2015) transformed three Chinese and two US soybean cultivars with an hpRNA construct harbouring SMV *HC-Pro* sequences. Transgenic SMV-resistant plants were obtained for each genotype. Coherently, SMV-induced seed coat mottling was absent in seeds derived from $T_2$ virus-inoculated plants. Importantly, this work shows the feasibility of significantly improving SMV resistance in multiple elite soybean cultivars.

Zhang and colleagues (2011) applied a refined hpRNA strategy to confer resistance to three viruses in soybean. They built a transgene capable of transcribing an RNA containing three short hairpins interspersed with single-stranded RNA sequences. Each hairpin harboured sequences from a different virus, in particular

SMV, the *Alfamovirus Alfalfa mosaic virus* (AMV) and the *Comovirus Bean pod mottle virus* (BPMV). Transgenic plants of three independent lines exhibited a strong systemic (but not local) resistance when simultaneously challenged with SMV, AMV and BPMV.

### 11.4.7 Cassava

The Cassava brown streak disease (CBSD), which is caused by the ipomoviruses *Ugandan Cassava brown streak virus* (UCBSV) and *Cassava brown streak virus* (CBSV), has recently been identified as one of the major threats to food security in Africa. CBSD causes corky necrotic lesions in the storage roots, with yield reduction of 30–85%. Importantly, no resistance genes have been found in the cassava genotypes traditionally used by farmers.

Yadav and colleagues (2011) showed that cassava plants transgenic for UCBSV *CP* sequences were highly resistant to UCBSV. In addition one line, accumulating a high level of UCBSV CP-derived siRNAs, was resistant to both UCBSV and CBSV in a field trial (Ogwok *et al.*, 2012).

In Africa, cassava production is also affected by the cassava mosaic disease (CMD), which is caused by nine different cassava-infecting begomoviruses. Vanderschuren and colleagues (2012) exploited a combination of natural CMD resistance and engineered CBSD resistance. 'TEM7', a farmer-preferred cassava landrace possessing the dominant *CMD2* resistance gene, was transformed with an hpRNA for CBSV *CP* sequences. Transgenic TEM7 scions grafted on CBSV- or UCBSV-infected rootstocks remained symptomless and did not accumulate detectable viruses. In addition, the resistance was substantially maintained when transgenic TEM7 scions were grafted onto rootstocks co-infected with CBSV and *East African cassava mosaic virus*, a *Begomovirus* causing CMD. Thus, the combination of natural and engineered resistance is a promising approach to combat complex viral diseases. Previous work by Vanderschuren *et al.* (2009) also showed that hpRNA technologies could profitably be applied to confer resistance to the *Begomovirus African cassava mosaic virus*.

### 11.4.8 Sugarbeet

Rhizomania, which is caused by the *Benyvirus Beet necrotic yellow vein virus* (BNYVV), is one of the most devastating sugarbeet diseases worldwide, with yield reduction up to 80% reported in some susceptible varieties. The natural vector of the virus is the soilborne obligate root-infecting plasmodiophorid *Polymyxa betae*, whose spores can retain infectivity for up to 15 years. Resistance genes *Rz1* and *Rz2* have been used to manage the disease. However, several observations suggested that the resistance could be overcome under severe disease pressure (McGrann *et al.*, 2009). An hpRNA targeting the BNYVV B-type replicase gene was used to transform sugarbeet (Lennefors *et al.*, 2006), and transgenic plants displayed high resistance to BNYVV A-, B- and P-types when inoculated with *P. betae*. Importantly, virus titres in the transgenic plants were significantly lower than in the resistant

'Holly' carrying *Rz1*; resistance was not overcome by co-infection with four other common sugarbeet-infecting viruses from the genera *Pomovirus*, *Polerovirus* and *Closterovirus* (Lennefors *et al.*, 2008).

### 11.4.9 Tomato

Among the viruses causing severe damage to tomato plants, of particular importance are those belonging to the genera *Begomovirus*, *Cucumovirus* and *Tospovirus*. Tomato yellow leaf curl disease, caused by a group of begomoviruses belonging to the *Tomato yellow leaf curl virus* (TYLCV) complex, is regarded as the most devastating threat to cultivated tomatoes. Fuentes and colleagues (2006) showed that an hpRNA directed against TYLCV *C1* gene sequences conferred useful levels of resistance to the virus. Field trials of F6 progenies of an immune line, as well as the F4 progenies of an additional tomato line obtained with the same construct, confirmed the capacity of this approach to confer high resistance to TYLCV (Fuentes *et al.*, 2016). However, both resistant lines showed slight developmental abnormalities, probably due to the unintentional targeting of host genes by C1-derived siRNAs.

hpRNA technology was also effective in conferring resistance to multiple tospoviruses in tomato (Peng *et al.*, 2014). Transgenic plants harbouring an hpRNA construct containing the conserved sequence region of the *Watermelon silver mottle virus* (WSMoV) *L* gene were highly resistant to WSMoV, *Groundnut ringspot virus* and *Tomato spotted wilt virus*.

### 11.4.10 Barley

One of the most economically significant and widespread viruses of cereals is the *Luteovirus Barley yellow dwarf virus*-PAV (BYDV-PAV). In barley, yield losses ranging from 5.6% to 21.1% have been reported and, in some cases, the quality of the grains of BYDV-infected barley plants is so poor that they can be used only for animal feed. In barley, the natural resistance gene *Yd2* has been used in breeding programmes, although its effectiveness depends on the plant genetic background. Wang *et al.* (2000) transformed barley with an hpRNA construct containing BYDV-PAV polymerase sequences, and two lines highly resistant to BYDV-PAV were obtained. As in nature BYDV-PAV infections can be associated with the *Cereal yellow dwarf virus*, plants were also co-inoculated with both viruses, and none lost resistance.

### 11.4.11 hpRNA reviewed

The above indicates that hpRNA is a powerful technology. However, although hpRNA constructs efficiently activate RNA silencing of VSOI, this does not always lead to useful levels of resistance (Shimizu *et al.*, 2011). There are different reasons for this: (i) low accessibility of target region; (ii) low efficiency of the siRNAs produced; or (iii) insufficient amounts of the hpRNA-derived siRNAs , which cannot efficiently halt virus replication. Accurate selection and validation of the viral sequences capable of conferring high level of resistance are clearly required (e.g. Sasaya *et al.*, 2014).

## 11.5 Artificial miRNAs and ta-siRNAs: New Tools for Conferring Virus Resistance

MicroRNA (miRNA) and trans-acting small interfering RNA (ta-siRNA) silencing pathways have been exploited recently for conferring virus resistance (Niu *et al.*, 2006; Singh *et al.*, 2015).

### 11.5.1 Artificial miRNAs

miRNAs differ from siRNAs in being generated from the successive processing of genome-coded longer single-stranded RNA precursors (the pri-miRNA and the pre-miRNA), characterized by having specific stem-loop structures. The mature miRNAs, which are 19–24 nt in length, are recruited to the RISC complex to downregulate their target mRNAs by degradation or translational repression in a sequence-specific manner. Importantly, it is possible to manipulate a pri-miRNA/pre-miRNA in such a way that the mature miRNA sequence is complementary to the desired target RNA (Tiwari *et al.*, 2014). These engineered miRNAs are known as artificial miRNAs (amiRNAs). The first evidence that amiRNAs can confer virus resistance came from Niu and colleagues (2006). Successive studies showed that amiRNA technology can confer resistance to different virus genera such as: (i) *Potyvirus*, *Potexvirus*, *Cucumovirus*, *Tymovirus* and *Tritimovirus* (positive-sense RNA viruses); (ii) *Tospovirus* (a negative-sense RNA virus); and (iii) *Mastrevirus* and *Begomovirus* (single-stranded DNA viruses). Although several of these studies were conducted on model plants, they provided valuable information for developing robust amiRNA-mediated resistance. In particular, more than one effective amiRNA, directed against different conserved regions of a viral genome, should be expressed at a high level in plants to reduce the likelihood that the virus can overcome resistance.

Two studies have shown that the expression of a polycistronic amiRNA (Fig. 11.2A) is a feasible approach to confer robust virus resistance in two important staple food crops, wheat and barley. Fahim and colleagues (2012) built a polycistronic amiRNA precursor targeting five distinct regions of the WSMV genome. Importantly, three transgenic lines for the amiRNA construct were immune to the virus. Similarly, Kis *et al.* (2015) showed that, through an accurate *in silico* and *in vivo* selection of the amiRNAs to be assembled into the polycistronic construct, it was possible to obtain transgenic barley plants fully resistant against the *Mastrevirus Wheat dwarf virus*. *In silico* selection refers to the bioinformatics-based selection of the best amiRNAs, while *in vivo* selection refers to their validation in transient assays to confirm the real efficacy of designed miRNAs. This screening is mandatory to increase the possibility of achieving a high level of resistance.

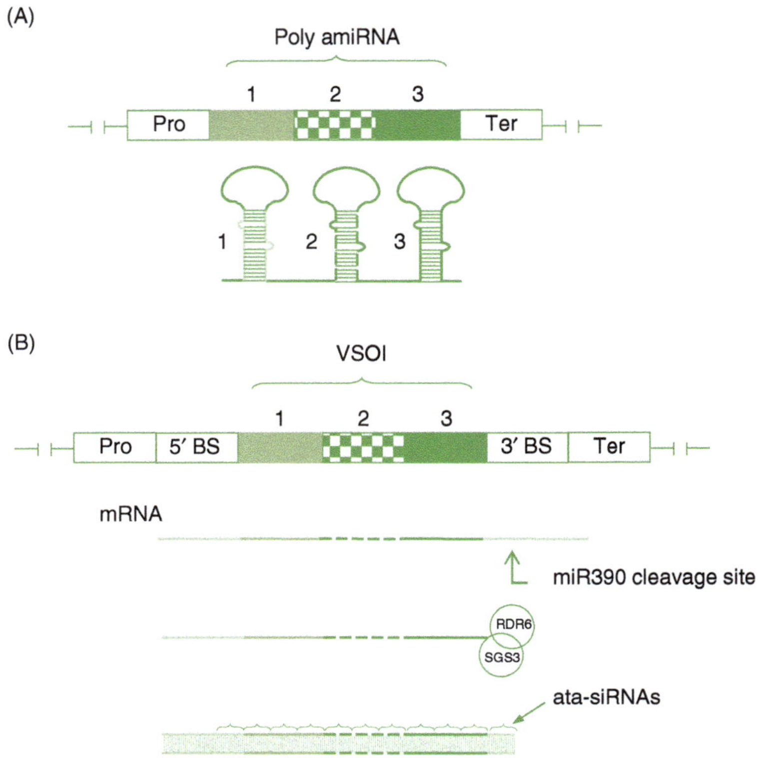


**Fig. 11.2.** Schematic representation of amiRNA and ata-siRNA constructs. A: Construct encoding a polycistronic amiRNA. B: TAS3-based ata-siRNA construct. VSOI: viral sequences of interest; Pro: promoter; Ter: terminator; 5′BS and 3′BS: 5′ and 3′ miR390 binding sites, respectively.

### 11.5.2 Artificial ta-siRNAs

Ta-siRNAs are a different group of plant siRNAs and play a significant role in regulating gene expression. Ta-siRNAs are the result of a hierarchic, sequential action of specific miRNAs, RDR6 and DCL4 on the non-coding transcripts of the *TAS* genes. The biogenesis of the ta-siRNAs starts with the recognition and cut of TAS transcripts by a specific miRNA, followed by the conversion of one of the cleavage TAS RNA products into a dsRNA molecule by the action of RDR6 and SGS3. The resulting dsRNA is then diced by DCL4 in successive steps starting from the end generated by the miRNA-guided cleavage, thus producing 21 nt phased ta-siRNAs (Fei *et al.*, 2013). It should be noted that, similar to the miRNAs, the *TAS* genes can be manipulated to produce artificial ta-siRNAs (ata-siRNAs), allowing them to be used for the potential targeting of any gene of interest (Montgomery *et al.*, 2008). Four *TAS* gene families have been described in *Arabidopsis*. In particular the *TAS3* family, which is targeted by a two-hit miR390-mediated mechanism, has been exploited recently to build engineered constructs expressing ata-siRNAs (Fig. 11.2B) directed against both DNA and RNA viruses. Singh *et al.* (2015) designed an ata-siRNA vector (TRiV) in which two miR390 binding sites (BSs) were separated by short sequences (200–250 bp) of the *AC2* or *AC4* genes of the *Begomovirus Tomato leaf curl New Delhi virus* (ToLCNDV). Transgenic tobacco plants harbouring TRiV-AC2 or TRiV-AC4 accumulated ata-siRNAs specific for the *AC2* and *AC4* sequences, respectively. Importantly, most of the transgenic plants remained symptomless and accumulated a low amount of viral DNA when challenged with ToLCNDV or *Tomato leaf curl Gujarat virus*.

A more refined strategy to confer resistance to both the *Potyvirus Turnip mosaic virus* (TuMV) and the *Cucumovirus Cucumber mosaic virus* (CMV) was adopted afterwards (Chen *et al.*, 2016). In this case six bioinformatically selected siRNAs, 21 nt in length and targeting different TuMV and CMV genes, were assembled in a phased order and cloned between miR390 BSs of the *Arabidopsis TAS3a* gene. Transgenic *Arabidopsis* plants accumulating CMV- and TuMV-specific ata-siRNAs were fully resistant when co-challenged with both viruses. This refined approach permits the expression of only the selected ata-siRNAs, so increasing the chances of efficiently and selectively silencing the intended viral genome. Moreover, the fact that TAS3 transcripts and the trigger miR390 are conserved in monocot and dicot plants should make this a tool of wide applicability.

## 11.6 RNA Silencing of Plant Susceptibility Genes: An Additional Route for Virus Resistance

As reported in the previous sections, RNA silencing against virus-derived sequences, pursued through different tools (hpRNA, amiRNA or ata-siRNA), is a feasible approach for introducing virus resistance traits in crops.

RNA silencing against susceptibility plant genes can also be exploited, in some instances, to confer virus resistance. In fact, as viruses are obligate intracellular parasites characterized by limited gene repertoires, they need to use and hijack host factors to establish a successful interaction. Thus knockout, knockdown or mutations of those plant genes facilitating pathogen infection (susceptibility genes) should, in principle, reduce the ability of the virus to cause disease.

The majority of host factors are indispensable for plant viability. However, those that – once knocked out – do not adversely affect plant growth and productivity can be exploited to confer resistance (Wang, 2015). In particular, susceptibility genes encoding the eukaryotic translation initiation factors 4E (eIF4E), 4G (eIF4G) or their isoforms, eIF(iso)4E and eIF(iso)4G; the TOM1 and TOM3 proteins; the small GTP-binding ARL8 protein; and the DEAD-box RNA helicase-like PpDDXL, among others, appear dispensable for plant growth in some species (van Schie and Takken, 2014).

In nature, virus resistance traits conferred by a mutated susceptibility gene usually become apparent in homozygosis and so are, from a genetic point of view, recessive characters. Interestingly, most of the natural recessive plant genes encode for eukaryotic translation initiation factors. Depending on the specific virus–host combination, mutants of *eIF4E*, *eIF(iso)4E*, *eIF4G* or *eIF(iso)4G* genes were shown to confer resistance to *Potyvirus*, *Sobemovirus*, *Carmovirus*, *Bymovirus* and *Waikavirus* (Sanfacon, 2015).

Recently, the *P. domestica eIF(iso)4E* gene was cloned and an eIF(iso)4E/hpRNA construct was introduced into plum plants. Transgenic plants silenced for the *eIF(iso)4E* were resistant to a PPV-D isolate (Wang *et al.*, 2013) and, based on previous results, these plum plants should be characterized by a broad-spectrum resistance to PPV.

The genes encoding TOM1 and TOM3 proteins are required for *Tobamovirus* multiplication in *Arabidopsis*. Importantly, these genes are conserved among a variety of plant species, suggesting that they could be exploited to confer *Tobamovirus* resistance in crops. In fact, transgenic tobacco plants harbouring hpRNA constructs for both *TOM1* and *TOM3* were highly resistant to several *Solanaceae*-infecting tobamoviruses such as *Tobacco mosaic virus*, *Tomato mosaic virus*, *Tobacco mild green mosaic virus* and *Pepper mottle mosaic virus* (Asano *et al.*, 2005).

Recently, Lapidot *et al.* (2015) have identified *Pelo* as the gene responsible for the TYLCV resistance conferred by the major recessive locus (*ty-5*) in the tomato line TY-172. *Pelo* encodes for a messenger RNA surveillance factor implicated in the ribosome recycling phase of protein synthesis. Notably, they showed that TYLCV-susceptible tomato plants transformed with an hpRNA construct capable of silencing the wild-type *Pelo* allele were resistant to TYLCV.

The RNA silencing approach directed against susceptibility plant genes shifts the virus-resistance trait from recessive to dominant, and this is important in permitting a ready transmission by crossing. This aspect is of relevance, especially when it is desirable to introduce the resistance trait into elite cultivars characterized by a long life cycle and a high degree of heterozygosis, such as *Prunus* species. Also, when more than one copy of a susceptibility gene is present per haploid genome, the RNA silencing is markedly superior to classical breeding in fixing recessive resistant traits. In fact, an hpRNA, amiRNA or ata-siRNA construct in hemizygosity can silence all copies of the susceptibility gene, thus conferring full resistance.

## 11.7 Aspects to Be Considered when Designing an RNA Silencing-mediated Resistance Strategy

In nature, mixed viral infections often occur. It has been shown that, in some cases, RNA silencing-mediated resistance can be overcome if the plant is previously infected with a different virus (Vassilakos, 2012). This phenomenon is the result of the ability of the VSR of the infecting virus to interfere with the RNA silencing pathway. Therefore, the potential impact of mixed viral infections should be taken into account when designing the RNA silencing strategy. However (as reported in sections 11.3 and 11.4), in several cases a previous infection with a different virus did not jeopardize resistance (Wang *et al.*, 2000; Lennefors *et al.*, 2008; Scorza *et al.*, 2013).

Another possible concern associated with siRNA-mediated resistance is its potential inhibition at low temperature (Vassilakos, 2012). Although a partial reduction of siRNA accumulation can be observed at low temperatures, this does not always impact hpRNA-mediated resistance (Di Nicola *et al.*, 2014). Also, amiRNA-mediated resistance was proved to be effective at low temperatures (Niu *et al.*, 2006; Kis *et al.*, 2015).

A potential drawback is the so-called 'off-target' effect. This phenomenon occurs as the outcome of sequence homology between some small RNAs derived from the transgene and mRNAs of the recipient plant. In particular the processing of long dsRNA, which generates a large pool of different siRNAs, can result in some instances in the unintentional downregulation of plant gene expression (e.g. Fuentes *et al.*, 2016). However, the expression of few selected ami- or ata-siRNAs (Kis *et al.*, 2015; Chen *et al.*, 2016) can prevent off-target effects, while still conferring resistance.

## 11.8 Concluding Remarks

Useful resistance has been obtained by gene silencing in many plants against a wide variety of viruses belonging to more than 20 different genera, and comprising single-stranded-DNA, single-stranded-RNA (both positive and negative) and double-stranded-RNA viruses. For the sake of brevity, we have only reported on selected examples of hpRNA-mediated resistance for the top ten crops ranked by production in tonnage; however, there are many other interesting applications in crops used for human feeding (e.g. banana, cucurbits and citrus). Resistant transgenic plants have been commercialized and are now used by farmers. Growing knowledge of silencing mechanisms and plant genomes allow the continuous improvement and refining of silencing-based tools, making them ever more specific in targeting virus sequences without affecting the expression of plant genes. In particular, amiRNA- and ata-siRNA-based silencing strategies (Carbonell *et al.*, 2014; Jacobs *et al.*, 2016), together with internet resources for amiRNA and ata-siRNA design (e.g. WMD3 http://wmd3.weigelworld.org/cgi-bin/webapp.cgi; P-SAMS http://p-sams.carringtonlab.org/) are expected to improve our ability to obtain transgenic plants with multiple virus resistance, with no detrimental effects on their agronomic traits. However, extensive field trials are required to validate the effectiveness of these new tools in conferring stable and durable virus resistance. From the point of view of biosafety, transgenic plants expressing short sequences of foreign RNAs are undoubtedly safer and more acceptable than those expressing foreign proteins.

## References

Abel, P.P., Nelson, R.S., De, B., Hoffmann, N., Rogers, S.G., Fraley, R.T. and Beachy, R.N. (1986) Delay of disease development in transgenic plants that express the *tobacco mosaic virus* coat protein gene. *Science* 232, 738–743.

Aragão, F.J., Nogueira, E.O., Tinoco, M.L.P. and Faria, J.C. (2013) Molecular characterization of the first commercial transgenic common bean immune to the *Bean golden mosaic virus*. *Journal of Biotechnology* 166, 42–50.

Arif, M., Azhar, U., Arshad, M., Zafar, Y., Mansoor, S. and Asad, S. (2012) Engineering broad-spectrum resistance against RNA viruses in potato. *Transgenic Research* 21, 303–311.

Asano, M., Satoh, R., Mochizuki, A., Tsuda, S., Yamanaka, T., Nishiguchi, M., Hirai, K., Meshi, T., Naito, S. and Ishikawa, M. (2005) Tobamovirus-resistant tobacco generated by RNA interference directed against host genes. *FEBS Letters* 579, 4479–4484.

Bai, Y., Guo, Z., Wang, X., Bai, D. and Zhang, W. (2009) Generation of double-virus-resistant marker-free transgenic potato plants. *Progress in Natural Science* 19, 543–548.

Bologna, N.G. and Voinnet, O. (2014) The diversity, biogenesis, and activities of endogenous silencing small RNAs in *Arabidopsis*. *Annual Review of Plant Biology* 65, 473–503.

Bucher, E., Lohuis, D., van Poppel, P.M., Geerts-Dimitriadou, C., Goldbach, R. and Prins, M. (2006) Multiple virus resistance at a high frequency using a single transgene construct. *Journal of General Virology* 87, 3697–3701.

Carbonell, A., Takeda, A., Fahlgren, N., Johnson, S.C., Cuperus, J.T. and Carrington, J.C. (2014) New generation of artificial microRNA and synthetic trans-acting small interfering RNA vectors for efficient gene silencing in *Arabidopsis*. *Plant Physiology* 165, 15–29.

Chen, L., Cheng, X., Cai, J., Zhan, L., Wu, X., Liu, Q. and Wu, X. (2016) Multiple virus resistance using artificial trans-acting siRNAs. *Journal of Virological Methods* 228, 16–20.

Chung, B.N., Yoon, J.Y. and Palukaitis, P. (2013) Engineered resistance in potato against *potato leafroll virus*, *potato virus A* and *potato virus Y*. *Virus Genes* 47, 86–92.

Csorba, T., Kontra, L. and Burgyán, J. (2015) Viral silencing suppressors: tools forged to fine-tune host-pathogen coexistence. *Virology* 479, 85–103.

Di Nicola, E., Tavazza, M., Lucioli, A., Salandri, L. and Ilardi, V. (2014) Robust RNA silencing-mediated resistance to *plum pox virus* under variable abiotic and biotic conditions. *Molecular Plant Pathology* 15, 841–847.

Elbert, A. and Nauen, R. (2000) Resistance of *Bemisia tabaci* (*Homoptera: Aleyrodidae*) to insecticides in southern Spain with special reference to neonicotinoids. *Pest Management Science* 56, 60–64.

English, J.J., Mueller, E. and Baulcombe, D.C. (1996) Suppression of virus accumulation in transgenic plants exhibiting silencing of nuclear genes. *The Plant Cell* 8, 179–188.

Fahim, M., Ayala-Navarrete, L., Millar, A.A. and Larkin, P.J. (2010) Hairpin RNA derived from viral NIa gene confers immunity to *wheat streak mosaic virus* infection in transgenic wheat plants. *Plant Biotechnology Journal* 8, 821–834.

Fahim, M., Millar, A.A., Wood, C.C. and Larkin, P.J. (2012) Resistance to *Wheat streak mosaic virus* generated by expression of an artificial polycistronic microRNA in wheat. *Plant Biotechnology Journal* 10, 150–163.

FAO (n.d.) FAOSTAT: Crops. Available at: http://faostat3.fao.org/browse/Q/QC/E, accessed 14 December 2016.

Fei, Q., Xia, R. and Meyers, B.C. (2013) Phased, secondary, small interfering RNAs in posttranscriptional regulatory networks. *The Plant Cell* 25, 2400–2415.

French, R. and Stenger, D.C. (2003) Evolution of *wheat streak mosaic virus*: dynamics of population growth within plants may explain limited variation. *Annual Review of Phytopathology* 41, 199–214.

Fuentes, A., Ramos, P.L., Fiallo, E., Callard, D., Sánchez, Y., Peral, R., Rodríguez, R. and Pujol, M. (2006) Intron–hairpin RNA derived from replication associated protein C1 gene confers immunity to *tomato yellow leaf curl virus* infection in transgenic tomato plants. *Transgenic Research* 15, 291–304.

Fuentes, A., Carlos, N., Ruíz, Y., Callard, D., Sánchez, Y., Ochagavía, M.E., Seguin, J., Malpica-López, N., Hohn, T., Lecca, M.R., Pérez, R., Doreste, V., Rehrauer, H., Farinelli, L., Pujol, M. and Pooggin, M.M. (2016) Field trial and molecular characterization of RNAi-transgenic tomato plants that exhibit resistance to *tomato yellow leaf curl geminivirus*. *Molecular Plant-Microbe Interactions* 29(3), 197–209.

Gao, L., Ding, X., Li, K., Liao, W., Zhong, Y., Ren, R., Liu, Z., Adhimoolam, K. and Zhi, H. (2015) Characterization of *Soybean mosaic virus* resistance derived from inverted repeat-SMV-HC-Pro genes in multiple soybean cultivars. *Theoretical and Applied Genetics* 128, 1489–1505.

Gonsalves, D. (2014) Hawaii's transgenic papaya story 1978–2012: a personal account. In: Ming, R. and Moore, P.H. (eds) *Genetics and Genomics of Papaya*. Springer, New York, pp. 115–142.

Guo, J., Gao, S., Lin, Q., Wang, H., Que, Y. and Xu, L. (2015) Transgenic sugarcane resistant to *sorghum mosaic virus* based on coat protein gene silencing by RNA interference. *BioMed Research International* 2015, ID 861907.

Jacobs, T.B., Lawler, N.J., LaFayette, P.R., Vodkin, L.O. and Parrott, W.A. (2016) Simple gene silencing using the trans-acting siRNA pathway. *Plant Biotechnology Journal* 14, 117–127.

Kis, A., Tholt, G., Ivanics, M., Várallyay, É., Jenes, B. and Havelda, Z. (2015) Polycistronic artificial miRNA mediated resistance to *wheat dwarf virus* in barley is highly efficient at low temperature. *Molecular Plant Pathology* 17(3), 427–437.

Lapidot, M., Karniel, U., Gelbart, D., Fogel, D., Evenor, D., Kutsher, Y., Makhbash, Z., Nahon, S., Shlomo, H., Chen, L., Reuveni, M. and Levin, I. (2015) A novel route controlling begomovirus resistance by the messenger RNA surveillance factor pelota. *PLoS Genetics* 11, e1005538.

Lennefors, B.L., Savenkov, E.I., Bensefelt, J., Wremerth-Weich, E., van Roggen, P., Tuvesson, S., Valkonen, J.P.T. and Gielen, J. (2006) dsRNA-mediated resistance to *beet necrotic yellow vein virus* infections in sugar beet (*Beta vulgaris* L. ssp. *vulgaris*). *Molecular Breeding* 18, 313–325.

Lennefors, B.L., van Roggen, P.M., Yndgaard, F., Savenkov, E.I. and Valkonen, J.P. (2008) Efficient dsRNA-mediated transgenic resistance to *Beet necrotic yellow vein virus* in sugar beets is not affected by other soilborne and aphid-transmitted viruses. *Transgenic Research* 17, 219–228.

Lindbo, J.A., Silva-Rosales, L. and Dougherty, W.G. (1993) Pathogen derived resistance to potyviruses: working, but why? *Seminars in Virology* 4, 369–379. DOI:10.1006/smvy.1993.1036

McGrann, G.R., Grimmer, M.K., Mutasa-Gottgens, E.S. and Stevens, M. (2009) Progress towards the understanding and control of sugar beet rhizomania disease. *Molecular Plant Pathology* 10, 129–141.

Missiou, A., Kalantidis, K., Boutla, A., Tzortzakaki, S., Tabler, M. and Tsagris, M. (2004) Generation of transgenic potato plants highly resistant to *potato virus Y* (PVY) through RNA silencing. *Molecular Breeding* 14, 185–197.

Montgomery, T.A., Yoo, S.J., Fahlgren, N., Gilbert, S.D., Howell, M.D., Sullivan, C.M., Alexander, A., Nguyen, G., Allen, E., Ahn, J.H. and Carrington, J.C. (2008) AGO1-miR173 complex initiates phased siRNA formation in plants. *Proceedings of the National Academy of Sciences* 105, 20055–20062.

Niu, Q.W., Lin, S.S., Reyes, J.L., Chen, K.C., Wu, H.W., Yeh, S.D. and Chua, N.H. (2006) Expression of artificial microRNAs in transgenic *Arabidopsis thaliana* confers virus resistance. *Nature Biotechnology* 24, 1420–1428.

Ogwok, E., Odipio, J., Halsey, M., Gaitán-Solís, E., Bua, A., Taylor, N.J., Fauquet, C.M. and Alicai, T. (2012) Transgenic RNA interference (RNAi)-derived field resistance to cassava brown streak disease. *Molecular Plant Pathology* 13, 1019–1031.

Palukaitis, P. (2012) Resistance to viruses of potato and their vectors. *The Plant Pathology Journal* 28, 248–258.

Park, H.M., Choi, M.S., Kwak, D.Y., Lee, B.C., Lee, J.H., Kim, M.K., Kim, Y.G., Shin, D.B., Park, S.K. and Kim, Y.H. (2012) Suppression of NS3 and MP is important for the stable inheritance of RNAi-mediated *rice stripe virus* (RSV) resistance obtained by targeting the fully complementary RSV-CP gene. *Molecules and Cells* 33, 43–51.

Peng, J.C., Chen, T.C., Raja, J.A., Yang, C.F., Chien, W.C., Lin, C.H., Liu, F.L., Wu, H.W. and Yeh, S.D. (2014) Broad-spectrum transgenic resistance against distinct tospovirus species at the genus level. *PloS One* 9, e96073.

Sanfacon, E. (2015) Plant translation factors and virus resistance. *Viruses* 7, 3392–3419.

Sanford, J.C. and Johnston, S.A. (1985) The concept of parasite-derived resistance—deriving resistance genes from the parasite's own genome. *Journal of Theoretical Biology* 113, 395–405.

Sasaya, T., Nakazono-Nagaoka, E., Saika, H., Aoki, H., Hiraguri, A., Netsu, O., Uehara-Ichiki, T., Onuki, M., Toki, S., Saito, K. and Yatou, O. (2014) Transgenic strategies to confer resistance against viruses in rice plants. *Frontiers in Microbiology* 4, 409.

Scorza, R., Callahan, A., Dardick, C., Ravelonandro, M., Polak, J., Malinowski, T., Zagrai, I., Cambra, M. and Kamenova, I. (2013) Genetic engineering of *plum pox virus* resistance: 'HoneySweet' plum—from concept to product. *Plant Cell, Tissue and Organ Culture (PCTOC)* 115, 1–12.

Shimizu, T., Nakazono-Nagaoka, E., Uehara-Ichiki, T., Sasaya, T. and Omura, T. (2011) Targeting specific genes for RNA interference is crucial to the development of strong resistance to *rice stripe virus*. *Plant Biotechnology Journal* 9, 503–512.

Singh, A., Taneja, J., Dasgupta, I. and Mukherjee, S.K. (2015) Development of plants resistant to tomato geminiviruses using artificial trans acting small interfering RNA. *Molecular Plant Pathology* 16, 724–734.

Smith, N.A., Singh, S.P., Wang, M.B., Stoutjesdijk, P.A., Green, A.G. and Waterhouse, P.M. (2000) Gene expression: total silencing by intron-spliced hairpin RNAs. *Nature* 407, 319–320.

Tiwari, M., Sharma, D. and Trivedi, P.K. (2014) Artificial microRNA mediated gene silencing in plants: progress and perspectives. *Plant Molecular Biology* 86, 1–18.

UN-DESA (n.d.) Transforming our world: the 2030 Agenda for Sustainable Development. Available at: https://sustainabledevelopment.un.org/post2015/transformingourworld, accessed 14 December 2016.

Vanderschuren, H., Alder, A., Zhang, P. and Gruissem, W. (2009) Dose-dependent RNAi-mediated geminivirus resistance in the tropical root crop cassava. *Plant Molecular Biology* 70, 265–272.

Vanderschuren, H., Moreno, I., Anjanappa, R.B., Zainuddin, I.M. and Gruissem, W. (2012) Exploiting the combination of natural and genetically engineered resistance to *cassava mosaic* and *cassava brown streak* viruses impacting cassava production in Africa. *PloS One* 7, e45277.

van Schie, C.C. and Takken, F.L. (2014) Susceptibility genes 101: how to be a good host. *Annual Review of Phytopathology* 52, 551–581.

Vassilakos, N. (2012) Stability of transgenic resistance against plant viruses. In: Çiftçi, Y.O. (ed.) *Transgenic Plants – Advances and Limitations*. InTech. Available at: https://www.intechopen.com/books/transgenic-plants-advances-and-limitations/stability-of-transgenic-resistance-against-plant-viruses, accessed 21 April 2017.

Wang, A. (2015) Dissecting the molecular network of virus-plant interactions: the complex roles of host factors. *Annual Review of Phytopathology* 53, 45–66.

Wang, M.B., Abbott, D.C. and Waterhouse, P.M. (2000) A single copy of a virus-derived transgene encoding hairpin RNA gives immunity to *barley yellow dwarf virus*. *Molecular Plant Pathology* 1, 347–356.

Wang, X., Kohalmi, S.E., Svircev, A., Wang, A., Sanfaçon, H. and Tian, L. (2013) Silencing of the host factor eIF (iso) 4E gene confers *plum pox virus* resistance in plum. *PLoS One* 8, e50627.

Waterhouse, P.M., Graham, M.W. and Wang, M.B. (1998) Virus resistance and gene silencing in plants can be induced by simultaneous expression of sense and antisense RNA. *Proceedings of the National Academy of Sciences* 95, 13959–13964.

WHO (n.d.) Health topics: pesticides. Available at: http://www.who.int/topics/pesticides/en/, accessed 14 December 2016.

Yadav, J.S., Ogwok, E., Wagaba, H., Patil, B.L., Bagewadi, B., Alicai, T., Gaitan-Solis, E., Taylor, N.J. and Fauquet, C.M. (2011) RNAi-mediated resistance to *Cassava brown streak Uganda virus* in transgenic cassava. *Molecular Plant Pathology* 12, 677–687.

Zhang, X., Sato, S., Ye, X., Dorrance, A.E., Morris, T.J., Clemente, T.E. and Qu, F. (2011) Robust RNAi-based resistance to mixed infection of three viruses in soybean plants expressing separate short hairpins from a single transgene. *Phytopathology* 101, 1264–1269.

Zhang, Z.Y., Wang, Y.G., Shen, X.J., Li, L., Zhou, S.F. and Li, W.C. (2013) RNA interference-mediated resistance to *maize dwarf mosaic virus*. *Plant Cell, Tissue and Organ Culture (PCTOC)* 113, 571–578.

# Index